The Evolution of Future Consciousness

The Nature and Historical Development of the Human Capacity to Think about the Future

by

Thomas Lombardo, Ph.D.

Bloomington, IN Milton Keynes, UK

authorHOUSE

157005213

AuthorHouse™
1663 Liberty Drive, Suite 200
Bloomington, IN 47403
www.authorhouse.com
Phone: 1-800-839-8640

AuthorHouse™ UK Ltd.
500 Avebury Boulevard
Central Milton Keynes, MK9 2BE
www.authorhouse.co.uk
Phone: 08001974150

First published by AuthorHouse 6/20/2006

ISBN: 1-4259-4446-9 (sc)

Library of Congress Control Number: 2006905329

Printed in the United States of America
Bloomington, Indiana

This book is printed on acid-free paper.

*To my Sweetheart and
Intellectual
Companion*

Jeanne

ACKNOWLEDGEMENTS

The beginning of this book and its companion volume, *Contemporary Futurist Thought*, first sprang into existence in my mind in 1992 while I was standing in a check-out line in a Safeway grocery store. I had just moved to Arizona and aside from being hired as the new chair of the psychology and philosophy departments at Rio Salado College I was also put in charge of "Integrative Studies" – the capstone course for the Associate degree offered at Rio Salado. Integrative studies could focus on whatever theme I chose to select, as long as the theme somehow pulled together the breadth of undergraduate courses, especially the sciences and humanities that students took in completing their Associate degrees. I had been thinking about different possible topics for the course but that day in Safeway I still hadn't decided on anything yet. As I was waiting in line in Safeway I started browsing through the paperback bookstand in check-out and noticed that Alvin Toffler, author of the highly popular book *Future Shock*, which I had read back in the 1970's, had a new book out titled *Powershift: Knowledge, Wealth, and Violence at the Edge of the 21st Century*. In a flash – it hit me – why not do the Integrative Studies course on the 21st Century – in fact, on the future? Most, if not all of the main areas of study in an undergraduate college program could be addressed and synthesized in the context of the future. Having had a long standing fascination with and love for science fiction since childhood, as well as studying evolution and the nature of time in college and post-graduate school, I felt I had some background and expertise on the topic of the future. Creating a course on the future, it seemed to me, would be an interesting and challenging endeavor, and I felt that such a course would be of great value for students. Shouldn't we all try to think about and understand the future? So I charged into the topic and in the following months read whatever books I could find on the future beginning with Toffler's *Powershift*, his earlier book *The Third Wave*, and John Naisbitt's *Megatrends* series, and constructed a course on the future.

Once I started teaching the course I realized that although there were many good books that dealt with selected aspects of the future, no one book (that I was aware of) covered all of its main dimensions. I was reading books on the future that dealt with science and technology, human society and culture, and economics and politics, but there were always major gaps in any individual reading. Hence, I started to write short articles for my students to supplement and integrate what they were reading. In fact, the first paper I wrote explained the value of thinking about the future - a short two-page piece titled "The Nature and Value of Thinking about the Future." This paper was the first reading I assigned in the class; to me it made sense to begin a course by asking what the value of studying the main topic of the course was. This beginning paper, over the course of a dozen years, grew into chapter one of *The Evolution of Future Consciousness*.

There were multiple sections of Integrative Studies offered at Rio Salado and early on I found a group of adjunct faculty who were interested in teaching sections of the course. Two of these faculty members, Robert Brem and Dr. Matt Bildhauer, were highly enthusiastic about the course and became very good friends in the coming years. We spent many hours in restaurants and coffee shops discussing and debating the future and related topics in history, philosophy, social studies, and science. I am grateful for their friendship, collegiality, and unwavering support as the course evolved in those early years. Matt, with a Ph.D. in philosophy, stimulated me into writing a supplemental paper for students on how thinking about the future developed in religion and philosophy through history, and this early paper grew into chapters three and four of *The Evolution of Future Consciousness*.

Eventually, the various readings for students grew and coalesced into an entire textbook on the future. In those first years I immersed myself in the literature on the future and encountered the works of Kevin Kelly, Walter Truett Anderson, Riane Eisler, Barbara Marx Hubbard, Hazel Henderson, and Francis Fukuyama, among others, and became a devout reader of *The Futurist* magazine. Pulling all this material together, I wrote *The Odyssey of the Future*, which covered all the main areas of future studies from science, technology, and space travel to culture, society and religion. After finishing the book, I gave a copy of it to the Chancellor of Maricopa Community Colleges, Dr. Paul Elsner. (Rio Salado is part of the Maricopa college system.) Dr. Elsner read the book and was extremely positive in his comments on it. In fact, in the following couple of years he supported the establishment of a Futures Institute at Rio Salado and funded the development of a website which I created on the study of the future. Through the subsequent years, Paul has become another good friend who emerged as a consequence of my work in the study of the future, and has remained highly encouraging in my ongoing writing every since.

Yet to really test the waters regarding the value and substance of my book, I sent a copy to Dr. Wendell Bell, former chair of sociology at Yale University and one of the best known academic futurists in the country. To my great satisfaction, Wendell was also very complimentary in his reaction to the book and encouraged me to publish it. Events in my life though took some unexpected twists and turns and the book was never published, but many of the elements and themes of my continuing work on the future were begun in that book.

Much of *Contemporary Futurist Thought* began in the writing of *The Odyssey of the Future*. In those early years, Wendell and Paul did much to validate my confidence in what I was doing since I very much respected the intelligence, wisdom, and scholarship of both of them. Additionally, during this time, I began a correspondence with Dr. Richard Slaughter, former President of the World Futures Studies Federation, and Richard accepted for publication my first article on the future. I thank Richard for validating my work, as well as providing a host of interesting writings of his own on the future that I have read over the years.

Also during this period I shared my writings, ideas, and enthusiasm on teaching the future with various colleagues at Rio Salado College. I want to especially thank Vernon Smith, Beatriz Cohen, and Larry Celaya, fellow faculty members, and my academic supervisor Vice-President Karen Mills, all of whom read sections of my early book and collaborated with me on work I was doing on futures education.

By the late 1990s, I began to attend the annual meetings of the World Future Society (WFS) and started to give presentations. I met Dr. Peter Bishop, faculty member in the Master's Program in Future Studies at the University of Houston at Clearlake, and Peter was also supportive of the academic work I was doing. At one of the earliest meetings of the WFS that I attended, I met a very energetic, intelligent, and extremely warm and friendly person on the shuttle from the airport to the convention hotel. His name was Jonathon Richter and over the last half a dozen years we have become great friends and professional collaborators, writing papers and giving presentations together. Jonathon and I continue to work on developing educational approaches to teaching the future and enhancing future consciousness. I also met Rick Smyre of Communities of the Future at one of the convention meetings and he has become another friend and colleague who I met because of my futurist work. Rick has also been very supportive of my writing and educational efforts.

Eventually Paul Elsner retired as Chancellor, the Futures Institute ended, the Integrative Studies course was dropped as a requirement in Maricopa, and I took down my website, but I continued to read and write on the future. I discovered new non-fiction writers such as Ray Kurzweil, Robert Nisbet, Lee Smolin, Sally Goerner, Peter Watson, and dove back into contemporary science fiction, totally enthralled and amazed by the recent novels of Dan Simmons, Stephen Baxter, Vernor Vinge, and Greg Bear. I am grateful to all these writers for stimulating my thinking on the future.

In completing these two books on the nature and development of future consciousness and futurist thought, I have explored history and the ideas of innumerable important figures from the past. I must mention, at the very least, Heraclitus, Aristotle, the Taoists, Leonardo da Vinci, Kepler, Spinoza, Leibnitz, Hegel, Nietzsche, and Darwin with whom I feel a strong sense of connection and resonance across time. All these visionaries of the past contributed to my understanding and study of the future.

Of special note in this regard, I am very appreciative to Antonio Damasio for providing directions to Spinoza's house and grave in his book *Looking for Spinoza*. On a cold, dark, and rainy day, my wife, Jeanne, and I searched out and found Spinoza's grave in The Hague last year, following Damasio's directions. The importance of finding Spinoza's home and gravestone is that Spinoza has always been one of my central guiding lights in my intellectual pursuits. Though Spinoza lived over three hundred years ago his cosmic vision, his forward-looking ideas on humanity, and the character of his life have been great sources of inspiration in thinking about many aspects of reality, including the future.

I would also like to thank two of my earliest teachers, before I discovered the future, James J. Gibson and J. T. Fraser, who tremendously informed and stimulated my young intellect on the nature of time, evolution, knowledge, and the human mind.

In the last few years, encouraged by Jeanne, I have been submitting articles for publication to the World Future Society, and have had quite a few accepted. In this process of submission and dialogue with editors, I have gotten to know Howard Didsbury, Ed Cornish (editor of *The Futurist* and former President of the World Future Society) and Timothy Mack (present President of the World Future Society). They have also all been very supportive of my work and my writings on the future. Thanks to all of them.

Finally, I want to thank my dear and wonderful wife Jeanne. Not only does Jeanne love to discuss ideas and philosophical topics, she also is my partner on my new website (www. odysseyofthefuture.net), contributing her poetry and occasional newsletters, and edited both books from beginning to end. I have dedicated *The Evolution of Future Consciousness* to her.

CONTENTS

INTRODUCTION

+‡==‡+

Are there any unique qualities that humans possess that make us special within the world of nature? Since the beginnings of recorded history, we have pondered this question. We have proposed many different answers, including abstract and logical reasoning, morality and ethics, an immortal soul, technology and civilization, a conscious sense of self-identity, and language. Humans also distinctively seem to require an overall sense of purpose and meaning in their lives. There has been debate, of course, over all of these possible answers. Do we really possess an immortal soul? Don't animals show many of these attributes, only to a lesser degree? Still, what if many of these qualities can be connected with one fundamental capacity, that even if present in animals is so vastly more developed in humans that it sets us apart from the rest of nature? What if technology, civilization, morals, self-identity, and purpose and meaning are all built upon a single quintessential foundation?

Following the lead of other writers, such as Anthony Reading in his book *Hope and Despair*, what I propose is that the human capacity for a highly expansive and complex sense of time, and in particular the future, is the foundation for much of what

1

makes us special within nature. Although our sense of time, of past and future, is built upon basic capacities found in animal psychology, we have greatly extended these abilities, and in so doing, have acquired the necessary mental powers for creating many of the unique features of human life, including ethics, technology, and civilization. The expansion of our temporal horizons has been the critical feature in our dominance among the species on earth. Furthermore, our continued survival and success will depend on developing this general ability even further in the future.

This book is about the human capacity to be conscious of the future, to create ideas, images, goals, and plans about the future, to think about these mental creations and use them in directing one's actions and one's life. I will first examine and describe the main psychological components of this general ability, which I refer to as "future consciousness," and consider how it is essential to normal human life. I will look at the intimate connection between future consciousness and ethics, purpose, and the conscious self. I will also look at the various benefits and value of future consciousness and outline ways to enhance it.

In the remaining chapters of the book I will describe the evolution and historical development of future consciousness, beginning with the emergence of life and sentient organisms; moving to our prehistoric ancestors and the appearance of humans; continuing through early mythic, religious, and philosophical thinking on the future; and finishing with modern beliefs and approaches to the future. Through this evolutionary and historical review, I will explain the connection between our emerging capacity to think about the future and the development of civilization, love and violence, science, and religion. My survey of the growth of future consciousness in this book will end at the beginning of the twentieth century. Completing my historical and theoretical overview, in a second book, *Contemporary Futurist Thought*, I will examine ideas and modes of thinking on the future over the last century up to the present time.

Through this study of the nature and evolution of future consciousness I will reveal how pervasive and central future consciousness is to human life and the human mind and how rich and varied human thinking on the future is.

CHAPTER ONE

⊹⇒⊶⇐⊹

THE PSYCHOLOGY AND VALUE OF FUTURE CONSCIOUSNESS

*"It is not the fruits of past success
but the living in and for the future
in which human intelligence proves itself."*

Friedrich von Hayek

Introduction

What are the main psychological processes involved in our awareness of the future? And what are the values associated with these mental capacities? In this opening chapter I describe the fundamental components and features and the functions and the benefits of future consciousness. I conclude the chapter with a discussion of why humans need to expand further their ability to imagine and think about the future.

As a starting point, let me provide a general definition and description of future consciousness. **Future consciousness** is part of our general awareness of time, our temporal consciousness of past, present, and future. Future consciousness includes the normal human capacities to anticipate, predict, and imagine the future, to have hopes and dreams about the future, and to set future goals and plans for the future. Future consciousness includes thinking about the future, evaluating different possibilities and choices, and having feelings, motives, and attitudes about tomorrow. Future consciousness includes the total set of ideas, visions, theories, and beliefs humans have about the future - the mental content of future consciousness. I define "future consciousness" as the total integrative set of psychological abilities, processes, and experiences humans use in understanding and dealing with the future.

Given this description of future consciousness, it should be evident that future consciousness is absolutely necessary for normal human psychological functioning. We would not be able to perform essential human tasks without an awareness of the future. Without the psychological capacities of anticipation, hope, goal setting, and planning we would be aimless, lost, mentally deficient, passive, and reactive.[1] We would not seem intelligent or for that matter even human without future consciousness.

The capacities of future consciousness, though, come in degrees. As we mature in life our awareness of both time and the future grows; our sense of time is very narrow in infancy and childhood. Even mature adults demonstrate great variability in being able to imagine possible futures, to set goals and plan, and to live with an "eye on tomorrow." Some people are more oriented to the present or immediate future; others are more "future oriented." Also our attitudes, modes of thinking, and frames of mind regarding the future can vary from the negative, counter-productive, or apathetic to the optimistic, positive, and active. Finally, there seems to be significant cultural variability in future consciousness. Some cultures are more future focused while others are more present or past focused.[2] In describing the main components of future consciousness and their importance in normal psychological functioning, I will also identify ways to improve these capacities and attitudes and demonstrate how enhancing our future consciousness benefits us in innumerable ways.

The Perceptual Awareness of Time

*"The past is consumed in the present and the present
is living only because it brings forth the future."*

James Joyce

Future consciousness is built upon the most fundamental of psychological processes; it is built upon the perceptual awareness of time. Though it is frequently stated or assumed that perceptual awareness has no sense of past or future or the passage of time, our sense of time is actually grounded in perception. Perceptual consciousness provides the beginnings of consciousness of time, of past, present, and future. Through perception we are aware of duration, stability, and change; of becoming and passing away; of patterns, rhythms, and forms of change; and of an experiential direction to time.

Basing my argument on the ideas of James J. Gibson, one of the most significant figures in the history of the psychology of perception, the perception of time is based upon the perception of events in the environment. We do not perceive empty or abstract time as such; rather we perceive dynamic events and the temporal relationships between events. We perceive motions and changes in the motion of objects; interactions and collisions between objects; changes in shapes and surfaces; sounds and sequences of sound; patterns of behavior in nature and animals; and natural and periodic rhythms. Further, the perception of time is relative rather than absolute. Just as the perception of space is relative, where the location and motion of objects are seen within the context of a spatial framework of the ground and surrounding objects, the perception of time is relative to a framework of temporal events. We perceive events and temporal relationships between events; shorter events are experienced in the context of longer events; before and after, longer and shorter durations, slower and faster, persistence and change, and the pattern and structure of changes in the environment are all relative qualities within our perception of time. As everyone can attest, the subjective experience of time clearly seems to depend on what is happening and how much is happening within our lives. Contrary to Kantian philosophy and the Newtonian idea of absolute objective time, Gibson argued that perceptual time is not some independent reality that flows within consciousness – there is no perceptual experience of empty time.[3] If nothing happened (or changed) in the environment, there would be no perception of time. The perception of time is grounded in the perception of events – of things happening and the relationships among these events.

Based on this idea that our fundamental mode of awareness of time is anchored to concrete events and the relationships among these events, I would propose that the entirety of our experience and understanding of time, even including higher forms of knowledge and thinking such as memory and anticipation, is relative to a framework of real, recollected, and imagined events. We have no sense of, nor can we imagine empty time. Psychological time is structured, filled, and delineated by events, real and imagined. Hence, expanding and enriching our consciousness of time requires building up our mental framework of experiences, ideas, and principles through which we understand and experience time. We can't just simply expand our consciousness of time without anchoring it to a mental framework of events and patterns in time. Learning history, reflecting on events in our own personal lives, studying contemporary trends and patterns of change, and exploring different possible futures, both personal and general, provide the substance and psychological framework required for developing our temporal consciousness. Our understanding and consciousness of time is contextual and relational and grows through adding detail and content and organizing the pieces into ever expanding and intricate maps of time in our minds.

For Gibson, the most fundamental distinction made in the perception of events is between relative persistence and relative change. The most basic experiences of time are seeing things change and seeing things stay the same. In fact, persistence and change are reciprocally distinguished in perception.

Perceptual persistence and change are relative, rather than absolute. Things are experienced as changing relative to things experienced as staying the same and things are experienced as persisting relative to things changing. Hence, at the most basic level, perception provides our mental anchor and framework for the reciprocal experiences of continuity and change.[4] Without the reciprocal experiences of persistence and change, there would be no sense of past or future, or the sense of a connection of past, present, and future.

If awareness of time is built on the perception of persistence and change, then there is always a sense of temporal duration within the perception of time - an experience of continuance. The experiences of persistence and change can only occur across a duration of time; neither persistence nor change can be defined at any one instant of time. Something persisting means continuing across some extent of time, and change means some transformation across some duration of time. Without the perception of persistence and change, and consequently duration, consciousness would be a set of disconnected and momentary experiences. But because the perception of time is built upon the relative experiences of persistence and change, it is not reducible to a set of disconnected instances.

Animals and humans perceive persisting objects, surfaces, and spatial forms as relatively stable. Without this fundamental sense of stability within the environment, adaptive and coordinated behavior would be impossible. Perception also reveals certain regular patterns of change, such as characteristic and repeatable forms of action in same species behavior and natural phenomena critical for survival. This overall awareness of a relatively stable world provides a sense of order to the flow of awareness and a framework in which to act. There is an order, pattern, and stability inherent in perceptual time that is necessary for ordered and patterned behavior.

Humans and animals also see dynamical transformations, such as objects breaking, moving, growing, or shrinking. In particular, regarding the perception of change, there is a clear awareness of things ending and new things beginning - of "becoming" and "passing away." The experience of change, of becoming and passing away, brings life and animation to our consciousness and provides the foundation for our awareness of the past and the future. Without experienced change – of becoming and passing away - there would be no discriminative sense to the notions of past and future and no sense of a passage of time. And also, to reinforce an earlier point, the sense of passage of time – of things retreating into the past and things emerging in the future – is relative and anchored to experienced events. The passage or flow of time is not an empty or content-less experience.

One popular theory of the perceptual awareness of time is that perception is limited to the immediate present. This theory assumes that there exists a conscious present which is an instant – a line, edge, or a point in conscious time. Within this theory the present has no duration. The futurist, Edward Cornish, for one, has argued for this theory of "duration-less present."[5] Anthony Reading, in his book *Hope and Despair*, also strongly argues that perception, in and of itself, does not yield an awareness of time, but only produces a series of "snapshots" of the present.[6] This presumed reality of an absolute immediate now separates what was - the past - from what will (may) be – the future. Within this model, perception is reducible to a set of disconnected conscious instances.

Yet as I have argued above, since perceptual time is built upon persistence and change, this view of perceptual time can not be correct. Persistence and change are temporal relationships and can not be defined within a single instant. Moreover, how could a state of consciousness possess no duration? Can the fundamental units of our consciousness of time possess no time? As Richard Morris argues, whatever the conscious "now" is, it is not an instant.[7] Further, as a critical source of evidence, the sensory organs and perceptual systems of both animals and humans do not react to instantaneous stimulus values, but rather relationships, temporal and spatial, between stimuli. In fact, the most critical dimension within physical stimuli for sensory receptors is change, and temporal change, obviously, is a property that can only exist across durations of time.[8]

Introspecting on human consciousness, as clearly evinced in perceptual awareness, we do not experience some momentary frozen "snapshot" of time or a sequence of such frozen instants; we experience flow, continuance, and duration. We experience temporal events or patterns that have duration and interconnect with each other in a nested framework of shorter and longer temporal events.[9]

The theory that perceptual awareness is limited to a hypothetical instantaneous present assumes that the conscious present can be clearly separated from our awareness of the past and the future, but there is no clear conscious dividing line between past, present, and future. Attempt to distinguish the present from the immediate past or the unfolding future. The present flows into the past in one direction and the future in the other.[10] Perceptual consciousness extends both into the past and into the future because we perceive change and persistence across time. Conscious time is durational and relational. We see "becoming" and "passing away" – the perceptual support for our consciousness of the future and the past. We also see objects and structures persisting – providing a sense of continuity of past, present, and future. The experience of time within perceptual consciousness is not of an instantaneous present, but rather of continuation and transformation across time.

It has been argued that human infant consciousness, presumably uncontaminated by memory or anticipation, appears limited to the immediate present. Infant consciousness is pure perception.[11] But the counter-argument is that infant consciousness is simply limited and restricted regarding the perception of persistence and change, rather than being totally devoid of any sense of duration and connection of events. The infant does not live in an instantaneous present or a chaos of disconnected moments. Once the sense-organs and perceptual systems mature and develop sufficient neuronal-synaptic connections, the human infant clearly shows the capacities to attend to change and interesting stable objects.

Another significant feature of our perceptual awareness of time is that we experience time as asymmetrical with a direction. Time metaphorically flows. We experience relative persistence, but we also experience succession, and the experience of succession goes in a direction. If one were to envision time as a line or sequence of events, then our experience of time moves in one direction across this line. It is sometimes said that we experience time always moving forward into the future and never backwards into the past. Time is always experienced as moving forward into what we call the future. This direction to temporal consciousness is often referred to as the **subjective arrow of time**.[12]

Without an experienced direction to time, there would be no way to distinguish between becoming and passing away, or more generally between past and future. The past is what "has been" relative to the present and the future is what "will be" relative to the present. What "will be" lies ahead on the arrow of time, and what "has been" lies behind. Past and future are a relative distinction defined by the direction of the experienced arrow of time, or stated in the converse past and future define the direction. Thus if one were to lose the sense of the past, one would necessarily lose the sense of the future, for the sense and experienced direction in time would disappear.

There are, of course, cyclical and persistent features to our experience of time, but there is an overall sense of linearity and directional flow. The future is to a degree different from the past. This asymmetry within perceptual experience clearly undercuts the notion that temporal consciousness is built upon momentary instants for there is no meaning that can be given to this experience of direction within the confines of an instant. A sense of direction or passage implies an experience of duration and an experienced change across this duration.

The temporal extent of perceptual consciousness may not be that extended in scope, but it is the experiential beginning and foundation upon which higher and more complex levels of temporal awareness, involving thinking and imagination, are built. As I have noted, it is a common belief that animals, human infants, and even many carefree adults live primarily in the "immediate here and now," but at best this is a relative distinction, for perceptual awareness, a capacity shared by animals, infants, and all adult humans, provides a window into extended time, of past, present, and future.

Emotion, Motivation, and Future Consciousness

Emotion, along with perception, is a second basic form of awareness that contributes to future consciousness. Emotion is a relatively constant feature of all human consciousness. Even though the intensity of experienced emotion varies across time, we are always feeling some emotion or set of emotions. All mammals, including humans, possess a clearly defined area of the brain, specifically what is referred to as the limbic system, which is responsible for producing a wide variety of emotions and basic motivational states, such as sexual arousal, fear, pleasure, aggression, and anxiety. This emotional area of the brain exists in rats, cats, rabbits, and humans and all other mammal species.[13] Long before the emergence of humans and the capacity for abstract and hypothetical thinking, animals exhibited emotional responses, such as fear and excitement to anticipated negative and positive events in the environment. The emotional life of animals is probably not as rich as that of humans,[14] but animals do have emotions, and often these emotions make reference to events beyond the present. Emotions, such as fear, hope, and anxiety clearly have a future focus – such emotions are anticipatory and are not simply reactions to the "here and now."[15]

Reading and other psychologists distinguish between emotions that have a present-focus and emotions that are "prospective" or anticipatory. Happiness and sadness, presumably, have more of a present focus, as emotional reactions to what is happening right now. Hope, as a positive prospective emotion, and fear, as a negative defensive prospective emotion, are future focused.[16]

One could argue that there is always an emotional dimension to future consciousness. When we anticipate what is to come, we have feelings as well as thoughts and images. Feelings or emotions provide the positive and negative color of future consciousness. The future is felt as bright or dark, exciting or depressing. Emotions fall into two general categories - pleasurable and painful. Pleasurable emotions about the future draw us; painful emotions about the future repel us. Reading goes so far as to argue that emotions provide the basis for assessing the value of different imagined futures and evaluating different goals. Without emotion, one future would be as good or bad as the next. It is through our emotional feelings about different possible futures that we determine what is desirable or preferable.[17] Hence, any comprehensive approach to the future and the development of future consciousness must address this dimension of the human mind. We feel the future as much as think it.

In his study of future consciousness, Reading highlights hope and despair (or depression), two of the primary prospective emotions. He defines hope as the energizing and pleasurable emotion connected with the anticipation of future goals and events that will enhance our well-being. Conversely, depression or despair is the painful and debilitating emotion connected with the loss of anticipation of positive future events or the anticipation of destructive future events.[18] Furthermore, hope also entails a positive realistic appraisal that one can achieve envisioned goals in the future, whereas depression or despair is associated with a sense of impotence about creating a better future. Hope, as an emotion, motivates people into action; despair de-motivates. Hope and despair come in degrees of intensity, with mania being the extreme form of hope and clinical severe depression with suicidal impulses being the extreme form of despair. Suicide is the abandonment of the future.

Based on these definitions of hope and despair, a couple important points should be highlighted. Although hope and despair are emotions, there is a cognitive dimension to both feelings. This cognitive dimension is thought. There are thoughts concerning one's capacity or lack thereof to realize future goals that are an integral part of the resulting emotions. Thoughts influence emotion. Conversely, emotion impacts cognition. For example, both depression and apathy (the lack of feelings about the future) depress thinking and imagination. In contrast, it is hope, an emotion, which energizes and stimulates higher levels of future consciousness; our capacity to imagine and think about future possibilities is severely hindered without the feeling of hope. Finally, we should note that hope motivates. As an emotion, it energizes people into planning and taking action to realize their goals. For Reading, hope is the engine and the mechanism that has driven the human species to progressively create the world that we live in. It is the foundation of the growth of civilization.

From the above discussion, we see that emotion and motivation overlap and interconnect. To use two other examples, fear and lust are emotions, but also motives. Motives are the causes of and reasons for behavior, and sometimes motives are emotions. When we are afraid, we run, freeze or even attack; when we feel lust, we approach and attempt to entice the object of desire toward us.

Motivation is another basic component of future consciousness. Motives make reference to the future in that motivated behavior, such as approach or avoidance, is directed either toward some desirable end in the future or the avoidance of some undesirable end in the future. Most human behavior is motivated, and generally we describe such motivated behavior as purposeful. Acting with purpose involves intentional behavior to achieve some anticipated end in the future, even if it is only the short-term future of the next few moments. Hence all purposeful or motivated action involves a form of future consciousness. This general capacity to act with purpose and regard to the future is a normal feature of all adult human minds.[19]

Motives often take the form of explicit conscious goals. Goals are psychological realities and involve mental images and thoughts ("cognitive representations"), invariably enriched with emotional color, about intended or desired states in the future. Action motivated by goals is referred to as goal-directed behavior. Goals can be either short-term or long-term. As the psychologists Karniol and Ross note, there is significant variability across individuals in setting long term versus short term goals. Some people limit themselves to setting mostly short term goals, while other people have the capacity to identify and pursue more long term goals as well. Goals can be either negative or positive; we could envision something desirable that we want to attain, or something aversive that we want to escape from or avoid. There is individual variability on this parameter as well; some people conceptualize the future more in terms of positive things to approach or realize; other people conceptualize future goals primarily in terms of things to avoid or escape from.[20] Although humans seem to show significant differences in the number, type, and intensity of goals in life, at the very least everyone demonstrates momentary goal-directed behaviors throughout the day.

Given the pervasiveness of goals and purpose in most of human behavior, the argument has been made in psychology that a person's conceptualization of the future is a fundamental determinant in explaining human action. Many psychologists, such as Freud and Skinner, have focused on how the past determines present human behavior (and there is clearly some truth in this position for goals do reflect past learning), but at a conscious level, when people act they are usually acting with the future in mind. Hence, one can argue that most human behavior is in fact determined by future consciousness - the sense of desired goals for the future.[21]

Goals reflect the influence of the environment, learning, social upbringing, inner biological needs and desires, emotions, and active and creative thought processes. Karniol and Ross emphasize that goals are a result of values; the goals that a person pursues depend upon that person's values, such as wealth, love, truth, or professional success. The sociologist and historian of the future, Frederick Polak, makes a related point regarding images of the future; according to him, the images of the future that a society creates (which in essence are envisioned ideals or goal states) strongly reflect the values of that society. Within the history of both psychology and philosophy there has been ongoing debate regarding the source of values (and in particular, moral values). Some argue that biological inheritance determines values, others argue that culture creates values. Still others, such as Reading, suggest that it is emotion and feelings of pain and pleasure that are the basis of values, while some contend that reason, at least, should be the foundation of values. Later, in this chapter I will examine the connection of ethics and future consciousness, but for the moment we should at least note that values and value driven motivation (including ethics and morals) play a central role in determining the goal content of future consciousness.[22]

As one measure of the development of future consciousness, we can ascertain the degree to which a person has identified, nurtured, and acted upon goals for the future, and in particular long term or novel, creative goals. One can enhance the development of goals and goal-directed behavior, and consequently future consciousness, through educational and therapeutic efforts (environmental influences) and self-directed introspective efforts to formulate, clarify, and expand an individual's personal goals. I will expand on this topic further throughout this chapter.

There is an interesting reciprocal connection between present psychological states and anticipated goals for the future. Positive and negative emotional states in the present influence the creation, development, and sustainability of goals for the future; happiness in the present amplifies and strengthens future goals; sadness weakens goals. Reciprocally, cultivating and maintaining positive goals for the future enhances present well-being, whereas the anticipation of disaster and misery in the future brings a person emotionally down in the present.[23]

Just as emotion has an impact on thinking and imagination, so does human motivation. In fact, the argument has been made that it is goals that set thinking in motion. Without motivation, we would not think – there would be no reason to.[24] As many philosophers and psychologists have argued throughout history, all thinking has an agenda – there is a goal behind all acts of thinking. To whatever degree this theory of thinking is correct, all human thinking has an inherent future focus – thinking serves the realization of goals.

In the next section I focus on human cognition and thinking, but in describing how emotion and motivation are important features of future consciousness, it has become clear that thinking and imagination are intimately connected with emotion and motivation. Regarding human emotions, as we have seen, hope and depression contain a cognitive component. Other emotions such as happiness, sadness, anxiety, and fear do so as well. Regarding human motivation, although goals are connected with basic biological needs and feelings of pleasure and pain, goals are also strongly influenced by thinking. We feel our goals but we also think out our goals. We articulate and rationalize our goals; we formulate our goals using concepts, beliefs, reasoning, and language. Reading, for one, strongly emphasizes the role of "cognitive representations" in the formation of goals.

Within the psychological study of motivation and emotion, one central finding over the last few decades is how strongly thinking affects both motivation and emotion. The cognitive theory of emotions states that thought and emotion are not totally distinct states of mind, and what we think, to a great degree, determines how we feel. Events in the world trigger interpretations, which in turn trigger emotions. If a person interprets an event using negative concepts, such as "awful," "disastrous," or "dangerous," the person will feel negative emotions such as anxiety, fear, or depression.[25] Another important factor is whether a person believes that he or she can do anything to influence the anticipated situation. If a person believes that he or she can influence the future and perhaps prevent some possible misfortune (which is a thought), then the person does not feel as depressed or fearful as a person who believes that he or she is powerless (which is also a thought).[26] (Recall that hope includes the thought or belief in individual power over the future, whereas depression includes that belief in powerless over the future.) So the emotions we have, which make reference to the future, are influenced by thoughts we have of these anticipated events and our beliefs about our own capacities to deal with these anticipated events.

Further, thoughts about the future and one's capacity to influence the future don't simply affect emotions; such thoughts also affect motivation and goal directed behavior. How we think and what we think about the world around us strongly influences what we desire and consequently what goals we strive for. As noted above thoughts create negative and positive interpretations of objects, events, and our goals for the future. In turn, our evaluative thoughts set certain behaviors in motion, determining what we avoid and what we seek.[27] Positive and uplifting interpretations of future goals and our abilities to realize these goals produce enthusiastic and tenacious action. On the other hand, depressed people, who have hopeless and helpless thoughts about themselves and the future, do not simply feel bad; they have marginal or negative goals, they do not act except to avoid, and often they have no desire to act in any constructive way. They are behaviorally frozen in the face of a fearful and hopeless future.

Hence positive and negative thoughts impact the fundamental emotional-motivational dimension of hope versus depression and fear. **Hopefulness** can be defined as having positive images, thoughts, and feelings of the future and one's abilities. Positive dreams for the future bring passion, enthusiasm, and excitement to life – they fuel and direct our motivational energy. Hope is essential for happiness and psychological health. Conversely, depression can be defined as a sense of **hopelessness**,[28] and is generated when individuals have negative images, thoughts, and feelings about their future, or no images at all, and feel helpless to change anything.

Depression, in fact, could be seen as a disorder of future consciousness – a state where future consciousness has collapsed into nihilism and negativity. Depression can be brought on through fear of anticipated negative events in the future. Just as significantly, apathy can arise when the future looks empty of promise. If people fail to imagine something, negative or positive, the energy goes out of life. When people have nothing to look forward to, they psychologically and physically wither and die.

Negativity and apathy about the future are often connected psychologically. A loss of mental imagery and energy often occurs because of fear of the future, either because of its perceived uncertainty or because the images have become too negative, anxiety producing, or even terrifying. We suppress, repress, or ignore what frightens us. Our conscious minds go blank. We become apathetic because we are negative and scared.

Just as thinking influences emotion, as I noted earlier, emotion influences thinking. Positive emotions, such as joy and happiness, enhance creative and opportunistic thinking and promote a win-win mindset and openness to others and the world. Negative emotions, such as anxiety and depression, produce more defensive and critical thinking, a win-lose mindset, and increased self-absorption in thoughts and images.[29] Negative emotionality seems to prime the human mind to seeing what can go wrong in the future, whereas positive emotionality does the reverse, facilitating thinking of what can go right and how to achieve it.

The futurist Michael Zey is especially concerned about the loss of positive images of the future in modern times, and the replacement of the positive with either negative visions or the total lack of any images. We have moved from hope to depression and apathy. Zey is not alone in this assessment of our contemporary depressive and apathetic mindsets regarding the future.[30] As Best and Kellner point out in their in-depth study of Postmodernism, the positive and hopeful images of the future generated in the period of the Enlightenment and the Industrial Revolution have come under critical attack over the last century, often being replaced by disappointment, despair, and nihilism regarding the promises and prospects of modern civilization.[31] The historian Robert Nisbet, in a similar vein, argues that in the last century, Westerners have lost faith in the positive image of progress.[32] Zey believes, as do many other futurists, that modern society should find new positive images of the future to create a renewed sense of hopefulness and counteract the prevalent sense of negativity and nihilism in our world.[33] In the last section of this chapter, I will discuss this issue further.

We should keep in mind though that negative images of the future per se do not necessarily cause apathy or depression. If we believe we are incapable of doing anything to alter these future negative events then we get depressed and exhibit a sense of **helplessness**.[34] Negative images could stimulate us into action in order to influence the future in a different, more positive direction. But, just as likely, negative images of the future may generate immobility or avoidance. A sense of helplessness, which is a state of mental and behavioral paralysis, is an essential feature of depression and nihilism.

Within the psychology of motivation, a common distinction is made between **approach** and **avoidance motivation**. Approach motivation produces behaviors that move toward a perceived positive object, whereas avoidance motivation produces behaviors that move away from a perceived negative object. People exhibiting positively motivated behaviors are approaching desirable objects; people exhibiting negatively motivated behaviors are avoiding aversive objects. Positive approach behaviors are associated with feelings of hope, enthusiasm, and elation; negative avoidance behaviors are associated with fear, anxiety, and depression. Even if negative imagery of the future can stimulate individuals into action, the behaviors are avoidance motivated. A life of attempting to avert or avoid disasters is a life governed by fear and anxiety. It makes much more sense from a psychological point of view to create and identify positive or desirable objects to move toward then to continually expend time and energy avoiding or escaping from perceived negative realities.[35] Mental health, in fact, can be defined as the degree to which one's behavior and personality centers around approach motivation and hope, and conversely mental illness can be defined as the degree to which a person's life revolves around fear, avoidance, and escape.[36] Mental health can therefore be described as a positive, approach-oriented mode of future consciousness, and mental illness can be described as a negative, avoidance-oriented mode of future consciousness.

Noelle Nelson and Wallace Wilkins are two psychologist/counselors who have looked at the emotional and motivational effects of beliefs about the future. Wilkins argues that the future is possibilities rather than certainties. We should focus on the positive future possibilities; such positive anticipations will increase the quality of life now. Wilkins argues that these positive anticipations, which need to be acted upon and not simply imagined, do not even need to be accurate. A positive mindset about where we are heading and the future consequences of our actions makes us feel better today.[37] Basically Wilkins is stating that an approach-motivational mindset about the future generates psychological well-being and mental health in the present. Reading makes a similar point regarding how hope creates a positive and pleasurable emotional state in the present.

Nelson distinguishes different types of beliefs about the future and their effects on mental health and personal success. According to Nelson, fear of the future produces negative emotional states and inaction. (A negative mindset about the future produces an emotionally unpleasant state of consciousness in the present.) But for Nelson, to uncritically think the future will be wonderful is unrealistic and will invariably lead to frustration and disappointment. Nelson, instead, argues for what she terms a "Winner" mindset about the future. A "Winner" mindset about the future involves acknowledging both the negative and positive possibilities of tomorrow, and believing that we have some power and choice in determining which possibilities are realized. Hence, perceiving the risks, but also seeing that one has some control over what will come to pass, generates good mental health and emotional well-being. As I previously noted, a sense of personal empowerment over the future positively affects one's emotional state. Conversely Nelson argues that believing that external forces beyond our control determine the future, or that the future is set, generates apathy and other negative emotional states. Thoughts, and particularly helpless thoughts about the future, instigate negative feelings about the future.[38]

In general, the particular ideas and beliefs individuals have about the future affect their emotional and motivational states. Zey and others are concerned about positive and negative beliefs about the future, but another important distinction noted in the above discussion is between realistic and unrealistic beliefs about the future. For Nelson, unrealistic beliefs lead to frustration and other negative emotional states. (Reading too includes in his definition of hope the factor of realistic beliefs regarding the achievement of goals.) Of course, given the uncertainty of the future, the issue of what is realistic and what is unrealistic is often debatable. Still, considered and thoughtful beliefs about the future stand a better chance of generating happiness and fulfillment in life.[39]

Despair and depression versus hope and enthusiasm about the future leads us into another important feature of motivation that impacts future consciousness. The famous personality psychologist, Abraham Maslow, separates all human motivation into the two general classes of security needs versus growth needs.[40] I would state that people exhibit the complementary or oppositional motives for security and adventure. Security motivation includes the needs for stability, certainty, and safety. The desire for adventure includes the needs for change, surprise, and risk. All people possess both sets of motives, yet it appears that there is considerable variation among people in the relative strength of each set of motives.[41] This variability in motivation impacts people's feelings, attitudes, and approaches to the future.

First, let us look at stability versus change. People psychologically require a degree of stability in their lives, as well as some degree of change. Too much stability and people become bored; too much change and people become anxious and confused. Stability provides mental order and identity in our lives; change provides excitement and a sense of growth and learning. Yet, some individuals appear highly inflexible and entrenched in old habits - they may revere tradition and the past. They are hooked on stability. Other individuals seem to embrace what is new and different, forever changing the wardrobes of their mind and their lifestyle. Cultures can show variability on this trait, as well. At an ideological level, one main area of disagreement in contemporary approaches to the future is over stability versus change. The popular contemporary writer Virginia Postrel refers to this ideological clash as between "stasis and dynamism."[42] Many people want the future to be like the past; many people want the future to be different.

Connected to the needs for stability and change, are the needs for both a level of certainty about life as well as a need for surprise, openness, and unpredictability. Again there are individual and collective differences in these two motives and this variability shows up in another fundamental clash in attitudes to the future. Some approaches to the future highlight certainty and commitment while other views acknowledge and even revel in the unpredictability of life. For example, the most popular position espoused in the World Future Society is that the future is possibilities and that this uncertainty about the future is a positive thing. Because the future is open, we have some choice, and consequently, self-empowerment over what will occur in the future. In contrast, fundamentalist religious views emphasize the certainty of the future. There are many religious systems which describe in considerable detail what the future will bring and followers of these belief systems are

24

absolutely convinced that these prophecies are unequivocally correct and will come to pass. Although all humans require some degree of certainty, from an epistemological perspective we are probably not justified in believing that we are certain about anything in the world, in particular the future, and from a psychological point of view the excessive need for certainty reflects insecurity, closed-mindedness, and a lack of critical thought.

People often refuse to anticipate or plan for the future much because the future is to some degree unpredictable, and where there is uncertainty, there is the possibility of frustration and disappointment. Because of the uncertainty of the future, for many people, to different degrees, the future is something to fear. People often retreat to the past or hold on to the present because it is more tangible, certain, and determinate. Peter Russell argues in *The White Hole in Time* that people have a psychological attachment to time, viewing their personal identity as bound up with what happens in time. Therefore, people fear time because they fear for their very survival. The future may change them or even terminate their existence.[43]

Finally, let us consider the polarity of safety versus risk. If the future is to a degree uncertain, to open one's mind toward the future, acknowledging the uncertainties, and yet set goals, plan, and act, involves realistic risk and consequently courage. We realize that what we hope for, what we plan for, what we attempt to accomplish may not come to pass. But, of course, where there is the possibility of failure, there is also the possibility of growth and success. Certainty provides mental safety as well as security, and people of course desire some level of safety or else they would feel excessively fearful and anxious. Yet, too great a concern for safety breeds defensiveness and withdrawal. In fact, a strong need for safety both reflects and further magnifies feelings of insecurity. When people avoid risk, it is because they are insecure. If they rigidly and desperately protect and try to preserve their stable and certain realities they further exacerbate their own insecurities.[44]

Howard Bloom takes a very interesting perspective on risk-taking. According to Bloom, humans, as well as various animals, form pecking orders or hierarchies of dominance. Further, humans and animals alike compete with each other to topple those individuals higher on the pecking order and move up in the social hierarchy. As a result, pecking orders are frequently in a state of change and the reshuffling of relative positions. Bloom contends that individuals or groups that are moving up in a pecking order are greater risk takers, while individuals or groups who are losing status and position are conservative and take few risks. Hence, the degree of risk taking is a sign of ongoing success or failure in one's movement within a pecking order. Bloom's argument would seem to connect with the idea that avoidance of risk reflects insecurity; those individuals or groups who are moving upward feel less insecure than those individuals or groups who are falling within a dominance hierarchy. Yet, if Bloom is correct in his general theory that life necessarily involves competition, then to become increasingly conservative and less risk-taking amplifies one's insecurity and decreases the chances of future success.[45]

In summary, our consciousness of the future and our ability to formulate and act on goals for the future is strongly influenced by our needs for security and adventure. Both sets of needs exist in all of us, but if our security needs become too powerful, we limit, if not totally repress, our openness to the future. Curling into a ball or burying one's head in the sand in the face of *the* unavoidable elements of risk, uncertainty, and change in life produces phobias, fears, paranoia, and depression. Of course, one shouldn't be foolhardy, Pollyannaish, impulsive, and thoughtless in life. The key is some sense of balance between security and adventure.

So, in summary, emotion and motivation frequently make reference to an anticipated future, a future that is colored and charged with human feelings, positive and negative images and goals, and cognitive interpretations of reality and the self that determine how we act and what, if anything, we strive for in the future. Further, our emotional states influence the type of thinking and motivation we engage in regarding the future. Positive emotional states are associated with approach motivation and constructive hopeful thinking; negative emotional states are associated with avoidance motivation and defensive thinking.

The Cognitive Dimension of Future Consciousness

The term "cognition" refers to all those psychological processes involved in the acquisition, storage, use, and creation of knowledge - cognition is knowing. Cognitive processes include perception, learning, memory, imagination, conceptual and abstract understanding, thinking, and language. Psychologists often distinguish between "lower" and "higher" cognitive abilities. Animals seem to clearly possess certain lower order cognitive abilities such as perception, learning, and rudimentary memory, but it is only the most advanced mammals, and in particular, humans, who demonstrate the higher cognitive abilities to think and use language.

Although since the beginnings of the study of psychology there has been ongoing debate and controversy over to what degree even higher mammals are able to think, to abstract, or to use language, it seems clear that relative to the rest of the animal kingdom, these capacities are greatly amplified in humans. It is a common argument that these significantly evolved higher cognitive abilities in humans have vastly expanded our capacity for both past and future consciousness.[46] Through thinking, imagination, and symbolic language humans can transcend the confines of perception and consider abstract, hypothetical, and imaginary realities; we can mentally represent to ourselves and communicate to others events distant in time, conceptualize and describe the grand panorama of history, and formulate plans of action that extend far into the future. Future consciousness, as well as historical consciousness, is immensely enriched through thought, imagination, and language.

Reading is one writer who emphasizes the importance of higher cognitive capacities in future consciousness. Within his theory, awareness of the future requires the capacity to create symbolic mental representations of the world. According to Reading, only humans can achieve this. (Humans only begin to show this capacity around two years of age.) Symbolic mental representations require the emergence of both the capacity for recall and the rudiments of human language. Both future and past consciousness begin to appear and expand when children apply their internal mental representations to organizing and making sense of the world. Future consciousness, in particular, is the ability to use internal mental representations to predict and understand the future.[47]

One important point that Reading highlights is the significant connection between learning and memory and the capacity for future awareness. Learning and memory are cognitive processes that involve the acquisition of knowledge through experience and interaction with the world. For Reading, learning and memory provide the foundation – the informational content – upon which mental representations of the world are created. Although what a human has learned and remembers derives from the past, mental representations based upon learning and memory, provide the knowledge we use to make predictions about the future. Through learning and memory, we acquire an understanding of the patterns and regularities of change in the world and apply this understanding to anticipating the future. Reading notes that recall emerges at approximately the same time as the capacity to predict the future, and that the area of the brain (the prefrontal cortex) involved in recall is the same area involved in making predictions.[48] Hence, to return to a couple of points made earlier, consciousness of the past and the future are intimately tied together, and increasing one's understanding of the past benefits one's ability to anticipate the future.

Although I believe that Reading draws too sharp a distinction between the capacities of perception (which he sees highly limited to the here and now) and higher forms of cognition involving thinking, language, and abstraction, I think that he is correct in emphasizing the importance of mental representations based upon memory in greatly expanding the human capacity for future consciousness. Reading, in fact, believes that the essence of human knowledge lies in the related capacities to form mental representations based on learning and being able to apply these representations to interacting and adapting to the world. Other writers on human psychology have also presented similar arguments, contending that human intelligence basically consists of the capacity to make predictions of the future based on the past.[49]

There is one important cautionary note that should be mentioned regarding the connection between memory and future consciousness. Although memory (and ideas of the past) may serve as a foundation for anticipating the future, future consciousness often extends beyond memory and the past. In fact, to believe that the future will be like the past is to remain stuck in the past. Experiences from the past, such as traumas and frustrations, can inhibit any new thinking about the future. Yet, one thing we learn from history is that there is always novelty and change; history does not entirely repeat itself. The future will not be the same as the past. As Karniol and Ross note, individuals at times will abandon, reject, or ignore the past in attempting to create a new and different reality for themselves in the future.[50] As I will discuss below, creativity is an essential component of future consciousness. Humans can create mental representations that include novel features that are not entirely simple reflections of memory and learning.

Putting the pieces together, the cognitive dimension of future consciousness, which begins with perception, is greatly amplified through learning and memory. Of special note, mental representations of reality and patterns of change begin to emerge as we observe and interact with the world around us. These mental representations go beyond perception since we can mentally recall and represent within our minds realities that presently do not exist. We can represent events, real and hypothetical, in the near and distant past and in the near and distant future. And although these mental representations may derive from learning and memory, we have the ability to create novel representations that transcend the past.

One key dimension of mental representations is imagination. Imagination is the capacity to create "perceptual like" conscious images and hypothetical realities in our mind without the appropriate physical stimuli being present. We can imagine colors, shapes, sounds, tactual sensations, and more complex perceptual realities sitting in the dark and in the silence without our senses being stimulated at all. Imagination transcends our present physical or perceptual reality and the relative immediate here and now.[51] Although the raw material of imagination involves perceptual experiences and memories of these experiences, imagination can go beyond what we have experienced. We can imagine realities we have never experienced, and this capacity is highly significant regarding future consciousness. When we imagine the future, we are creating perceptual like representations of events that, at least in some respects, we have never encountered.

Although imagination need not pertain to the future – we may engage in pure fantasy or we may have images of the past – we often do create images of the future. I will refer to this ability to visually imagine the future as **"visual foresight"** or simply "visioning." All adult humans possess the capacity of visual foresight. We could not intentionally act on conscious future goals unless we possessed some minimal ability to envision the future – seeing in our "mind's eye" our goals and aspirations and extending our consciousness beyond the here and now.

Although the ability to imagine goals is a normal psychological capacity, it varies significantly among individuals. We may not engage and evolve this capacity as much as we can and should. For one thing, we may envision only very short-term goals or we may routinely direct our lives with habitual goals that are rarely questioned or altered. We may put little effort into attempting to imagine alternative goals from those we have followed in the past. We may not think much about changing our goals. Reading refers to such habitual goal setting as "passive expectation" and contends that it is motivated by security needs.[52] If we simply imagine the same goals over and over again and act on them, we could say that our lives are stuck in the past.

Also, independent of goal-setting, the capacities of imagination and in particular, imagining different scenarios for the future varies greatly among people. Some people have trouble thinking of what might happen tomorrow, let alone years or decades down the line; some people can imagine only one possible future; other people, and I want to highlight science fiction writers as a case in point, can imagine all kinds of richly defined alternative futures often extending centuries, if not thousands or millions of years, into the future.

There are various benefits associated with developing our capacity for foresight. In imagining the future, we are asked to think hypothetically, to visualize possibilities, to repeatedly pose the question "What if?" Developing our visual foresight amplifies our powers to envision new and more complex goals - goals that reach further out into the future. Even if we simply engage in speculative visioning about the future, with no thought as to what practical or personal relevance such visioning entails, we nourish our imaginative powers. We expand the universe of our mind. Further, by enriching our minds with new possibilities and expanding the psychological space in which we think, we increase our mental and potential behavioral freedom. Exercising foresight builds the power of imagination and enhancing the power of imagination expands the human mind.

In using our imagination to envision and consider future possibilities, creativity often comes into play. Human creativity is the production of novel ideas, inventions, and behaviors. Although imagination is built upon memories of real perceptual experiences, imagination can be creative. We can combine together various elements of experience into novel images and scenarios. There are many ways to nourish creativity in people;[53] but exercising and developing foresight is definitely one of the best. Imagining possible futures is not completely bounded by the perceived constraints of the past and present. What is possible and impossible in the future? Imagining possible futures means breaking out of mental sets.

Another feature of future consciousness that is connected with imagination and creativity is **possibility thinking**. As noted above, envisioning the future is not limited to one possible reality. Although many people envision only one possibility when they imagine the future, perhaps because they believe in fate or fatalistic determinism, or have a great need for security, one can with minimal effort entertain all different types of possible futures. Are we headed toward an amazing new world or are we headed toward disaster? Will computers de-humanize us, or will they bring us together? Will we travel into space? Will we colonize the stars? Or will we remain earthbound? Will life disappoint us, or will we realize our dreams? We may be limited to one definite present reality or past, but when we turn our mind to the future, the universe unfurls into many possible trajectories. We deal with multiple and alternative possibilities, rather than singular, definite facts. Possibility thinking, which is only limited by our creative imagination, facilitates open-mindedness and mental flexibility.

In contemporary future studies literature possibility thinking is often referred to as **scenario building**. Through brainstorming and various collaborative activities, futurists attempt to create different detailed scenarios for the future and consider the implications of each scenario, good and bad.[54] Also, as noted above, science fiction provides a rich array of alternative possibilities for the future, and science fiction writers, like futurists, consider the positive and negative aspects of these different visions.

In considering possibility thinking and scenario building, it is important to note that other higher cognitive processes come into play besides imagination. As Reading points out, a key element behind the cognitive power of mental representations in humans is symbolic consciousness. Through the learning

32

of language we develop a symbolic system for representing and thinking about the world. This system allows us to form abstractions and conceptualize various complex relationships about reality. When we think about the future using mental representations we not only create images but we also formulate hypothetical descriptions and interpretations of the future using this symbolic system of language. We engage in internal dialogue and symbolically think about what we are envisioning in the future. We use abstractions, represented through symbolic language, in organizing and making sense out of the future.[55] Thinking about the future therefore involves, at the very least, a combination of imagination and symbolic consciousness.

The capacity for thinking is one of the key capacities of the human mind. Throughout history various definitions have been proposed regarding the nature of thinking. Thinking appears to be an internal or mental process involving sequences of images, symbolic representations, and most generally ideas. Thinking has been described as an internal dialogue and a form of information processing that goes on in the mind. Thinking seems to involve the use of abstractions and concepts. Through thinking we attempt to understand, to solve problems, to make decisions, and to plan. Our thinking processes can be either relatively creative or relatively habitual. There are different types of thinking, for example: Analysis, where we mentally divide something into its component parts; synthesis, where we combine together and organize a set of ideas; and reasoning, where we draw conclusions from premises and assumptions. Having looked at imagination, foresight, and symbolic consciousness, let us examine some other key dimensions of thinking that are involved in future consciousness.

Critical thinking is the principled evaluation of ideas and beliefs based on standards of reason. Logic, analysis, criticism, and self-reflection are all aspects of critical thinking. Although critical thinking can be applied to almost all aspects of human life, and is therefore not exclusively linked to future consciousness, it is a necessary component in the thoughtful and rational consideration of the future. Everyone engages in critical thinking – we all scrutinize and assess the validity and credibility of our ideas and the ideas of others - but as with other cognitive skills, there is great variability in how well the skill is practiced. Critical thinking is cognitive skill that can be improved[56] and improving critical thinking skills clearly facilitates the development of future consciousness. Improving critical thinking as it applies to the future enhances the rationality and realism of future consciousness.

When we think about different goals and options, to various degrees we assess the realistic probabilities of reaching different goals. We assess the pro's and con's of different goals and the risks involved in realizing our various objectives. To whatever degree we evaluate and compare different possibilities for the future, we are engaging in critical thinking. When we analyze and clarify our ideas on the future, when we self-critique, and when we consider the logic of our speculations – these are all examples of the role of critical thinking in future consciousness. Within futurist thinking, the distinction is often made between "possible, probable, and preferable futures."[57] After imagining possible futures, futurists would argue that we should evaluate and compare these different possibilities, considering which ones are most probable and which are most preferable. Such cognitive activities are examples of critical thinking.

Open-mindedness is a relative quality of thinking and is another significant cognitive dimension of future consciousness. Critical thinking and open-mindedness are connected processes and mutually support each other within future consciousness. Open-mindedness, in fact, is an essential element of critical thinking. From a critical thinking perspective, nothing is taken for granted. Ideas are not simply accepted as unequivocally true or dismissed as unequivocally false. Critical thinking, in fact, has been defined as the opposite of being closed-minded. Closed-mindedness is authoritarian, dogmatic, and egocentric – highly protective of any perceived threats to the legitimacy of a professed belief system.[58] Critical thinking necessarily involves comparing different points of view in order to judge their relative validity and consequently supports open-mindedness and works against close-mindedness. One cannot engage in critical thinking if one entertains only a single point of view.

Two other higher cognitive skills and forms of thinking that are components of future consciousness are **problem solving and decision making**. Again, although each of these thinking skills can be applied to areas of life that do not directly pertain to the future, both skills are often invoked in thinking about the future. Further, both skills can be improved, thus facilitating the development of future consciousness.

Humans engage in problem solving all the time. Problem solving is a form of thinking where some challenge, puzzle, question, or difficulty presents itself and a solution or answer needs to be identified. Problem solving has been studied extensively by psychologists, and although people often rely on critical thinking to solve problems, there are also strong components of insight and intuition that come into play in human problem solving. People show various degrees of creativity as well as closed-mindedness, bias, and habitual thinking in solving problems.[59]

As George Santayana said, "Life is not a spectacle or a feast; it is a predicament." When we turn our minds to the future, either at the personal level or the global level, we are confronted with a set of problems in need of solutions. How do we save enough money for retirement? How do we fulfill our life dreams and personal aspirations? How do we learn to live together peacefully without war and violence? How do we control and manage our resources and our environment in order to provide for everyone, yet without destroying and depleting nature? The arena of future consciousness is populated with problems, puzzles, and questions. We can become paralyzed and overpowered by the problems facing us, and thus mentally retreat from attempting to understand and solve these myriad problems, or we may attempt to tackle them, hoping to find reasonable solutions.

Problem solving is therefore an integral part of future consciousness because the future is filled with problems and challenges. If we face the future realistically, we confront problems, and if we attempt to develop a constructive attitude toward the future we need to work at the successful solution to innumerable problems. Expanding our problem solving abilities facilitates the development of a realistic and constructive mode of future consciousness.

Decision making is also an essential component of future consciousness. When we make decisions, we are making choices among various alternative goals and courses of action. As noted above, future consciousness opens up various possibilities for tomorrow, but we do not act on all these possibilities. Decision making is the process of selecting among different possibilities. As noted above, people apply, to various degrees, critical thinking in evaluating these different possibilities. But people also use intuition, hunches, and gut feelings in making decisions. Regardless of what methods are used, making decisions and acting on these choices, is clearly a future oriented skill. Conversely, the inability to make decisions – to stand immobile in the face of different opportunities and choices – is a deficiency in future consciousness. If I can not make choices and act on these choices, I cannot move into the future. People who suffer from depression often can neither make decisions nor act on them.

Both problem solving and decision making are selection processes. We are faced with different possibilities and need to identify the best solution – the best course of action. Imagine what the human mind would be like if we did not possess the abilities to solve problems or make decisions. We would stay stuck in previous habits of behavior and instinctual responses – a clear impoverishment of future consciousness, as well as a lack of self-initiative and creativity. Problem solving and decision making probably evolved in humans as ways to more intelligently and constructively deal with the future.[60] They are cognitive processes that address the conscious universe of possibilities.

Another major component in thinking about the future is **planning**. Goal setting often leads to planning – the identification of actions and steps needed to reach a goal. Planning is a higher cognitive skill involving conditional and hypothetical thinking. Conditional thinking involves "if...then" sequences of thought – if I do X, then what do I believe will follow? Hypothetical thinking involves considering possibilities rather than actual realities. In planning, possible actions and anticipated results are imagined and considered. Consequently, planning also involves linear thinking, the process of thinking through a series of ideas where one idea follows from the previous idea. Planning involves identifying a sequence of steps and anticipating the consequences of each step along the way. The execution of a plan is also a linear process, where actions proceed in a series of steps. A plan may involve only one step and its anticipated result, or many steps, with many sub-goals along the way. Planning and the execution of plans give human consciousness and human behavior a focused and linear direction.

Although planning and the execution of plans are both linear processes, people may develop plans by working forwards in their minds from their present state to the desired end state, working backwards from the desired state to their present state, or some combination of the two. Also, in the execution of plans, people may initiate a series of steps and at any place along the way, based on feedback from the environment, revise their plan, and go back to the beginning and modify their actions.[61]

As with goal setting and foresight, planning how to reach a goal is a cognitive skill that by definition is a form of thinking about the future. It involves the identification of steps necessary to achieve a desired end in the future. The identified steps in a plan all refer to future events. As with all the other cognitive processes identified above, people demonstrate various levels of competency at planning, but this ability can also be enhanced through practice, learning, and education. Clearly it is a skill that develops in people as they mature from infancy to adulthood – infants seem incapable of planning. Over the last few decades numerous educational programs have been developed to teach individuals and organizations "strategic planning."[62]

In general, the capacity to plan is a cognitive skill that is necessary in order for us to function in a complex world – most of the important goals of life require some degree of planning. Most behaviors in human life require sequences of actions with multiple steps and short and long-term goals. Imagine an individual who couldn't plan – his sphere of consciousness and action would appear reduced to the immediate present – to simple reactions to present stimuli.[63] There would be innumerable types of behaviors closed off to him or her. Hence, learning to plan is necessary in order to function and survive in our world – without this capacity we would not quite seem human.

Planning is a form of future consciousness that is of obvious value to human life, but the capacities for creativity and cognitive flexibility need to be brought into the picture to get a more balanced and accurate view of the nature and benefits of planning. Planning and the implementation of plans can vary in degree of rigidity versus flexibility.

Plans can be very precise and definite in the formulation of steps and goals. Sometimes it is important to tenaciously stick to a plan, where exactitude of implementation is critical or where personal tendencies to waver or give up need to be countered.

Yet a fundamental problem with rigid planning and implementation is that it does not acknowledge the unpredictability of the future. If everything goes as anticipated and is entirely predictable, there is no problem, but reality rarely behaves in a totally predictable fashion. There are often unintended or unanticipated consequences to the actions we take.

The uncertainty and surprises of life imply that rigid plans will often fail. Further, rigid planning and implementation can be seen as reflecting a need to excessively control life, or as indicative of a simple lack of creative imagination. Rigid planning and inflexible action often reflect habitual modes of thinking and behavior as a way to preserve the past rather than to create a future.[64]

At the opposite end, plans can be general and flexible. One can sketch out some main steps in the plan and avoid getting too specific. One can revise plans based on the contingencies of life. In fact, flexibility is often an essential part of planning - to consider various possible challenges and to be prepared for them with alternative strategies if the situation warrants it. Because there is uncertainty in life, staying open and flexible in planning is often critical to success. Creative imagination and thinking in planning and re-planning open possibilities to the mind and support the flexibility necessary in life. Flexible planning acknowledges the adventure of the future.

The value of articulating some level of specificity in planning should not be minimized or discounted though. Even if plans get repeatedly revised along the way, if plans aren't formulated, usually nothing much happens. Of course, there is chance and luck in life, but "chance favors the prepared mind." As the futurist Wendell Bell argues "Failing to plan is planning to fail."[65] Planning provides a cognitive sense of focus, direction, and emotional impetus. Planning for the future is proactive and purposeful, rather than reactive or passive. Plans turn a person into a creator of the future rather than a victim of it.

The philosophical study of knowledge will help to further clarify this point regarding flexibility versus specificity in the nature of planning. Within philosophical epistemology the two main theories of knowledge in Western intellectual history are **empiricism** and **rationalism**. Empiricism is the view that knowledge comes through observation, experience, or sense perception. Rationalism is the view that knowledge comes through reason and logic.[66] If one followed a rationalist approach to planning, one would think out all the steps ahead of time and then act. If one followed an empiricist approach, one would experiment and try out different possibilities, and through a process of trial and error move toward one's goal. Everyday human knowledge invariably involves some combination of the two – of reasoning and trial and error observation. When we engage in planning and goal directed behavior, we also show varying degrees of both reasoning and observation. If we were to think plans out, without any testing of our ideas, and then attempt to follow through on our plans regardless of consequences, we would be demonstrating extreme rigidity and the need to control. If we did not think out any plan of action and simply felt our way through life, we would be totally passive and reactive. Empiricism and rationalism need to be combined in planning and acting.

On one hand, life seems to be a combination of reasoning, valid foresight, and personal control, and on the other hand, accidental occurrences, trial and error, and surprises, both good and bad. Hence the future involves the dual elements of adventure and risk, and determinism and personal control. Because of this combination of uncertainty and control, planning for the future should involve a balance of focused and determined action and openness and flexibility. We are neither pawns to a capricious fate nor gods totally in control; we are somewhere in between. The key is a cognitive balance.

The concept of cognitive balance also applies to the dual processes of imagination and reasoning in thinking about the future. When we plan we articulate a series of steps in our mind – we engage in linear reasoning. When we imagine a future, we have a vision. Linear thinking is analytic – a distinctive series of steps is identified. When we have a vision, we often have the vision all at once – we have a holistic insight. Some would argue that visioning is actually more powerful and effective than linear analytical planning in approaching the future.[67] One of the powers shared by both religious mythology and science fiction is the creation of vivid and compelling visions of the future. But reasoning and imagination often support and enrich each other – we think about our images and intuitions, attempting to evaluate and clarify them and we often work out a plan that is inspired by a vision. The two processes of analytic reasoning and holistic visioning have distinctive features and distinctive strengths. Both cognitive processes are essential elements of futurist thinking; both processes contribute to the mental representations of the future that we create. As mentioned above, when people engage in problem solving and decision making about the future, they often use intuition and insight as much as critical and logical thinking.

Finally, in reviewing cognitive processes associated with future consciousness, the general capacity of **wisdom** should be included. Wisdom is the capacity to apply general knowledge gained in the past to challenging and novel situations in the present. Wisdom can also be defined as being able to grasp the big picture of reality and use this knowledge for the betterment of life. Wisdom is connected with numerous other psychological abilities. It integrates and utilizes the capacities for critical thinking, creativity, problem solving, and decision making and is connected with the virtues of courage and humility. The significance of wisdom regarding future consciousness is that "wisdom connects the heritage and lessons of the past with the thoughtfulness, openness, and creativity needed for the future. Wisdom involves an expansive synthesis of temporal consciousness and combats the excessive narrow presentism of today."[68]

As can be seen, future consciousness in humans is intimately connected with a variety of higher cognitive processes and skills. The power, complexity, and expansiveness of future consciousness are greatly amplified through the higher cognitive capacities of thinking, language, creativity, insight, planning, and wisdom. Although all these cognitive capacities are normal abilities in the mature adult mind, people show great variability in the execution of these skills and consequently variability in the level of development of future consciousness. We can enhance our capacity for future consciousness by developing these general cognitive skills, or working in the reciprocal direction, by consciously focusing on the future, through imagining multiple possibilities, thinking out new goals, critically evaluating probabilities, or planning courses of action, we can strengthen all these general cognitive capacities of the human mind.

Summary: Cognitive Processes
of Future Consciousness

Empirical
Observation
and Perception

The ability to perceive the
environment and understand
observable facts or patterns of facts.

Memory and
Learning

The acquisition of knowledge and
development of mental representations
based upon perception and interaction
with the environment.

Imagination

The ability to create mental images
and hypothetical "perceptual like"
realities in our minds.

Foresight

The ability to imagine or envision the
future.

Goal Setting

The ability to identify and
conceptualize goals of action.

Possibility
Thinking

The ability to imagine and
conceptualize multiple or alternative
hypothetical future realities.

Scenario
Building

The activity of imagining and
describing detailed, complex, and
realistic hypothetical future realities.

Language and
Symbolism

The ability to represent reality and
communicate through a symbolic
system of abstractions.

Thinking	A conscious mental activity of information processing and the creation and manipulation of ideas, often involving an internal dialogue.
Critical Thinking	The ability to apply principles of sound and valid reasoning to logical inference, the comparison and evaluation of different points of view, and the development and expression of theories and hypotheses; thinking about thinking—the opposite of egocentric thinking.
Open-Mindedness and Creativity	The ability to be flexible, to evaluate with fairness other points of view besides one's own view, and to be receptive to ideas that are different from standard beliefs. The production of novel ideas, inventions, and behaviors.
Problem Solving	A form of thinking where some solution or answer to a question, problem, or challenge is identified and successfully enacted.
Decision Making	The ability to make a choice among alternative goals and courses of action and follow through on the choice.
Planning	The ability to construct a hypothetical series of connected actions that lead to the realization of an identified goal.
Hypothetical Thinking	The ability to imagine, think about, and evaluate possibilities.

Holistic Insight   The ability to understand the "big
                      picture"—to see how the details of
                      a situation fit together—frequently
                      experienced in a rather sudden flash of
                      comprehension.

Wisdom   Integrative and expansive temporal
                      consciousness applied to present and
                      future challenges and problems

The Holistic Nature of Future Consciousness

Perception, emotion, motivation, purposeful behavior, learning, and higher cognitive abilities all contribute to future consciousness. Our consciousness of the future is multi-dimensional. Further, these psychological processes interact in creating and supporting future consciousness. The components of future consciousness form an interdependent and interactive whole. As one important example of interaction among different psychological processes involved in future consciousness, I have already described some of the ways that thinking in general, and thinking about the future in particular, both influences and is influenced by motivation and emotion.

Frederik Polak, in his classic study of the historical development of *The Image of the Future*, emphasizes the multi-dimensional quality of awareness of the future. As he states, many psychological elements contribute to thinking about the future, including reason, faith, emotion, intuition, and imagination. Further, the power and draw of images of the future created by various cultures throughout history is not entirely due to rational considerations. Aesthetic, emotional, and spiritual factors also contribute to the influence futurist images have on people's behavior.[69]

The holistic nature of future consciousness reflects the holistic nature of the human psyche. The fundamental dimensions of human psychology, including thought, emotion,

44

motivation, behavior, and self-identity, form a reciprocal system. Each capacity is a relatively distinct psychological reality, yet they are all clearly interdependent and form a unified whole.

The noted psychologist Albert Bandura, in his theory of **reciprocal determinism,** captures many of these psychological interdependencies.[70] He believes that the fundamental variables of human psychology – mind, behavior, and the environment – interact in a circular and mutually dependent fashion. Mind (through thoughts, emotions, and desires) initiates behavior, which in turn affects the environment, which in turn has an effect back on our thoughts and emotions through the perception of what happens in the environment when we act. Humans continually modify their thoughts, emotions, or motives as a consequence of observing the results of their actions in the environment. Because thoughts and emotions guide behavior, which in turn affects the environment, the mind has an influence on the environment, and reciprocally, the environment, through its effect on perception, has an effect back on the mind.

Reading adopts a similar model in describing the relationship between hope and action. Hope directs and energizes future oriented behavior toward the realization of some goal. Based on feedback from the environment, a person may modify his behavior, change his thinking, or even alter the goal. If the result of future oriented behavior leads to the realization of a future goal, or some intermediate step toward the goal, a person's sense of hope increases. If future oriented behavior does not lead to either the goal or a step toward the goal, hope and motivation decline. If efforts toward realizing a future goal are repeatedly frustrated and a person concludes that there is no realistic hope toward realizing the goal, depression and despair increase. Reading describes this whole process in terms of circular feedback loops, where mind, behavior, and environmental consequences impact each other, either raising or lowering a person's level of hope.[71]

At the most general level, the person and the environment form a reciprocal whole.[72] The person and the environment mutually affect and determine each other. When we engage the environment in attempting to create a positive future for ourselves, the future unfolds as a result of a reciprocal interaction between ourselves and the world around us. The actual results of our plans reflect the input of the environment as well as our states of mind and behavior. We are neither a passive victim of the environment nor totally in control of it. It is more accurate to say that we are participatory in the creation of our future and that the future will involve a co-evolution of both ourselves and the world around us. What happens in our lives is an interaction effect.

Since how we approach the future affects all aspects of the human mind, theories of the future should be evaluated regarding their overall effect on human psychology. For example, even if technological visions of the future promise increased economic productivity and intellectual augmentation, what impact would such a future have on human personality, human emotion, and social interaction? When we assess the various methods for envisioning and thinking about the future, such as science fiction, myth, religion, science, or reason, we should consider how different approaches engage the fundamental dimensions of the human mind. Because the whole human mind participates in the creation and experience of the future, any viable approach to the future needs to address all the fundamental psychological dimensions.

The Self and Future Consciousness

Self-identity – who we are – necessarily contains an element of future consciousness. People cognitively represent their personal identities as "narratives" or "stories." To use an expression of the neurophysiologist Antonio Damasio, we understand and describe ourselves in terms of an "autobiographical self." The object of self-consciousness is not a static thing; rather we are an ongoing story we tell ourselves. Interestingly, we are both the narrator and the main character in this story; we are both subject and object; the creator and the thing created.[73] The narrative of the self includes both the past and the future - of significant events and themes that summarize our life journey to the present, as well as hopes and aspirations for our future. The narrative of the self gives temporal coherence to our lives and our consciousness, connecting and relating past, present, and future.[74] We have a sense of how the main events in our life are related through the narrative of the self. How have we come to where we are and where is this ongoing journey heading in the future? As the creator of this narrative, we interpret and sculpture the meaning and substance of the journey and consequently who we are.

Reading highlights the important connection between the self and one's conception of the future. According to Reading, the narrative of the self, involving both past and future, gives consciousness sequential coherence. Without the narrative of the self, consciousness would be fragmented. In fact, he argues that true consciousness only arises in humans who can conceptualize and mentally represent a narrative self. Animals and human infants possess "sentience" but not true consciousness. Hence, for Reading, all true consciousness involves an ongoing sense of self-consciousness, and since self-consciousness is narrative in form, all consciousness involves an integrated consciousness of both past and future. Adding to Bernard Baars' model that consciousness is like a theatre, Reading contends that consciousness is more like an interactive TV, where the self is both what is being monitored as well as manipulating what is coming up on the screen. Further, because as humans we have a conscious sense of self, we also have a sense of choice and free will. We see ourselves as agents able to make choices among different possible futures. Consequently, the conscious experience of choice and free will ultimately depends upon future consciousness; the narrative self entails a sense of future consciousness and without a narrative self there is no sense of choice. Also, the idea of having choices clearly involves a sense of the future for choices refer to different possible actions in the future.[75]

One particular theme that is highly significant in our self-narratives is the relative strength of **optimism** and **pessimism** in how we view the story of our lives. The psychologist Martin Seligman, who has studied optimism and pessimism extensively, argues that the belief that one can positively affect the future is critical to optimistic thinking. Seligman defines optimism as a way of thinking involving the beliefs that misfortunes are relatively short-lived, limited in their effect, and due to external circumstances. Pessimists not only have negative images about the future, they believe that they cannot positively affect any change in what is to come. They believe that they are doomed to failure. They feel hopeless and helpless. Seligman defines pessimism as involving the beliefs that misfortunes have long-term and pervasive effects and are the fault of the individual. Following a cognitive theory of motivation and emotion, Seligman contends that the emotional state of depression is primarily due to pessimistic thinking. One other general point that he brings up is that both optimism and pessimism are self-fulfilling prophecies. Consequently, each mode of thinking gets reinforced since it tends to lead to the very results it anticipates. The particular attitude, whether it is optimism or pessimism, forms a reciprocal loop with the evolution of one's personal environment. Beliefs lead to behaviors that produce environmental effects which confirm and strengthen the beliefs. (Reading describes this self-reinforcing process as the "snowball" effect of hope and depression.) Seligman sees optimism and pessimism as "habits of thought" which obey the laws of reinforcement – they are reinforced through confirmation. In his mind, based upon a great deal of accumulated experimental evidence, these habits of thought can be changed through re-learning, education, and training.[76]

Optimism and pessimism come in degrees and the same person can possess elements of both attitudes in their assessment of themselves and the possibilities of their lives.

Polak argues that cultures show combinations of optimism and pessimism in their images of the future. Further, the degree to which a person or a culture exhibits an optimistic or pessimistic attitude depends upon fluctuating factors both within the person or culture and within the environment. A string of successes or failures can raise or lower the levels of optimism and pessimism. Still, as Seligman argues, optimism and pessimism depend upon certain inner habitual ways of thinking and interpreting reality; hence, a general optimist will interpret success or failure differently than a pessimist and react differently. Seligman and Reading argue that optimism and pessimism have both positive and negative values associated with each attitude; optimism energizes and is self-fulfilling, but pessimism attunes people to the possibilities of danger. Reading sees the need for balance and flexibility in optimism and pessimism within an individual.[77]

In this review of optimism and pessimism, we see again that beliefs about the self are an essential element in a person's attitude toward the future. Pessimists believe they are impotent and feel depressed about the future; optimists believe they have power to positively affect the future and feel hopeful about the future. It should also be noted that such beliefs and emotions pertaining to the self affect behavior. Optimists act as if they believe they are going to succeed; pessimists act as if they believe they are going to fail.

Bandura has studied the beliefs people have about their own "self-efficacy" and its affect upon behavior.[78] "Self-efficacy" is defined as the belief in one's ability to achieve one's goals. People show different degrees of "self-efficacy." A person with low self-efficacy believes he is relatively powerless with respect to the future, whereas a person with high self-efficacy believes he has a high level of control or influence on the future. High self-efficacy is the opposite of perceived helplessness. People with high self-efficacy set realistic goals and persist in achieving these goals. People with low self-efficacy set unrealistic or minimal goals and are very likely to give up as soon as challenges or difficulties arise.

Within many forms of psychotherapy, including cognitive therapy, the goal is to help people to see that there are alternatives to the negative future scenarios they foresee for themselves and that they have some power to change the direction of their lives. Opening the mind to the possibilities of tomorrow and raising one's perceived self-efficacy in realizing positive possibilities, in essence, is what psychotherapy is all about – it is a form of changing and expanding one's future consciousness. Psychotherapy often involves helping clients to set new goals, articulate plans, and monitor follow through on these plans, all forms of future consciousness. Psychotherapy also works toward helping people to think differently about themselves – to see themselves as more capable or open to change than they previously believed. Psychotherapy facilitates changes in a person's self-narrative.

Acting on the future proactively alters one's self-identity. The future is a challenge, necessarily involving an element of risk and uncertainty, and when people meet challenges rather than running from them, they increase their self-confidence and self-esteem. Expanded foresight, goal setting, planning, and goal-directed behavior give a person a sense of increased control and self-empowerment. Developing one's capacity to think about the future – to identify and seize opportunities and tackle challenges - improves one's self-image and self-confidence.

Howard Bloom expresses a popular sentiment that people need challenges in life in order to feel happy. He would also add the related factors of a sense of control over one's life and a determined commitment to goals as essential to happiness.[79] As psychological research has shown, stress in life seems to have more to do with feeling out of control than having too much to do. Hence, following this argument, psychological well-being depends upon a person's self-image and actions with respect to the future. Embracing the challenges of the future with a sense of self-efficacy and commitment to success generates happiness.

Another important connection between the self and future consciousness pertains to approach versus avoidance motivation. As Karniol and Ross report, within self-discrepancy theory it is hypothesized that individuals behave in accordance with two different types of visions of a future self. The first vision is referred to as the "ideal-self" which includes positive hopes about a future self that a person attempts to approach and realize, such as career and personal goals. The second type of future self is the "ought self" which includes future states that a person wants to avoid, such as unethical or criminal ways of life. The "ideal self" consists of "do's" and the "ought self" consists of "don'ts". This distinction corresponds roughly to Freud's idea of the ego-ideal (the "ideal self") and the conscience (the "ought self"). Karniol and Ross report that each type of future self seems to have different effects on motivation, performance, and behavior. Just as people can view the future as filled with positive outcomes or negative ones, people can see their future selves in a similar manner.[80]

The evolutionary biologist John Stewart has also examined the connection between future consciousness and self-identity.[81] Stewart, who comes at the issue of psychological development from an evolutionary rather than therapeutic perspective, suggests that individuals can become increasingly self-evolutionary. According to Stewart, since people are self-conscious, they possess the ability to assess and evaluate their beliefs, attitudes, and habits and consider to what degree their personal traits are of benefit to them for the future. As Stewart notes, many personal traits may have had value in the past but may not have value for the future. (For example, Walter Truett Anderson notes that at one time, having a set character was seen as a psychological strength, but with increasing change in our world a more flexible self - a multifaceted self - might be of more benefit.[82]) For Stewart, self-evolving individuals assess the future value of different possibilities of their own psychological make-up and attempt to direct their psychological growth toward qualities that will serve them best in the future. In essence, Stewart is suggesting that people can and should increasingly apply possibility thinking to their own selves and evaluate which possibilities seem most desirable for the future - the identification of "preferable selves" - and attempt to modify their personal identities accordingly.

Stewart argues that throughout natural history life has evolved an increasing capacity to adapt to time frames farther and farther out into the future. In general, he believes that evolution moves from adapting to local and immediate concerns (the here and now) to adapting to more expansive spatial and temporal parameters. He sees the most primitive life forms and states of consciousness as present focused, whereas human evolution and the rise of modern civilization have generated longer and wider spheres of temporal consciousness. Throughout evolution and history humans see increasingly farther out in space and time.[83] Interestingly, this same evolutionary trend is mirrored in the psychological development of individual humans. Infants are described as exceedingly egocentric, functioning primarily in the immediate here and now; as we mature we broaden our perspective in both space and time.[84] Stewart sees the future evolution of the self as following a similar pattern of increasingly expansive perspectives - the self will take longer and longer views of its own development and evolution.

People undoubtedly show significant variation in terms of the extent of their temporal and spatial consciousness. But as I have argued, all normally functioning human adults are conscious of the future to some degree and how they see the future impacts their psychological states in many ways. Following Stewart and the ideas of many futurists, expanding the range of our temporal consciousness regarding our own self-identity into the longer-term future has many beneficial effects.

As Reading notes, people search for both an explanation of their existence and a purpose for their life.[85] In essence, we want to connect our identity to a past and to a future. Our lives and our personal identities will acquire greater coherence and meaning the more we see ourselves in the context of greater expanses of time, both past and future. A self that focuses on relatively immediate needs and short-term results is correspondingly a more fragmented identity – a series of relatively disconnected episodes and states of consciousness without any overall direction or integration. In fact, it is one of the great struggles in life to keep ourselves focused on our goals and fundamental priorities without being distracted or lost within the chaos of the present. Taking a long term view of personal identity and our lives and working toward the realization of this vision brings greater coherence to the self.

The Integrative Dimension of Future Consciousness

There is an integrative dimension of future consciousness. Raising our awareness of the future teaches us that nothing exists in isolation, and that we must learn to look at the whole and not just some individual part or slice of reality. In the academic and professional worlds, a particular topic or discipline is usually studied and practiced in isolation of other aspects of life, such as in biology, business, medicine, law, psychology, or economics. The academic and professional worlds breed specialization.[86] Yet in life all these different areas interact with one other. Changes in human psychology and mental health affect economics and vice versa. Biological and medical advances affect society and social growth. Just as the human mind involves the reciprocal interaction of many psychological processes, our world is a vast reciprocal network of institutions, cultures and belief systems, and social and technological forces. Our future world will be the result of many interactive variables. In imagining the future, we are stimulated to different degrees into both **integrative** and **interactive thinking**.

Science fiction, which creates complex and realistic scenarios about the future, is one avenue for understanding the interactive quality of the evolution of the future. A good science fiction novel must discuss the environment, social institutions, belief systems and transformed cultures, as well as advances in technology and science and consider how all these variables interact and intermesh to produce a future reality.

Yet to move in the opposite direction in time, one of the most dramatic and effective ways to see the interactive quality of all the dimensions of human life is through the study of history. History, in fact, has much to offer in understanding the future and when people really begin to ponder the possibilities of the future they almost always begin to think more about history. As noted earlier, future and past consciousness are interconnected. Patterns and trends, which are a basis for making predictions about the future, are revealed through history. The full scope of human existence is revealed through the reading of general histories of humankind. History expands a person's temporal consciousness. History reveals both relative persistence and change – it reveals temporal patterns. Finally, history provides innumerable examples of the causative relationships across time – how events trigger other events and so forth. But in particular, because history expands our temporal awareness and reveals the richness and scope of change through time, we begin to see more clearly how the different elements of human life actually are interdependent. We find that varied and multiple aspects of human life have changed across time and we see that the changes are due to the effects all components have on each other. History shows a dynamic interactive tapestry – a tapestry that we can take as one central source of information for thinking about the future.[87]

Though thinking about the future expands our consciousness, it also brings the abstract down to earth; it personalizes our world. If science, technology, business, or world ecology changes, what impact will such transformations have on our lives? We often fail to notice in the present how big events set the tone of our existence. Yet if we were to alter the fundamental parameters of our civilization, it would be much more apparent how we are connected to the grander scheme of things. Imagine a world transformed and then imagine oneself in it. Significant worldwide developments in the future, such as the pervasive application of genetic technology to medicine, reproduction, and food production will impact our personal lives in numerous ways.[88] Thinking about how the world may change in the future connects us to our world.

Expanding our future consciousness stimulates us into considering how changes in our world are going to impact us and what we need to do to take advantage of these changes. We may be motivated to become more flexible and creative.[89] If one follows the evolutionary logic of Stewart, ever-increasing flexibility has been a fundamental direction in life throughout the history of evolution. Stewart believes that the more we extend our perspective out into the future and the more we self-evolve in anticipation of the future, the better our chances for survival as a species.[90] As change accelerates, this "evolutionary imperative" will become increasingly necessary in the years and decades ahead.

Ethics, Values, and Virtues

Future consciousness is intimately connected with values and ethical thinking. As Wendell Bell has argued, part of futurist thinking is ethical in nature. When we imagine a better world for tomorrow, we are moved to consider and define what is "good." What is the good life? What makes a good society? These questions are ethical in nature. When we consider alternative possibilities, each of which could be realized depending on our actions today, we are considering different choices for tomorrow. And when we ponder which choice to make, the basis of our decision making is values, and in particular, ethical values. Not all human values are ethical or moral - values are ideals that could apply to any aspect of life. Ethics pertains to those values that are concerned with human conduct, human character, and human affairs in general. Still when we consider preferable futures, ethics invariably is part of the picture, and hence thinking about preferable futures is a form of ethical thinking, for what is preferable regarding human behavior and human life is defined in terms of some set of ethical values.

The future is a testing ground for our values. Through thinking about the future, we might choose to revise our values. We might decide that if we were to pursue some present direction based on some value, e.g., "it's good to have many children," that the future consequences would be quite undesirable. We may need to rethink the realistic and future implications of our values.[91] Can we imagine societies or individual lives built on different ethical principles? Would we find them desirable or realistic? Within the context of the future, we can discuss the present and our contemporary values. We can envision ideal future worlds and compare them to our present world, observing where we fall short of such ideals and how we could go about improving our world.

In the most general sense all ethical decision making is a form of future consciousness. Ethics provides principles to guide us in making choices. The capacity to see that we have choices entails that we are aware of different possibilities in the future and that we can select which of these possibilities is most desirable.[92] When we make decisions and choices based upon our ethics and values, we are considering what course of action to take in the future. Also when we think ethically, we consider consequences and consequences are anticipated events in the future. How will one's actions impact the future? As noted earlier, all purposeful behavior involves future consciousness, and behavior motivated by ethics is a form of purposeful behavior. One of the most influential theories of ethics in Western philosophy is the Consequentialist Theory, developed by Jeremy Bentham and John Stuart Mill, which defines the ethical value of actions in terms of the future benefit and costs of actions.[93] From this philosophical viewpoint, ethics unequivocally pertains to the future.

Ethics also comes into future consciousness when we consider future human generations. As I noted earlier, the ongoing evolution of future consciousness is part of a more general process of expanding consciousness from egocentricity and the here and now to ever widening spheres of awareness in space and time. The farther out we think in time, the more other humans come into the picture. What responsibilities do we have toward our children and their lives tomorrow? These are ethical questions. We need to consider possible future developments in society, business, and science in order to help them become better prepared as adults. In fact, whenever we think about how to raise our children we are obviously thinking about the future, as well as thinking about ethics.

What about the future generations of all humanity? Should we consider the future consequences of our present actions?[94] Are we not ethically responsible to all of our descendants? For example, should we be squandering our present resources and robbing future humans of the same opportunities for a good life that we have enjoyed?[95]

As one final point on ethics and future consciousness, I would argue that the evolution of future consciousness should be guided by a set of **virtues**, which in effect are ethical character traits.[96] In presenting this argument I am following the futurist idea, clearly articulated by Wendell Bell, that "preferable futures" is a critical dimension of futurist thinking.[97] Evolving our future consciousness must be guided by values and not simply predictions and probabilities. But I am also arguing that when we think about preferable futures, we need to particularly focus on ourselves. How should we guide ourselves into the future? What should we aspire toward in the future? Virtues define ideal or preferable character traits within the individual. Connecting back to the previous discussion on the self, and in particular, the "ideal self", virtues provide the ethical direction for the evolution of the self.

Following Seligman's work on key character strengths and how such traits are connected to human happiness, I would propose that our contemporary psychological and social reality can be significantly improved through a focused exercise and development of a core set of character virtues.[98] The idea that the "good life" can be achieved through the internalization of character virtues goes back at least as far as Aristotle. For Aristotle, a life of virtue not only creates happiness in the individual but equally contributes to the well being of the community.[99] Virtues are not simply self-centered or self-serving. Virtues are connected with values, in that a virtue is a value lived and internalized into the character of a person. If truth is a value, honesty and forthrightness are the corresponding virtues.

Some of the most significant and central virtues that will both improve our present psycho-social reality and expand our future consciousness are courage, self-responsibility, the love of wisdom and thinking, and evolutionary transcendence. Courage is necessary for dealing with uncertainty and change; self-responsibility supports self-initiative; the love of wisdom and thinking opens and enriches our minds, and as noted earlier, wisdom provides a creative guidance system for good choices in the face of novelty and complexity; and evolutionary transcendence provides us with optimism and hope for tomorrow and pulls us out of egocentric thinking and concerns.[100]

Virtues provide a value-structured way to approach the future. Instead of simply asking what might be, we ask what should be. Instead of simply looking at external forces that may shape the future, we look at ourselves and ask how we can help shape the future. Instead of simply considering an ideal future scenario, we focus on what might be an ideal future self. We can not have a better world, unless we have better human beings.

Philosophy, Cosmic Consciousness, and the Future

The future is the most cosmic, mind-expanding, and philosophically enlightening topic the human mind can entertain. Will we travel into space and find new and strange worlds that possess life and intelligence? Will we evolve machines that are smarter and wiser than we are? What great wonders lie beyond the horizon of tomorrow? Will we achieve a social utopia? How might humanity, biologically or psychologically, be transformed? Will we transcend our present biological bodies? What new revelations and achievements, technologically, scientifically, and even spiritually will emerge? Is humanity a stepping-stone on the journey of life and mind within the cosmos? What unbelievable realities will evolve in the universe and will we participate in their creation? The growth of future consciousness facilitates the growth of cosmic consciousness.

What we think is impossible today may become possible tomorrow. As H.G. Wells stated we may be only at "the beginning of the beginning." We may be only just awakening. What, indeed, can we say is impossible forever and always? Throughout history many of humankind's most cherished and deepest beliefs have been contradicted and transformed. What are the boundaries and the potentialities of reality? What are the limits to life, intelligence, technology, and civilization – to truth, beauty, and the good? And what may lie beyond these human categories of existence? The vast reaches and mysteries of the future seem to extend without end, challenging our intellect and expanding our consciousness and imagination. The future may be a reality of infinite mental space.

The growth of future consciousness expands one's philosophical mindset. Many of the classic philosophical questions, concepts, and themes are being reconsidered and debated within the context of futurist thinking. The great mysteries of life are taking on new meanings as we consider the far-reaching possibilities of tomorrow. Biotechnology and cyborgization are leading to redefinitions of life and the nature of being human. Artificial intelligence and computer technology are stimulating new theories and ideas regarding the nature of consciousness and mind. Given the incredible possibilities of virtual reality and the growing infusion of electronic simulations into all aspects of human life, what are we to think about the nature and possibilities of reality? Discussions of globalization and cultural pluralism are stimulating new ideas in ethical and political theory. Immortality, a topic traditionally reserved for religion, is another issue seriously being considered in futurist thought.[101]

Spirituality and theology also get re-examined in the context of the future. The existence of God has been addressed and re-conceptualized in the far-reaching speculations of futurists and cosmologists such as Frank Tipler.[102] As the great futurist science fiction writer and philosopher Olaf Stapledon clearly saw, the future *can* be seen as a great cosmic or spiritual journey.[103] The future may have deep spiritual and metaphysical significance. The future is not simply an issue of technological and scientific possibilities but, perhaps more so, an issue of the possibilities of mind, spirit, and reality.

Although philosophical or spiritual issues could be seen as having a timeless or eternal quality, the great questions of humanity are, to a great degree, time-bound. Philosophy and theology have evolved through history, and there is no reason to think that they will not evolve further in the future. The big questions and the big answers get redefined through time. Sometimes the big questions get answered, as is the case with science often developing ways to address questions that previously seemed beyond human understanding. A dramatic case in point is how recent advances in physical cosmology are beginning to answer the question of the origin of the universe, or simply put, "Why is there something rather than nothing?"[104] Besides old questions being answered, undoubtedly new mysteries and ideas will emerge, contingent upon the challenges and perplexities of the world of tomorrow. In thinking about the future, we can at times get a glimpse of what these new insights and issues might be.

When we think about the future we engage in "possibility thinking" - we stretch and expand the universe of ideas, vistas, and realities. We entertain the unthinkable and make it thinkable. All philosophical theories and spiritual perspectives are founded upon certain beliefs and assumptions regarding the nature of reality, human existence, and human knowledge. Yet possibility thinking stimulated by the simple question "What if?" opens the human mind to considering alternative visions of reality and existence. One clear case of such mind-expanding thinking concerns the emergence of a world-wide collective mind and consciousness as technology wires and connects humanity together into some kind of global brain.[105] In such a "possibility universe" what would be the meaning of the individual human self? Would there be some new kind of self woven into the personal identities of humans?[106] Would there be such a thing as a collective or global consciousness? How could this be? What would the idea of individual freedom mean? What would the idea of a society mean?

The Contemporary Transformation and the Challenge to Future Consciousness

"The central question of our time is
what to do about the future.
And that question creates a deep divide."

Virginia Postrel

Turning from the cosmic and the spiritual to the other end of the philosophical continuum, future consciousness is intimately tied to the concrete and pragmatic concerns of life. The future is the most important and pressing practical issue of our time. It is of great importance both for humanity as a whole, as well as for each of us individually.

As innumerable writers have pointed out, and almost everyone can see by observing the world around them, we live in a period of significant, if not monumental, change. Socially, technologically, and psychologically, humanity is being transformed. The contemporary transformation of humanity is pervasive, fundamental, and multifaceted. We are in the midst of a "paradigm shift" for all of human civilization.[107] As various futurist authors have stated, we live at a "turning point," a time of "future shock" and "creative chaos."[108] We live in turbulence. Our universe is bubbling and we are in the pot.

We see around us multiple trajectories of change – going forwards, backwards, and into the unknown - at times reinforcing and amplifying one other, at times in violent opposition to each other. Furthermore, these various changes are open to different interpretations; what some writers and social commentators see as a progressive change, other writers see as regressive and vice versa. To name just some of the most significant trends and developments, we could include globalization; ever increasing world population and the emergence of giant mega-cities around the world; international terrorism; the accelerative growth and ubiquitous spread of technology; the transformation of the natural environment; potential mass extinctions; the resurgence of tribalism; ever-increasing consumerism and consumption; the feminist movement and other human rights movements; the rise of religious fundamentalism; computerization, post-industrialism, and the new information economy; and the great postmodernist disillusionment with modernity and the Western vision of progress.[109]

Not only do we live in an age of fundamental change, to coin an expression, we live in the "Age of Velocity." History reveals change, and recent history seems to demonstrate that change is accelerating.[110] As Walter Anderson, James Gleick, and other writers on contemporary trends have pointed out, the exponential curve of growth is one of the defining symbols of our time - everything seems to be speeding up.[111] According to Gleick, time is compressing - more and more is happening in a day - in a week - in a year. One could say that the future is coming at us more rapidly than ever before - the flow of the river of time is speeding up. As Milan Kundera observes, "Speed is the form of ecstasy the technical revolution has bestowed on man."[112]

Because of the complexity, ambiguity, speed, and pervasiveness of these different changes in our world, there is uncertainty, fear, and conflict. The uncertainty of our times, and consequently the uncertainty of the future, is amplified by the diverse set of perspectives and theories offered to explain the contemporary transformation. There are competing views and philosophies of what is happening around us, what is right and what is wrong about these changes, where we are heading, and where we should be heading.

We are being bombarded with information, choices, belief systems, and possibilities. Many believe we are going through an epochal re-organization and evolution of human civilization, unparalleled since the time of the Industrial Revolution. Some believe modern Western civilization is crumbling.[113] Some see advancing technology as a boom, ready to finally realize humanity's oldest dreams of a paradise on earth;[114] some see human technology as a scourge on the earth, about to collapse under its own hubris and wastefulness and perhaps bring the whole earthly ecosystem down with it.[115] There is the clash, reflective of an ongoing disagreement running back through human history, over stability and conservatism versus change and innovation. Should we anchor ourselves to the past and to tradition, or should we embrace what is new and different?[116] All these perspectives vie for our attention and allegiance. There is, in fact, an ongoing struggle over the future, as Postrel notes, embodied in these various philosophies and perspectives, occurring both in the public arena and in our individual minds.[117] The future is undecided, confusing, and open to debate. Humanity is in a battle over the future. Our minds are in a battle over what to believe and what to do.

Because of the great turmoil and historical significance associated with our age, it is absolutely necessary to bring some mental order and understanding to the present chaos and complexity of our times, to see the main pieces of the puzzle and the pattern of it all, to evaluate the different viewpoints, theories, and possibilities, and to gain some sense of direction for tomorrow. In particular, given the diverse views regarding the future, it is important to find a balance between openness to different ideas and principled standards for evaluating these various ideas. It is essential to develop an informed, proactive, comprehensive, and integrative perspective on what is happening around us and where all these changes may lead. From the previous review of psychological dimensions of future consciousness, it is clear that we need to develop our critical thinking capacities, our decision and problem solving abilities, our sense of realistic optimism and self-efficacy, an evolving sense of self, and various virtues and character strengths to flourish and achieve direction in our complex, fast paced world. Perhaps most of all we need wisdom.

Yet we face a challenge. According to many writers, our conscious sense of the future is narrowing and weakening. We are becoming lost and forlorn in an overpowering present. Howard Didsbury believes that our increasing need for immediate gratification, supported by modern technology and its conveniences, is diminishing our sense of the future and the importance we place on it.[118] Stephen Bertman, coining the term "hyperculture," contends that the fast-paced modern world is destroying both historical consciousness, and our sense of the future. We quickly forget the past and don't have time to think about tomorrow.[119] As James Gleick says, "We live in the buzz."[120] Stewart Brand goes so far as to state that "Civilization is revving itself into a pathological short attention span."[121] The historian Robert Nisbet argues that the contemporary "Cult of the Present" – an intentional focusing on the now as all that really matters – is destroying both the past and the future.[122]

In the opening section, I argued that humans would become disoriented and dysfunctional without future consciousness. Given the increasing complexity of our modern world and the accelerating pace of change, it would appear highly maladaptive and paradoxical for humans to be losing their sense of the future. Given the momentous changes occurring around us, future consciousness is something that should be expanding rather than shrinking if we are to flourish or even survive as a species in the future.

What seems paradoxical about this presumed deterioration of both a sense of the future and the past is that it runs against evolutionary and historical trends. Future consciousness has evolved throughout the history of life. As a general evolutionary trend, which seems mirrored in the cognitive development of children (ontogeny recapitulating phylogeny), awareness and adaptive functioning becomes increasingly less egocentric and more expansive in both space and time throughout the history of life. There is a developmental movement in evolution from the "here and now" to the "there and then." The "mind's eye" sees out further and further.

In particular, awareness of the future has dramatically increased with the emergence of humans and our very large brains that are capable of both enhanced memory and foresight. Further, with the advent of human civilization and written culture, human awareness of time, both of the past and the future, went through a significant jump forward.[123] We began to keep written records of the past, accumulating knowledge to plan better for the future. Creating another great step forward, with the dawn of science our knowledge of the vast extent of time and "the life of the cosmos"[124] has grown immensely and we understand the intricacies and details of our human history and the history of the earth better than ever before.

Hence, although there are various arguments and social and psychological indications that future (and past) consciousness is regressing in contemporary times, there are contrary and more general indications that our consciousness of both space and time has been steadily growing throughout history. So which is it? Is future consciousness in humanity expanding or diminishing? The answer is probably both.

Consider, as writers such as Reading and Christopher Lasch in his *Culture of Narcissism* point out, that in times of uncertainty or pessimism over the future, people tend to focus more on the present and short term pleasures. Also people become more self-centered and cynical and retreat to the past or into the occult. If Postmodernism destroyed the credibility of the Western vision of progress, as many have argued, then the "morale" (or social equivalent of hope) in the general population has deteriorated; at least in the West, there is no common purpose or sense of direction. There are too many different possibilities, both good and bad; there is no single unifying credible image of the future. Also, growing up and living in a period of flux and uncertainty, without a sense of trust and stability, seems to retard the development of future consciousness.[125] Finally, as Polak points out, if a culture does not regularly renew and revitalize its image of the future, its population will regress to focusing more on the present. Hence, those very social factors that make it imperative that we evolve our future consciousness are contributing to its loss as well as to a loss of optimism.

While our general knowledge of past and future keeps accumulating through advances in science, history, and other disciplines, we may not always be using this knowledge in our everyday lives. In spite of its fast pace, we may have entered a depressive or disorganized period in human history, having lost our hope and vision in a positive future. There are clearly many forward looking organizations and social movements around the world, but perhaps the general population, in modernized countries, is caught in the madness of the present, while those in undeveloped and poor countries struggle with basic survival needs and the impoverishment of their uprooted lives. All in all, we are caught between time and Timbuktu; our horizons keep expanding, yet we feel increasingly lost amidst the complexity of it all. To whatever degree and in whatever ways our temporal consciousness is narrowing we need to understand why, and find ways to turn the process around and expand our minds outward into time and the future.

It is my belief that in the long run future consciousness in humans will continue to evolve. In agreement with Stewart and other writers, I think evolution favors a continued expansion in temporal consciousness. If future consciousness in humans doesn't grow we are doomed. Perhaps then a superior type of mind will pick up the gauntlet.[126] There are, though, as noted above, contemporary trends and social movements showing a heightened, rather than regressive sense of the future.[127] Perhaps what we need (and indeed what will happen) is a pervasive and global "paradigm shift" of the first order that will emerge out of some type of synthesis of these forward looking movements.

From a practical standpoint, one thing is certain. The future is the only reality that we can actually do anything about. Though there is much to be learned about the past that clearly is of great benefit to us, until time travel comes along, we can do nothing about the past. The past is both set and gone. Though it is often said that all we have is the present, the present, at best, is transitory and perpetually flowing into the future. The future is a vast arena of possibilities – the present is here and gone. Unless one is fatalistic and believes in pre-determination the future is the only arena of existence over which we have any practical influence or control. To whatever degree we can guide the future in a constructive and informed way we will benefit ourselves, and all others, who are affected by our actions.

We all can become better informed on the topic of the future. Becoming better informed on the future involves learning about present trends and future possibilities and the main theories, challenges, and controversies regarding the future.

We all can improve our thinking capacities about the future. Enhancing one's thinking capacities about the future involves improving one's abilities to imagine new possibilities for tomorrow, to more thoughtfully compare and evaluate different possibilities and alternative theories of the future, to synthesize information and ideas on the future, and to more clearly define values and preferred directions for tomorrow. We can expand our capacities for wisdom and integrative thought. These various capacities need to be applied both to the world around us and self-reflectively to ourselves

We all can improve our abilities to constructively guide and create the future. Improving one's ability to create and guide the future means becoming better at applying futurist understanding and thinking to the active and often inventive process of participating in the development of the future. It means connecting thought with behavior – theory with application. Constructively influencing the future also involves motivational, attitudinal, and emotional factors, for we need to be inspired, optimistic, and visionary if we are to successfully direct the evolution of our lives.

Summary and Conclusion

In summary and conclusion, it would be helpful to list the various benefits connected with the development of future consciousness and some of the different ways to further develop future consciousness that I have described within this chapter.

Benefits of Enhancing Future Consciousness

- Improves imagination, creativity, and flexibility.
- Fosters mental health.
- Improves higher-order cognitive abilities, especially planning, problem solving, critical thinking skills, and integrative understanding.
- Raises self-consciousness.
- Expands mental and behavioral freedom.
- Expands consciousness.
- Can work against depression, fear, apathy, and perceived helplessness.
- Gives meaning, purpose, and hope to life.
- Brings greater self-control over one's life.
- Brings greater coherence to the self.
- Is highly adaptive, especially in a world of rapid change – maximizes the chances of survival and thriving in the future.
- Facilitates the development of courage and wisdom.
- Can improve ethical thinking and character.
- Expands philosophical understanding and cosmic consciousness.

Ways to Develop Future Consciousness

- Challenge existing habitual beliefs about the future.
- Brainstorm on alternative visions and beliefs about the future.
- Become familiar with many diverse visions of the future, both from the sciences and the humanities.

- Challenge existing habitual beliefs about one's ability to influence the future.
- Clarify and assess your life plans and goals and imaginatively and critically consider alternative possibilities.
- Clarify and assess your self-narrative and imaginatively and critically consider alternative views of who you are, what you can accomplish, and where you are heading.
- Learn about history and especially long term trends that are continuing in the present.
- Learn to tolerate better, if not appreciate, the uncertainties and adventure of life - be willing to take calculated risks at times - don't be ruled by security and safety.
- Learn the psychological practices and techniques for enhancing optimism.
- Learn the psychological practices and techniques for enhancing thinking skills, visualization and imagination, and creativity.

In conclusion, future consciousness involves a fundamental set of normal psychological abilities and attitudes. Throughout our history, future consciousness has evolved, and there are many ways to facilitate its further development. It is of great value, personally and collectively, to pursue this continued evolution.

Whether one is abstract or concrete, ethereal or pragmatic, cosmic or personally focused in attitude and inclination, future consciousness energizes, enriches, and benefits the total human mind. The future is both a very practical concern and a mind-expanding cosmic adventure in creativity and imagination.

Future consciousness includes visual foresight, goal setting, planning, critical thinking, decision making, problem solving, ethics, and character virtues, and, most broadly, the multi-faceted capacity of wisdom. Future consciousness is also connected to the emotional, motivational, and personal dimensions of the human mind. Our psychological health – our sense of hope, optimism, adventure, and self-efficacy – is intimately connected with future consciousness. Future consciousness is holistic and impacts all aspects of human psychology.

The study of the future fosters intellectual synthesis and the development of higher cognitive processes; it helps us think about and understand the grand and interactive scheme of things. Future consciousness connects us to our world and all of humanity, present and future generations included, both cognitively and ethically. Future consciousness transforms us philosophically and spiritually. Given the scope and speed of change in our contemporary world, it is absolutely necessary that we continue to develop our future consciousness.

There are many different reasons why we should further evolve our future consciousness. It is an issue of survival, of mental health, and of transformation and transcendence.

References

[1] Damasio, Antonio *The Feeling of What Happens: Body and Emotion in the Making of Consciousness*. Orlando, Florida: Harcourt Brace, 1999.

[2] Reading, *Anthony Hope and Despair: How Perceptions of the Future Shape Human Behavior*. Baltimore, Maryland: The John Hopkins University Press, 2004, Pages 6-7, 9-10, 109-115, and 120.

[3] Gibson, James J. *The Ecological Approach to Visual Perception.* Boston: Houghton Mifflin, 1979; Lombardo, Thomas *The Reciprocity of Perceiver and Environment: The Evolution of James J. Gibson's Ecological Psychology*. Hillsdale, NJ: Lawrence Erlbaum Associates, 1987, Pages 301-302.

[4] Although Reading argues that perception does not produce an awareness of time, he does acknowledge that the experience of passage and change requires the experience of continuance or persistence. See Reading, Anthony, 2004, Page 120.

[5] Cornish, Edward "How We Can Anticipate the Future" *The Futurist*, July-August, 2001

[6] Reading, Anthony, 2004, Pages 40 – 41, 119.

[7] Morris, Richard *Time's Arrows: Scientific Attitudes Toward Time*. New York: Touchstone, 1986.

[8] Carlson, Neil *Physiology of Behavior*. 3rd Edition. Boston: Allyn and Bacon, Inc., 1986, Chapters 6 and 7.

[9] Reading would argue on this point that it is short-term or working memory that provides the mental "glue" that unites together the perceptual snapshots into a continual and flowing experience. Yet, his argument assumes that perception is a set of instantaneous snapshots, but perception is grounded in relationships rather than instantaneous and absolute values. See Reading, Anthony, 2004, Page 57.

[10] Johnson, Marcia and Sherman, Steven "Constructing and Reconstructing the Past and the Future in the Present" in Higgins, E.T. and Sorrentino, R. M. (Ed.) *Motivation and Cognition: Foundations of Social Behavior* Vol. II. New York: Guilford Press, 1990.

[11] Reading, Anthony, 2004, Pages 62 – 63.

[12] Morris, Richard, 1986.

[13] Wade, Carole, and Tarvis, Carol *Psychology*, 7th Edition. Upper Saddle River, NJ: Prentice Hall, 2003, Chapters Four and Eleven.

[14] Reading, Anthony, 2004, Page 83.

[15] Seligman, Martin *Authentic Happiness: Using the New Positive Psychology to Realize Your Potential for Lasting Fulfillment*. New York: The Free Press, 2002; Damasio, Antonio *Looking for Spinoza: Joy, Sorrow, and the Feeling Brain*. Orlando, Florida: Harcourt, Inc., 2003.

[16] Reading, Anthony, 2004, Page 76; Karniol, Rachel and Ross, Michael "The Motivational Impact of Temporal Focus: Thinking About the Future and the Past" *Annual Review of Psychology*, Vol. 47, 1996.

[17] Reading, Anthony, 2004, Pages 44-45, 74.

[18] Reading, Anthony, 2004, Pages 3-13, 150-160.

[19] The capacity for purposeful behavior directed toward anticipated future goals can be significantly impaired in brain damaged individuals. See Damasio, Antonio, 1999.

[20] Karniol, Rachel and Ross, 1996, Pages 1, 6-7.

[21] Karniol, Rachel and Ross, 1996, Page 1-2.

[22] Karniol, Rachel and Ross, 1996, Page 5; Polak, Frederik *The Image of the Future*. Abridged Edition by Elise Boulding. Amsterdam: Elsevier Scientific Publishing Company, 1973, Pages 9-10; Solomon, Robert *The Big Questions: A Short Introduction to Philosophy*. 6th Ed. Orlando, Florida: Harcourt College Publishers, 2002, Chapter 8; Wilson, E.O. "The Biological Basis of Morality" *The Atlantic Monthly*, April, 1998; Shermer, Michael *The Science of Good and Evil*. New York: Times Books, 2004; Jones, Dan, "The Moral Maze" *New Scientist*. November 26 - December 2, 2005, Pages 34 - 37.

[23] Karniol, Rachel and Ross, 1996, Page 2.

[24] Karniol, Rachel and Ross, 1996, Page 3.

[25] Ellis, Albert and Harper, Robert *A New Guide to Rational Living*. North Hollywood, CA: Wilshire Book Company, 1976; Seligman, Martin *Learned Optimism: How to Change Your Mind and Your Life*. New York: Pocket Books, 1998; Wade, Carole, and Tavris, Carol, 2003, Chapter Eleven.

[26] Seligman, Martin, 1998.

[27] Hergenhahn, B.R. and Olson, Matthew *An Introduction to Theories of Personality*. 6th Edition. Upper Saddle River, NJ: Prentice Hall, 2003. Especially see Chapter Eleven.

[28] Seligman, Martin, 1998.

[29] Seligman, Martin, 2002, Chapter Three.

[30] Zey, Michael G. *The Future Factor: The Five Forces Transforming Our Lives and Shaping Human Destiny*. New York: McGraw-Hill, 2000; Brand, Stewart *The Clock of the Long Now: Time and Responsibility*. New York: Basic Books, 1999.

[31] Best, Steven and Kellner, Douglas *The Postmodern Turn*. New York: The Guilford Press, 1997.

[32] Nisbet, Robert *History of the Idea of Progress*. New Brunswick: Transaction Publishers, 1994.

[33] Hubbard, Barbara Marx *Conscious Evolution: Awakening the Power of Our Social Potential*. Novato, CA: New World Library, 1998.

[34] Seligman, Martin, 1998.

[35] Hergenhahn, B.R. and Olson, Matthew, 2003. Especially see Chapter Ten.

[36] Maslow, Abraham *Toward a Psychology of Being*. New York: D. Van Nostrand Co., 1968.

[37] Wilkins, Wallace "The Art of Strategic Anticipation: Investing in Your Positive Futures" *The Futurist*, March-April, 2001.

[38] Nelson, Noelle "Beliefs About the Future" *The Futurist*, January-February, 2000.

[39] Ellis, Albert and Harper, Robert, 1976.

[40] Maslow, Abraham, 1968; Hergenhahn, B.R. and Olson, Matthew, 2003, Chapter Fifteen.

[41] Reading, Anthony, 2004, Page 3.

[42] Postrel, Virginia *The Future and Its Enemies: The Growing Conflict Over Creativity, Enterprise, and Progress*. New York: Touchstone, 1999.

[43] Russell, Peter *The White Hole in Time: Our Future Evolution and the Meaning of Now*. New York: HarperCollins, 1992.

[44] Maslow, Abraham, 1968.

[45] Bloom, Howard *The Lucifer Principle: A Scientific Expedition into the Forces of History*. New York: The Atlantic Monthly Press, 1995, Pages 299-303.

[46] Fraser, J. T. *Time, the Familiar Stranger*. Redmond, Washington: Tempus, 1987; Diamond, Jared *The Third Chimpanzee: The Evolution and Future of the Human Animal*. New York: HarperPernnial, 1992; Shlain, Leonard *Sex, Time, and Power: How Women's Sexuality Shaped Human Evolution*. New York: Viking, 2003; Calvin, William *A Brief History of the Mind: From Apes to Intellect and Beyond*. New York: Oxford University Press, 2004.

[47] Reading, Anthony, 2004, Pages 24-26, 40-42, 55-58

[48] Reading, Anthony, 2004, Pages 50-58.

[49] Dennett, Daniel C. *Consciousness Explained*. Boston: Little, Brown, and Co., 1991; Hawkins, Jeff *On Intelligence*. New York: Times Books, 2004.

[50] Karniol, Rachel and Ross, 1996, Page 8.

[51] Recall that perception is not completely limited to the immediate here and now and that the idea of an absolute now - distinct from past and future - is highly doubtful.

[52] Reading, Anthony, 2004, Page 2.

[53] Csikszentmihalyi, Mihaly *Creativity: Flow and the Psychology of Discovery and Invention*. New York: HarperCollins, 1996.

[54] Bell, Wendell *Foundations of Future Studies: Human Science for a New Era*. Volume I. New Brunswick: Transactions Publishers, 1997.

[55] Reading, Anthony, 2004, Pages 100-106.

[56] Paul, Richard *Critical Thinking: What Every Person Needs to Survive in a Rapidly Changing World*. Rohnert Park, CA: Foundation for Critical Thinking, 1993; The Critical Thinking Community - http://www.criticalthinking.org/.

[57] Bell, Wendell "Making People Responsible: The Possible, the Probable, and the Preferable", *American Behavioral Scientist*, Vol. 42, No.3, November-December, 1998.

[58] Critical Thinking Community - Taking Charge of the Human Mind - http://www.criticalthinking.org/resources/tgs/taking-charge-of-the-human-mind.shtml.

[59] Wade, Carole, and Tarvis, Carol, 2003, Chapter Nine; Myers, David *Psychology: Seventh Edition in Modules*. New York: Worth Publishers, 2004, Module 28.

[60] Calvin, William, 2004.

[61] Karniol, Rachel and Ross, 1996, Page 7.

[62] Bell, Wendell, Volume I, 1997.

[63] Again see Damasio, Antonio, 1999, in this case on brain damaged individuals who do not seem able to plan.

[64] Postrel, Virginia, 1999.

[65] Bell, Wendell, Volume I, 1997.

[66] Tarnas, Richard *The Passion of the Western Mind: Understanding the Ideas that have Shaped Our World View.* New York: Ballantine, 1991; Solomon, Robert, 2002, Chapter Five.

[67] Quinn, Daniel *Beyond Civilization: Humanity's Next Great Adventure*. New York: Three Rivers Press, 1999.

[68] Lombardo, Thomas "The Pursuit of Wisdom and the Future of Education" Odyssey of the Future - http://www.odysseyofthefuture.net/pdf_files/Readings/Pursuit_of_Wisdom.pdf; Lombardo, Thomas and Richter, Jonathon "Evolving Future Consciousness through the Pursuit of Virtue" in *Thinking Creatively in Turbulent Times*. Didsbury, Howard (Ed.) World Future Society, Bethesda, Maryland, 2004.

[69] Polak, Frederik, 1973, Pages 13, 22.

[70] Hergenhahn, B.R. and Olson, Matthew, 2003, Chapter Eleven.

[71] Reading, Anthony, 2004, Pages 17-20.

[72] Lombardo, Thomas, 1987, Chapter One.

[73] Damasio, Antonio, 1999, Pages 17-18, 134-143, 172-176.

[74] Reading, Anthony, 2004, Page 31.

[75] Reading, Anthony, 2004, Pages 60-70; Baars, Bernard J. *In the Theatre of Consciousness: The Workplace of the Mind*. New York: Oxford University Press, 1997.

[76] Seligman, Martin, 1998, Pages 4-7, 76-82, 107-115, 291-292.

[77] Polak, Frederik, 1973, Page 17; Reading, Anthony, 2004, Page 10.

[78] Hergenhahn, B.R. and Olson, Matthew, 2003, Chapter Eleven.

[79] Bloom, Howard, 1995, Pages 311.

[80] Karniol, Rachel and Ross, 1996, Page 8.

[81] Stewart, John *Evolution's Arrow: The Direction of Evolution and the Future of Humanity*. Canberra, Australia: The Chapman Press, 2000.

[82] Anderson, Walter Truett *The Future of the Self*. New York: Putnam, 1997.

[83] Shlain, Leonard, 2003.

[84] Wade, Carole, and Tarvis, Carol, 2003. See especially Chapter Fourteen.

[85] Reading, Anthony, 2004, Page 136.

[86] Wilson, E.O. *Consilience: The Unity of Knowledge*. New York: Alfred A. Knopf, 1998.

[87] Molitor, Graham T.T. "Trends and Forecasts for the Next Millennium" *The Futurist*, August-September, 1998; Molitor, Graham T.T. "The Next 1000 Years: The "Big Five" Engines of Economic Growth" in Didsbury, Howard F. (Ed.) *Frontiers of the 21st Century: Prelude to the New Millennium*. Bethesda, Maryland: World Future Society, 1999.

[88] Anderson, Walter Truett *Evolution Isn't What It Used To Be: The Augmented Animal and the Whole Wired World*. New York: W. H. Freeman and Company, 1996; Naisbitt, John *High Tech - High Touch: Technology and our Accelerated Search for Meaning*. London: Nicholas Brealey Publishing, 2001; Stock, Gregory *Redesigning Humans: Our Inevitable Genetic Future*. Boston: Houghton Mifflin Company, 2002.

[89] Russell, Peter, 1992.

[90] Stewart, John, 2000.

[91] Bell, Wendell, Vol. I, 1997.

[92] Reading, Anthony, 2004, Page 69.

[93] Solomon, Robert, 2002, Chapter Eight.

[94] Meadows, Dennis, Meadows, Donella, and Randers, Jorgen *Beyond the Limits*. Toronto: McClelland & Stewart, 1992; Slaughter, Richard "Futures Concepts" in Slaughter, Richard (Ed.) *The Knowledge Base of Future Studies*. Volume I. Hawthorn, Victoria, Australia: DDM Media Group, 1996.

[95] Mellert, Robert B. "Do We Owe Anything to Future Generations?" *The Futurist*, December, 1982.

[96] Lombardo, Thomas and Richter, Jonathon, 2004.

[97] Bell, Wendell, Vol.I, 1997; Bell, Wendell, 1998.

[98] Seligman, Martin, 2002.

[99] Solomon, Robert, 2002, Chapter Eight.

[100] Lombardo, Thomas and Richter, Jonathon, 2004.

[101] Stock, Gregory *Metaman: The Merging of Humans and Machines into a Global Superorganism*. New York: Simon and Schuster, 1993; Kurzweil, Ray *The Age of Spiritual Machines: When Computers Exceed Human Intelligence*. New York: Penguin Books, 1999; Kurzweil, Ray and Grossman, Terry *Fantastic Voyage: Live Long Enough to Live Forever*. U.S.A: Rodale, 2004.

[102] Tipler, Frank *The Physics of Immortality: Modern Cosmology, God, and the Resurrection of the Dead*. New York: Doubleday, 1994.

[103] Stapledon, Olaf *Last and First Men and Star Maker*. New York: Dover Publications, 1931, 1937. See also Wilber, Ken *A Brief History of*

Everything. Boston: Shambhala, 1996; Tipler, Frank, 1994; Hubbard, Barbara Marx, 1998.

[104] Smolin, Lee *The Life of the Cosmos.* Oxford: Oxford University Press, 1997; Prigogine, Ilya *The End of Certainty: Time, Chaos, and the New Laws of Nature.* New York: The Free Press, 1997; Hawking, Stephen *The Universe in a Nutshell.* New York: Bantam Books, 2001.

[105] Stock, Gregory, 1993.

[106] Kurzweil, Ray, 1999.

[107] Best, Steven and Kellner, Douglas, 1997; Cornish, Edward *Futuring: The Exploration of the Future.* Bethesda, Maryland: World Future Society, 2004.

[108] Capra, Fritjof *The Turning Point.* New York: Bantam, 1983; Rucker, Rudy, Sirius, R.U., and Queen Mu, *Mondo 2000.* New York: Harper Collins, 1992; Toffler, Alvin *Future Shock.* New York: Bantam, 1971; Toffler, Alvin *The Third Wave.* New York: Bantam, 1980; Toffler, Alvin *Power Shift: Knowledge, Wealth, and Violence at the Edge of the Twenty-First Century.* New York: Bantam, 1990.

[109] Glenn, Jerome and Gordon, Theodore *2004 State of the Future.* American Council for the United Nations University, 2004; Christian, David *Maps of Time: An Introduction to Big History.* Berkeley, CA: University of California Press, 2004, Chapter Fourteen; Best, Steven and Kellner, Douglas, 1997.

[110] Kurzweil, Ray, 1999; Christian, David, 2004, Chapter Fourteen.

[111] Gleick, James *Faster: The Acceleration of Just About Everything.* New York: Pantheon Books, 1999; Anderson, Walter Truett *All Connected Now: Life in the First Global Civilization.* Boulder; Westview Press, 2001; Gitlin, Todd *Media Unlimited: How the Torrent of Images and Sounds Overwhelms Our Lives.* New York: Metropolitan Books, 2001.

[112] Gleick, James, 1999.

[113] Berman, Morris *The Twilight of American Culture.* New York: W. W. Norton, 2000.

[114] Kaku, Michio *Visions: How Science Will Revolutionize the 21st Century.* New York: Anchor Books, 1997; Zey, Michael G. *Seizing the Future: How the Coming Revolution in Science, Technology, and Industry Will Expand the Frontiers of Human Potential and Reshape the Planet.* New York: Simon and Schuster, 1994.

[115] See for example Walter Anderson's discussion of the Deep Ecology movement in Anderson, Walter Truett, 1996.

[116] Reading, Anthony, 2004, Page 147.

[117] Postrel, Virginia, 1999; Zey, Michael G., 2000.

[118] Didsbury, Howard F. "The Death of the Future in a Hedonistic Society" in Didsbury, Howard F. (Ed.) *Frontiers of the 21st Century: Prelude to the New Millennium.* Bethesda, Maryland: World Future Society, 1999.

[119] Bertman, Stephen "Cultural Amnesia: A Threat to Our Future", *The Futurist*, January-February, 2001.

[120] Gleick, James, 1999.

[121] Brand, Stewart *The Clock of the Long Now: Time and Responsibility*. New York: Basic Books, 1999.

[122] Nisbet, Robert, 1994.

[123] See Shlain, Leonard, 2003 for a theoretical explanation of how primitive humans first became conscious of extended time. And see Fraser, J. T., 1987 for a general discussion of the evolution of temporal consciousness throughout the history of humanity.

[124] I take this expression from Smolin, Lee, 1997.

[125] Reading, Anthony, 2004, Pages 7, 126, 140-141, 148, 171-172; Lasch, Christopher *The Culture of Narcissism*. New York: Warner Books, 1979.

[126] Vinge, Vernor "The Coming Technological Singularity: How to Survive in the Post-Human Era" *Vision-21: Interdisciplinary Science and Engineering in the Era of Cyberspace NASA-CP-10129*, 1993 - http://www.ugcs.caltech.edu/~phoenix/vinge/vinge-sing.html; Kurzweil, Ray, 1999.

[127] Ray, Paul and Anderson, Sherry *The Cultural Creatives: How 50 Million People are Changing the World*. New York: Three Rivers Press, 2000; Hubbard, Barbara Marx, 1998.

Chapter Two

The Origins of Future Consciousness

In this chapter I describe the beginnings of future consciousness and how future consciousness has progressively evolved throughout the history of life and prehistoric humanity. I explain how the emergence and development of future consciousness was driven by survival needs and evolutionary forces and how future consciousness facilitates and further intensifies the evolutionary process. Evolution generated future consciousness, but future consciousness, expressed through culture, language, thinking, and technology, in turn has speeded up the process of evolution.

One key principle emerges in this historical – evolutionary survey. It is the principle of reciprocity. Throughout the chapter, I describe several important reciprocities relevant to the evolution of life, mind, and human society. I examine the reciprocal evolution of self and culture, genetics and culture, and male and female reproductive behavior. As a general conclusion, I argue that our evolutionary heritage and social-psychological make-up is a network of reciprocities. All these reciprocities have directly contributed to the evolution of future consciousness.

Life and the Environment

*"...the entire history of life on this planet could
be conceived as a striving by life-forms to
attain an ever-greater appreciation
of the vectors of space and time."*

Leonard Shlain

Following the views of John Stewart and Leonard Shlain, among other contemporary writers, it seems that throughout the history of life temporal and spatial sensitivity has expanded from the relatively momentary here and now to increasing vistas of space and time.[1] As life became more complex, adjustment and awareness evolved to more complex and expansive patterns in the environment. Specifically following Stewart on this point, evolution has driven the growth of future consciousness because adaptability and survivability are served by increasing sensitivity and awareness of the future. The farther out in time (or for that matter in space) one can "see" the more knowledgeable and capable one becomes in dealing with the twists and turns and variations in the environment. If everything stayed the same – in every direction in space and time – there would be no need to see beyond the "here and now" – but the world is filled with differences and changes extending outward in space and time.

All life is dependent on the environment in the sense that living forms utilize, and in fact, require various resources and physical conditions in their environment in order to perpetuate their existence. In attunement with the environment, living forms possess sets of abilities that allow them to seek out, identify, and use the resources of their world. Hence, life fits and adjusts to its environment.

The environment of life though is a complex and multifaceted reality. There is an intricate and highly organized spatial and temporal structure to the environment. The temporal structure of the environment involves various natural rhythms and periodicities, relative constancies, and often abrupt and drastic changes that occur in the world. There is a multitudinous array of animal behavior patterns within the environment. All these environmental events and temporal structures produce patterns of physical stimulation. There are complex temporal patterns of sound, pressure, and light connected with mating opportunities, food, shelter, protection, and danger.

All life shows some degree of adjustment and resonance with the temporal patterns of the environment. There are innumerable bio-rhythms built into the fabric of life - circadian, infradian (less than once a day), and ultradian (more than once a day) - which are in resonance with environmental patterns and temporal cycles.

First, let us consider the genetic foundations of environmental adaptability. From the simplest life forms, such as bacteria, which first emerged on the earth billions of years ago, all life possesses a common genetic foundation. As James Watson and Francis Crick discovered in the 1950's, the molecular code for all life on earth is embodied within the same complex molecule, DNA.[2] Differences among species are to a great degree due to variations within the DNA code. The DNA code of a particular species roughly determines a set of bodily structures and physiological, biochemical, and behavioral processes that allow the life form to successfully deal with its environment.[3] Because the genetic structure of a species is a product of natural selection due to the environment, the genetic make-up of a life form supports a set of inherited capacities that are adapted and attuned to the conditions of the environment, both its dangers and its necessities.

Although adaptability to the environment is built upon a genetic foundation, as life in its evolutionary history became more complex, other factors came into play. Multi-cellular life forms, including the first animals, dramatically appeared on the scene in great numbers and varieties during the Cambrian Explosion around 570 million years ago.[4] With the emergence of animals and complex nervous systems, distal sense organs, and muscular systems for locomotion and manipulation, perceptual and behavioral capacities were significantly enhanced. Animals can see motion, the speed and direction of the approach of predators, the receptive behavior signals of potential mates, and a host of other dynamical processes and events significant for their survival. And animals can respond with appropriate and complex behaviors to these patterns of information and environmental events. Animals show attunement in their perception and behavior with temporal patterns such as the seasons, day-night cycles, fertility rhythms, and lunar cycles.

Although the basic structure of an animal nervous system is genetically determined, this genetically endowed foundation allows for memory and anticipation based upon individual learning. Animals with nervous systems can go beyond inherited skills and capacities.[5] Through learning, existing behaviors can be modified or new behaviors can be acquired. (The nervous system, in fact, is transformed at a synaptic and biochemical level – at the very least – in conjunction with learning.) All animals demonstrate some capacity for learning, and therefore in some sense possess the ability to remember. The less complex the nervous system, the less flexibility the animal will demonstrate to learn and modify behavior.

Adaptation to the environment based on learning has a distinct advantage over adaptation simply due to genetic inheritance.[6] Genetic variation in animals only occurs across generations due to natural selection. Learning introduces flexibility during an individual lifetime – animals can modify their behavior during their lifetime; they are not rigidly constrained by a pre-determined set of inherited dispositions and behaviors. One could of course argue that the mutability of nervous systems that are able to learn is due to a certain type of genetic make-up that supports this capacity, but the specific learned associations and responses are a result of the unique interactions with the environment during an animal's individual life.

Because animals can learn they can also anticipate. Having encountered either specific dangers or resources before in certain environmental conditions, animals learn to move in the direction of what is valuable to them, and away from what is dangerous, before they directly sense the salient object or event. They demonstrate anticipatory behavior based on past learning. It is often argued that animals live in the "immediate here and now."[7] Yet, animals clearly show responsiveness to the anticipated future and a sense of having learned from the past even if their sense of time is limited. In general, the sense of the future and the past is enriched through the effects of learning. In particular, learning introduces increased flexibility in dealing with change.

Consequently, it has been argued by biologists such as Stewart that the genetic capacity for learning and increasing flexibility would be naturally selected for in the evolution of animals.[8] More flexible animals stand a better chance of reproducing. As a general trend, as animal life has evolved, nervous systems have become more complex and animals have become increasingly flexible and capable of learning. The capacity to adjust to change has evolved through time.

The evolution of life has also been a collective process. The array of different life forms, at any given period of time, has always existed in a network of interdependencies. A critical part of the environment of life is other life forms. Life needs life in order to survive and flourish. Through both competition and cooperation life evolves collectively or reciprocally. The evolution of predators stimulates the evolution of prey and vice versa. Symbiotic and parasitic relationships continue to emerge and evolve throughout history. As Harold Morowitz states in his panoramic review of the history of evolution, *The Emergence of Everything*, new emergent forms or properties in nature co-evolve.[9]

The social and interactive dimension of life and its evolution are emphasized in Howard Bloom's writings.[10] Bloom points out that even bacteria mutually influence each other through the sharing of genetic information. Bacteria exchange DNA and can modify their physiology and behavior in response to environmental changes through this process. As Bloom puts it, bacteria are collective learning machines.

According to Bloom, the capacity for individual learning, which emerges with animals, opens the door to a whole new mechanism for acquiring information about the past. Bloom argues that, with the development of the capacity for learning, social learning comes as well. Animals can learn through modeling and imitation of other members of their species. Members of a species can share learning and information with each other. Knowledge among animals is a social phenomenon - a group can learn and pass on information to new members. Offspring can learn from their parents or other more experienced members of their social unit. Information and learning can be passed on across multiple generations, building upon itself.[11]

In describing the evolution of humans in the next section, one significant trend that greatly contributes to the evolution of future consciousness is this ever growing capacity among our prehistoric ancestors to collectively share information and pass new learning on to offspring. The foundations of culture are built upon this ability. Culture in humans emerges, as Bloom would say, as a "collective learning machine".

The Prehistoric Evolution of Humans

"We are not fallen angels but risen apes."

William Calvin

The evolution of future consciousness in humans has been driven by adaptive challenges to life and is intimately connected to fundamental patterns of living. The development of tool making, coordinated hunting, male-female bonding, representational art, child rearing, and culture all have contributed to the expansion of human future consciousness. Much of what makes us unique, biologically, psychologically, and socially, is associated with our expanded and complex sense of time and in particular the future.[12]

In beginning the story of our ancestry it is important to keep in mind, as William Calvin notes, that "we are not fallen angels but risen apes". In spite of numerous mythic and religious stories of our having once lived in a more pure and elevated state (the Myth of the Golden Age), or the idea that humans began as non-material spirits or souls that were then placed in physical bodies, the overwhelming evidence indicates that humankind evolved through a series of stages from more primitive primates. We are evolutionarily and genetically connected with all of life (we all share DNA as a common genetic code), and in particular, we are close genetic cousins to that group of existing primates we call "apes".[13] As Desmond Morris aptly described us, we are "the naked ape".[14]

The evolutionary perspective on humans not only provides the most factually grounded explanation of our origins and nature, but also gives us a sense of hope and progression. Whatever our failings or limitations throughout history, and there seem to be many, our evolutionary story is generally one of advancement and achievement. Our depth and range of consciousness, our capacities for science, literature, and art, our technologies, our evolution of morals and cultural values, our creative abilities, our intricate social systems, and our vast capacities for learning and the acquisition of knowledge are all evolutionary advances progressively achieved across the long trajectory of our history. To view our species as having "fallen from grace" is depressing and factually in error; to see our history as progressive is elevating and factually correct.

To begin the saga of our evolution, our genetic ancestors, the primates, appeared after the extinction of the dinosaurs approximately 60 million years ago. On the primate evolutionary line, apes and monkeys diverged around 20 million years ago. Aside from developing hands with opposable thumbs for grasping, the primate - ape evolutionary line also showed increasing behavioral plasticity, greater maternal care of young, and an increasing brain/body ratio.[15] All these general trends continued in the evolution of humans and contributed to the ongoing development of future consciousness.

Humans, chimpanzees, and gorillas all evolved or branched off of the common ancestral line of great apes. Gorillas branched off first, approximately 12 million years ago. Based on genetic evidence, the chimpanzee and human evolutionary lines diverged approximately 7 million years ago. Chimpanzee DNA is 98.6 % identical to human DNA. Chimps are our closest living genetic relatives, and humans (and not gorillas or any other ape) are the closest genetic relatives to chimps. Genetically, we cluster with the chimps. Since there are two different present species of chimpanzees, the common chimp and the bonobo chimp, we are, as Jared Diamond has argued, the "third chimpanzee".[16]

Our relationship with the common and bonobo chimps is fascinating for we seem to combine psychological and behavioral features of both species. The common chimp can be very aggressive, and the males will show extreme group violence, attacking and ferociously killing other chimps that are not part of their social group, or other vulnerable animals of prey. The bonobos exhibit much less violence and engage in a great deal of sexual behavior as a way to apparently reduce aggressive tendencies and reinforce bonding among males and females. (As animal and human research has repeatedly demonstrated, hugging and other forms of affectionate physical contact reduces violent behavior.[17]) In general, bonobos show much more intra-species affectionate behavior than do common chimps. The common chimps have a more male dominated social order whereas the bonobos are more female and maternal dominant in their social order. Interestingly, humans appear to reflect and combine both the "killer" and "lover" dispositions of our closest relatives.[18] In fact, throughout written history and theories of human nature, these two general tendencies (to fight and kill versus to love) have frequently been conceptualized as the good and evil sides of humans. As we will see, sex and violence, as well as female versus male dominance, are significant themes in the evolution of future consciousness.

One important idea regarding our ancient genetic heritage that connects sex and violence is the hypothesis that male and female humans have evolved along different paths, the male in particular being selected for increasing violence because it increases male reproductive success. The anthropologist Michael Ghiglieri, in his book *The Dark Side of Man*, argues that hominid male evolution has been significantly shaped by sexual selection – that is the preferential selection of those behaviors that lead to securing female mates and eliminating male competition. According to Ghiglieri, it is violent behavior among males that gets selected for because it is the strongest and most intimidating males that get the female mates.[19] Hominid males learned to fight in order to make love – a rather paradoxical combination of traits to say the least. Another argument - in fact a rather popular one – connecting sex and aggression in males is the idea that our male ancestors became increasingly ferocious and effective hunters in order to win and maintain the commitment of female partners. Males killed animals of prey to get meat for females as a point of bargaining for sex and love.[20] If either or both of these hypotheses is correct, it is important to note that aggression and violence in our male human ancestors served future focused goals – sex and a committed partner.

Archeological evidence indicates that during the period of seven million to one million years ago a branching series of hominid life forms lived throughout areas of Africa and at several points migrated up into Asia and the Middle East.[21] The term "hominid" refers to all those bipedal apes that progressively emerged in the evolutionary line that separated from the chimpanzee line. Modern humans are hominids, and in fact, the only surviving member of this genetic group. In the past, often multiple hominid species co-existed in the same areas. Our ancestry is not some simple unitary line of descent, but a transforming family of various genetic cousins, a "meandering" and "multifaceted evolution". At times our hominid predecessors lived relatively peacefully together, but at other times engaged in competition, if not violent antagonism.[22]

During the earliest period of evolutionary branching around five to four million years ago, a bipedal posture and mode of locomotion emerged, freeing the hands to carry objects (including food and baby hominids) and eventually create tools. Erect posture and locomotion probably first evolved in adaptation to the move from a jungle-forest environment to the relatively open savannah, but there is ongoing debate as to the exact reasons for the change to bipedalism.[23] The earliest erect hominids were the *Ardipethecus* and *Australopithecus* genus, of which there was a variety of species that lived throughout Africa beginning around 4.5 million years ago.[24] The famous skeleton, Lucy, is an *Australopithecus afarenis* who lived 3.2 million years ago and had a brain the size of a chimp, approximately 400 cubic centimeters.[25] Modern humans are probably descended from one of the species lines of *Australopithecus,* but many of the other lines of *Australopithecus* died out. *Australopithecus,* who lived around 2.5 million years ago, probably made the first crude tools.

Tools are most strongly associated with the appearance of *Homo habilis* ("handy man") between 2.5 and 2 million years ago.[26] There is significant variation in the cranial size of habiline fossils, but there is an overall and quite significant trend toward increasing brain size in the habiline line. *Homo habilis* had a significantly bigger brain than *Australopithecus.* In fact, by around 1.8 million years ago hominid brain size had doubled to around 800 cubic centimeters.[27] According to Calvin, with the appearance of *Homo habilis,* meat consumption in our ancestors significantly increased (presumably to feed our bigger brain). For Peter Watson, it was the emergence of stone tools that enabled these early hominids to eat meat, providing a way to butcher animals and get at muscles and internal organs.[28]

It seems clear that the earliest stone tools were used to obtain or prepare food, both animal and vegetable. Various types of primitive tools were created through chipping and flaking particular types of stones and minerals, a process that involved both a high level of manual dexterity and thoughtful planning.[29] Chipping away at a rock to form an instrument for the intended future purpose of killing and skinning of animals, or whatever other uses these first tools served, indicates a clear awareness of the future, as well as planning for the future. A tool is made to serve a future purpose – the act of creating a tool is not an end in itself. In fact, archeological evidence seems to indicate that early hominids used tools to create other tools – chipping one stone with another – which would indicate a multiple step planning process.[30]

Another aspect of early tool making that demonstrates future consciousness is the fact that sometimes tools were made at places distant from where animals were killed or butchered. *Homo habilis* seems to have had the capacity or foresight to make tools ahead of time in one location and then bring the tools to another spot (up to ten miles away) in anticipation of finding animals and butchering them.[31]

These early stone tools were also connected with future consciousness in still another way – in this case a form of **social future consciousness**. Calvin suggests that social instincts, specifically regarding increased cooperation and sharing, evolved or developed with the emergence of *Homo habilis*.[32] These early hominid hunters, cooperatively working together to find meat, either through killing prey or scavenging, brought the meat back to the social group rather than consuming it on the spot. Further, they also appear to have engaged in cooperative butchering as is indicated by evidence at archaeological sites from the period. This securing of and then butchering meat for later consumption reflects delayed gratification and sharing. It is a social form of future oriented behavior that chimps appear totally incapable of doing.[33]

All told, these social and tool making capacities of *Homo habilis* would appear to demonstrate the ability to imagine and create mental maps. The creation of such mental images also seemed to be based on past learning; Homo habilis located and remembered places that animal prey frequented and would return to these hunting spots with tools in anticipation of finding meat for consumption.[34]

The culmination of the trend towards increasing brain size during the early tool making period was the emergence of *Homo ergaster* in Africa around 1.8 million years ago. The first hominid migration out of Africa appears to have been around 1.7 million years ago when *Homo erectus* – a slightly modified descendent of *Homo ergaster* - spread across the Middle East, Asia, and eventually as far as China and Indonesia. *Homo erectus* first learned to control and use fire, probably cooked some foods, developed more sophisticated and standardized tools than *Homo habilis,* probably engaged in body painting, collected crystals, pebbles, and shells, presumably for aesthetics and reasons of social status, and did not become extinct, at least in Asia, till less than one hundred thousand years ago.[35]

Gaining control over fire is an especially noteworthy accomplishment for this ability clearly distinguished *Homo erectus* from the rest of the animal world. Fire was no longer something simply to fear – it became a powerful tool of future consciousness that could be used for a variety of purposes. The conquest of fire is often listed as one of the critical events in the history and evolution of humans.[36]

There are indications that *Homo erectus* was significantly more socially advanced than earlier hominids. Ghiglieri proposes, based on a review of archeological and fossil evidence, that *Homo erectus* possessed a rudimentary form of language and culture. (Fossil evidence, in fact, indicates that Broca's area – the part of the brain involved in speech production in modern humans – was even present in *Homo habilis* brains.[37]) Ghiglieri defines being human as having self-awareness and using culture as a primary means of coping with the environment. Ghiglieri believes that *Homo erectus* had these qualities.[38] Culture depends upon socially transmitted ideas from the past, and thus *Homo erectus* would therefore have developed a rudimentary form of historical consciousness. It is important to also note that, if Ghiglieri is correct, the dual dimensions of increasing self-awareness and social awareness emerged together.

It has been argued that the emergence of human consciousness depends upon the development of two defining conceptual distinctions – the abilities to distinguish the self from the non-self and the past from the future.[39] Both of these distinctions are reciprocities; the ideas of self and non-self and past and future are interdependent and defined relative to each other. It can be debated whether animals do or do not have some rudimentary sense of self and other or past and future, but if we agree with the arguments of Ghiglieri, then *Homo erectus* possessed something approximating modern human consciousness. *Homo erectus* had crossed the line separating humans and the human mind from the rest of the animal kingdom.

Ghiglieri also believes that *Homo erectus* developed monogamous female - male relationships, with the male making an extended time commitment toward the raising of children. There is debate over the point in our evolutionary history at which we developed monogamy as a primary form of male-female bonding, but monogamy does represent a significant jump forward in future consciousness in that it indicates a conscious choice against impulsive sexual gratification with multiple partners. Monogamy means commitment and commitment involves reference to the future.

Howard Bloom, who emphasizes in his book *Global Brain*, the collective and social dimensions of life, adaptation, and learning, describes the first migration of *Homo erectus* out of Africa as the collective human mind going global, spreading its primitive culture, and technology across much of the Eastern Hemisphere.[40] Various populations of hominids during this first migration and later ones probably exchanged artifacts and inter-bred with each other. For quite some time, at least since 1.8 million years ago, we have been a burgeoning global species with varying degrees of awareness of other people and other cultures spread across other distant lands. We are creatures that form ever expanding and increasingly complex social networks.

A second significant jump in brain size occurred around 500 thousand years ago.[41] Brain size shot up another four to five hundred cubic centimeters. Connected with this second major increase in brain size is the emergence of hominids closely related to and including the first "archaic" examples of our species, *Homo sapiens*. These earliest representatives of our species first appeared in Africa.

In this most recent surge in the growth of the brain, the frontal cortex of the cerebrum at the top of the brain, in particular, expanded in size considerably. This is significant since, as revealed through modern neurological research, the prefrontal area of the frontal cortex (the large most forward section of the frontal cortex) is the part of the brain most strongly involved in future oriented decision making and purposeful behavior. The prefrontal area appears to be responsible for the temporal organization of thinking and behavior, planning and goal setting, self-initiation, and the consideration of alternative actions and consequences of behavior.[42] The frontal cortex is also strongly associated with heightened self-awareness in humans. This evolutionary surge in the growth of the frontal cortex and prefrontal area, demarcating our emergence as a species, would seem to indicate that it is our neural capacity for complex and expanded future consciousness that most strongly distinguishes our species.[43]

There are though a variety of explanations that have been offered regarding what instigated the relatively rapid growth in brain size in hominids over the entire history of the genus line during the last few million years.[44] The neurophysiologist William Calvin has hypothesized that the dramatic spurt of neurological growth in hominid evolution was triggered by sudden and frequent climate changes.[45] During the period of 2.5 million years to 500 thousand years ago there were frequent and sudden climate changes associated with the waxing and waning of innumerable Ice Ages. These climatic changes produced significant environmental changes in Africa, the home of our hominid ancestors, including decreasing rainfall and the periodic shrinking of forest and jungle. Surviving through repeated, unpredictable, and rapid change became a distinctive strength of our ancestral line – it appears that developing much bigger brains was connected with this capacity for dealing with change. As Calvin points out, tools as well as the first spurt in increasing brain size appear when the Ice Ages begin.[46]

Another explanation is that increasing brain size was connected with tool making and enhanced manual dexterity. Watson documents this theory of the evolution of the hominid brain.[47] The evolution of the brain and the development of tools occurred interdependently; increases in brain size stimulated advancements in tools which in turn triggered further increases in brain size. Within this theory, the human brain and technology form a co-evolutionary reciprocal whole. Yet there is debate over whether evolutionary jumps in brain size are closely correlated with significant improvements in the quality of tools.[48]

A third explanation, also connected with increasing cognitive and intelligence abilities, is the **"social intelligence"** theory. More complex cooperative behaviors were needed as hunting evolved in humans. Also, as social and, in particular, family units became more complex, our hominid ancestors needed more brain power to predict and influence the behavior of one another. As hominids became increasingly complex in their social interactions and organizations, a higher level of social intelligence was needed. Hence, we have bigger brains because our complex social relationships demand high levels of intelligence.[49] In this case, society and the hominid/human brain form a positive feedback loop of reciprocal evolution.

There is another popular theory, the **"social display"** or **"social mirror"** theory, supported by the anthropologist Charles Whitehead and others, which proposes that increasing brain size is most strongly connected to the rapid increase in forms of gesture, personal expression, mimicking, song and dance, ritual, play, and ceremony that have emerged in our evolution. This theory does not emphasize so much the importance of increasing intelligence and cognitive capacities as it does the heightened capacity in humans to express and represent their feelings, attitudes, personality, and motives. The psychologist Merlin Donald attributes the significant advances in social organization made by *Homo erectus* to the emergence of "mimetic" thinking and behavior.[50]

Social display theory is connected with the popular sociological theory that the self is a social construction. Through display we teach each other, and in particular the young, about the nature and make-up of our psychological states. What is private is first made public. Children develop a concept of the self by being taught through display the myriad intricacies of human behavior. Children learn about emotional, motivational, attitudinal, and cognitive states of the self through having such states expressed and demonstrated by adults. The child learns to mirror the social representation of the self. We are the most self-conscious animals on the earth, and this heightened self-awareness is a product of the complex social displays we broadcast to each other and then internalize.[51]

We should recall that Ghiglieri identified culture and self-awareness as the two defining features of being human. Social display theory connects the two factors together. What I would suggest is that the self and culture constitutes a significant reciprocity in the social-psychological make-up of humans. Although social display theory may be correct in that the group teaches its youth about the nature of the human self, individuals do not all turn out the same – we are not all carbon copies of some cultural template. Our individualized selves impact back on culture, contributing new and unique elements into it. Self and culture form a reciprocal loop, each influencing the evolution of the other.

An idea from Bloom helps to understand this reciprocity of self and culture. Bloom argues that within any social group there are **"conformity enforcers"** and **"diversity generators"**, providing for both cohesion and experimental variety in its repository of knowledge and behaviors.[52] These dual forces are analogous to the dual processes of genetic replication (producing uniformity) and genetic mutation (producing variety) in biology. Culture is one of the most powerful "conformity enforcers" within human groups bringing unity of purpose and identity to a people, whereas individual selves are diversity generators, bringing experimental variety into the group. Interestingly, as revealed through archaeological evidence, as human culture evolved, more inventiveness and creativity in artifacts shows up as well. Conformity and diversity work in opposition to each other, but these processes also work in reciprocity. It is no coincidence that humans possess both highly developed cultures and highly developed individualized selves.

Another noteworthy factor to consider in understanding the reciprocal evolution of the self and culture is the development of parental care in hominid history. Increasing parental care provides more opportunity for imitation and the learning of culture and for the development of the self. Ghiglieri argues that *Homo erectus* evolved a more committed male-female bonding relationship to improve the quality of child rearing. As a general trend observed in nature, mammals more than reptiles, and in turn, primates more than other mammals, spend more time raising their offspring.[53] As our hominid line evolved, more time was spent in caring for the young. Through this process of increasing parental care, both the transmission of culture and the intensification of self-awareness were facilitated.

As one final theory to consider regarding the dramatic increase in the size of the human brain, let us return to another hypothesis of William Calvin as presented in his book *Cerebral Symphony*.[54] Calvin identifies the execution of actions in anticipation of future events, such as the throwing of projectiles toward where we believe a running animal of prey will be in the immediate future, as a key perceptual-motor capacity that evolved in humans. This is a distinctive strength of the human brain - its capacity to predict the future even if it is simply the immediate future. Literally, we are very good at "seeing ahead".

I have already discussed the general hypothesis that what clearly distinguishes humans from other animals is our highly developed capacity for future consciousness. What I would like to introduce now is the neurological theory that the human brain is fundamentally a mechanism for making continual predictions about the future. As argued by writers such as Daniel Dennett and Jeff Hawkins, the human brain is continually generating predictions about what is going to happen in the future, from the short term to the long term. For Hawkins, human intelligence is nothing but the skill in making predictions. Although throughout the history of psychology, it has been emphasized that what distinguishes humans is our capacity for learning and memory – as great recorders of the past – the view being described here takes the opposite approach; what distinguishes the human brain is the highly developed capacity to predict. In fact, to drive the point home, if we consider the evolution of brains in animals, it is clearly more important that animals anticipate what is going to happen than to remember what has happened in the past. If we examine the neurological circuitry of the animal or human brain, sensory nerves do not simply convey information from sense organs to the brain, but rather, the brain, through numerous neural pathways running down the sensory nerves and the motor nerves that control adjustments in the sense organs, modulate sensory input; even basic perception and behavior is a forward looking process. Brains search and explore in anticipation of what is going to happen.[55] Hence, for whatever reasons that it became increasingly important, the recent surge in the evolution of the human brain involved a dramatic growth in the basic neural capacity for anticipating or predicting the future.

Whatever the reasons for increasing brain size, and there were probably several based on both archeological and genetic evidence, our modern species, *Homo Sapiens*, appeared first in Africa around 150,000 years ago.[56] The brains and body structure of these humans were basically identical to those of modern humans (but see below for a possible noteworthy difference).

After migrating into the Middle East, Asia, and eventually Europe, they co-existed for quite sometime with their genetic cousins, *Homo neanderthalenis*. Neanderthals were shorter and more solidly built than *Homo sapiens* and actually had a slightly bigger brain.[57] What is particularly fascinating is that archeological evidence indicates very little difference in tools and artifacts in their early years of co-existence between these two related species.

Neanderthals and modern *Homo sapiens* are probably related through a common ancestor, *Homo heidelbergensis (or archaic Homo sapiens)*, that lived throughout Africa and Eurasia approximately four hundred thousand years ago.[58] Neanderthals lived in Europe and Western Asia and appear to have been specially adapted to the rigors and climatic challenges of the Ice Ages. They were probably predominately meat eaters, made sustained hunting treks with both children and females, buried their dead, skinned animals for clothing, and had some level of spoken language, but did not show much variation or change in their material culture, referred to as the Mousterian culture, for most of the time of their existence from three hundred to twenty-eight thousand years ago.[59] Yet during the last ten thousand years of their existence they began to exhibit real advances in material culture, including representational art, after apparent contact with modern *Homo sapiens* in Western Europe, thus producing what is referred to as the distinctive Châtelperronian culture. But contact with *Homo sapiens* was probably the eventual undoing of the Neanderthals for chances are that they were out-competed by the superior culture and way of life of *Homo sapiens*.[60]

The demise of the Neanderthals was connected with something of great importance that happened in human history around forty thousand years ago. As noted above, *Homo sapiens* first appear around one hundred and fifty thousand years ago in Africa. Our species spread up into the Middle East sometime after that time, but did not distinguish itself in any significant way from the Neanderthals who also lived in that region. Our brains and bodies were basically the same as today, but we showed no indication of real material or technological superiority. Following Diamond, even if he is somewhat exaggerating the point, we were still more animal than human at least in our behavior and accomplishments.[61] Yet based on the most recent thinking on this matter, sometime around fifty thousand years ago, a relatively small group of genetically linked *Homo sapiens* came out of Africa carrying with them a distinctly different and highly more advanced material culture. This group of *Homo sapiens* first spread across Eurasia and then Australia, and eventually the entire globe, wiping out all other existing hominids in their way, as well as driving to extinction many Ice Age mammals due to their highly efficient hunting techniques and weaponry. This relatively sudden and momentous advance in culture, abilities, and behavior is referred to as **"The Great Awakening"** or "The Great Leap Forward", or as William Calvin calls it, "The Mind's Big Bang".[62]

The Great Awakening, Culture, and the Discovery of Death

"In the beginning was the image"

Leonard Shlain

Whereas prior to forty to fifty thousand years ago, *Homo sapiens* demonstrated little inventiveness in tools, worked with limited and local materials and resources, showed minimal variation in artifacts across different regions, and few examples of representational art – most artifacts were utilitarian – beginning with the Aurignacian cultural period in Western Europe (40,000 to 28,000 BP) things dramatically changed. Cave paintings, engravings, sculptures, body adornments, musical instruments, new multi-pieced weapons, ceramics, and weaving appeared in great variety and numbers. Also, long distance trading of materials and unique local cultures emerged. Further, cultural evolution went into high gear, with new distinctive cultures developing in relatively quick succession to each other. The Late Stone Age or Upper Paleolithic Age (40,000 to 11,000 BP) witnessed an explosion in human inventiveness.[63]

There are a variety of explanations for what instigated this acceleration in creativity and change. The emergence of modern language, the rise of patriarchy, and the psychological discovery of personal death have all been proposed as instigators of the Great Awakening. It has also been argued that the Great Awakening is more apparent than real. Throughout Africa, prior to the Great Awakening, there is piecemeal evidence for most of the significant advances connected with Aurignacian culture. When the final wave of migration of modern humans came out of Africa around 50,000 years ago they brought with them all the elements of Aurignacian culture that had been more slowly acquired over the previous one to two hundred thousand years.[64]

Randall White, the historian of prehistoric art, takes the view, however, that at least regarding the multifarious forms of art that emerged in Western Europe around 40,000 years ago, the cultural jump was relatively sudden and pronounced. Further, he makes the basic evolutionary point that representational art must have had a significant adaptive benefit. He suggests that perhaps its emergence was connected with contact and competition with the Neanderthals, though it should be recalled that Neanderthals and *Homo sapiens* coexisted for approximately one hundred thousand years prior to the Great Awakening.

But to follow White's argument that art had some important adaptive value, he points out that representational art provided a new medium or "space" in which to abstract or isolate features of the natural world and re-present these features where they could be rearranged and combined in new ways. That is, representational art provided a public and material "mental working space" in which to think in terms of images, icons, and symbols. This medium or new virtual reality, in which to represent information, possessed much more flexibility and openness than the natural world. For example, there are art objects and drawings that are **"therianthropic"**, where animals and humans are combined into single figures. Sequential time and motion are also represented through drawings of horses or other animals in successive body positions. Part of a whole complete object in the natural world could be separated and abstracted from the whole and represented as standing for the whole (referred to as "metonymy"). Representational art, by producing this vastly enlarged mental space in which to think, would have provided our ancestors with much greater cognitive power than that of any co-existing hominids, or for that matter, of any other animals who "think" only within the confines of the perceptual world. Art opened up a new universe of possibilities.

It is interesting that this development of a mental space for abstract, combinational, and possibility thinking parallels a similar process that presumably took place, according to social display theory, in the evolution of the self; in both cases a mental reality was initially expressed and developed in public. Over time, the public realm and private mental realm have intertwined into a reciprocal feedback loop, with inner realities manifesting outer expressions and outer expressions instigating further developments in inner realities. We draw and we write to "see" what we think and what we can imagine, but in turn, what we think and imagine provides stimulation and instigation for what we express within the public world. The arguments from White and Whitehead are that the public arena first instigated developments in the inner mental reality. In our present time, the development of computers, which provides a further enhancement of a public space in which to think and imagine, is probably instigating a new level of development in our inner mental capacities and reality.

The idea that representational art provided a new medium or space in which to think connects with an important argument presented by William Calvin regarding our cognitive evolution. Modern human thought, and its expressions through language, music, mathematics, and art, possesses complex and contextualized structure. Our thoughts are frequently not single ideas, but organized arrangements of ideas exhibiting various internal relationships and references. Language possesses syntax and grammar which provides a structure for the arrangement of words, logic identifies rules of implication and reasoning, planning involves the arrangement of steps in sequential order, narration places events in temporal and causal sequences, and musical composition involves a host of principles for harmony and development. Modern humans think in complex Gestalts – in particular, possessing sequential order and relationships.

Sometime in our evolution this capacity to operate in complex mental spaces, involving framing, nesting, and arranging of ideas within ideas developed. Although Calvin does not give precise dates, since it is extremely difficult at this point in time to precisely determine the details of what was going on or not going on in our ancestors' minds, he does suggest that complex human thought emerged just before or coincident with the Great Awakening.[65]

If representational art provided a medium in which to juxtapose, arrange, abstract, and recombine features of the external world, then it very well could have supplied the "mental space" in which to develop complex and modern thought. It was the medium that created the new message.

Another converging line of thinking on this cognitive jump concerns human language. To recall, Ghiglieri contends that *Homo erectus* had some level of language capacity, and White clearly believes that Neanderthals possessed language. Yet, according to Diamond, it was the emergence of modern language with complex syntax that instigated the Great Awakening.[66] Language is of course a prime example of a structured, contextualized, and rule governed capacity.

Other writers, such as Reading, also see the emergence of language as responsible for the Great Awakening. Reading believes that language provides a symbolic system for representing reality that allows humans to transcend the here and now and engage in abstract and hypothetical thinking. Further, it supports the complex sequential pattern of human thinking.[67] Language is the foundation for human future consciousness. Others have in fact made the argument that the emergence of language during the Great Awakening is what led to the emergence of representational art.[68]

I think that Calvin is on the mark though in arguing that what is fundamental is the complex form of thinking that appears in modern humans. Language is one example of this evolved cognitive capacity, but then so is representational art. Archeological evidence would indicate that music, another form of complex sequential behavior, may have emerged around the same time. Perhaps all these types of behaviors appeared relatively close together because of a general cognitive jump in the capacity to represent and organize "ideas" in complex arrangements.

The significance White places on representational art is that it is relatively permanent and publicly visible - providing a "tablet" to "read" from and a "canvas" on which to tinker, embellish, and create. It has even been hypothesized that cave art, along with other artifacts, was a "tribal encyclopedia" which recorded important information that members of a tribe needed to learn in order to function in the world. Hence, although language is usually cited as the one symbolic system that allows humans to plan out sequences of behavior ahead of time, there is evidence to support the idea that representational art also served the function of not only recording significant events and ideas but developing plans for the future as well.[69]

Complex thought, and its manifestations in art, music, and language provides a possibility space in which the mind can work. It is structured and anchored in symbols, images, and rules, but it opens up an arena of mental freedom. After the Great Awakening, humans became much more creative, rather than stuck in traditional or repetitive ways of life that lasted for tens and hundreds of thousands of years. As I mentioned earlier, perceptual consciousness involves a contextual structure for experiencing the flow of time and the organization of space. Complex thought provided a mental structure in which to represent time, as well as other aspects of reality, in a more powerful and expansive way than through the more primitive processes of perception and emotion.

Prehistoric representational art was connected with the development of cognition and consciousness, but what were the motives or reasons behind creating it? As noted above, it may have served the functions of record keeping and representing plans, but there are other explanations that have been offered as well. Two popular and related explanations are that 1) the art objects were totems embodying or representing spiritual or animal powers or 2) that the art served the function of sympathetic magic; by drawing animals this would bring success in hunting the animals. The latter explanation is clearly an example of future consciousness – the drawing presumably causes a future event to occur. But the paintings and drawings do not correspond very well with the animals that were hunted by the people who created the art, and very rarely are there explicit depictions of animals actually being hunted. A third, recently popular explanation is that the art was an expression of **shamanism** – the paintings or sculptures provided access to and perhaps power over a spiritual world.

White believes that prehistoric art probably served many purposes, including all of those listed above. As another function, jewelry and body adornment probably signified social status. Of special significance to the evolution of temporal consciousness, it has been noted in recent studies of cave paintings that the art on the walls does not appear to be random but arranged into coherent wholes. The different drawings and engravings fit together. It has been suggested that the collection of art in a particular cave form **"mythograms"**, that is, stories told in pictures.[70] This is highly significant for it implies that prehistoric humans were representing temporal sequences or narratives tens of thousands of years ago, and interestingly in the form of images. Again, the medium provided a mental space in which to organize and articulate a complex structure of thought – in this case the story – a temporal structure. Our first recorded stories, and perhaps myths, were "picture books". The image and the corresponding human capacity to imagine and visualize

103

has been a powerful dimension within temporal consciousness throughout the existence of our species. In fact, it may be critical to our unique and advanced mental abilities. It may have begun on the walls of caves.

Prehistoric art and cultural periods evolved and transformed during the Upper Paleolithic Age. The Aurignacian period was followed by the Gravettian period (28,000 to 22, 000 BP) in Western Europe. During the Gravettian period there was a large increase in human representations and musical instruments, and new materials and techniques emerged. An utterly fantastic and compelling polished ivory sculpture of the bust of a woman (the "hooded lady") was produced during this time and the woman clearly was not in the same style as the numerous "Venus" sculptures that were to follow in the next period.[71] The Gravettian period, in turn, was followed by the Solutrean period (22,000 to 18,000 BP), and then the Magdalenian period (18,000 to 11,000 BP). In each case there was a significant cultural transformation, with new materials, new motifs, new styles, and new types of objects appearing on the scene.

More art and artifacts have been uncovered from the Magdalenian period than all other periods combined. Although there are some examples dated from the Gravettian period, a profusion of "Venus" (fertility) sculptures appeared during Magdalenian period. Though there is debate on this point, these Venus figurines appear to highlight the sexual features of women and it has been argued that these sculptures reflect a mother Goddess religion that dates back tens of thousands of years. The woman was worshipped as the source and giver of new life.[72] Also, although abstract and geometrical designs have been found in earlier periods, even predating the Great Awakening, but there was a huge increase in visual abstraction during the Magdalenian period. White argues that such designs and symbols must have had cultural meaning and rules behind their use and placing – they were not "gratuitous decorations." The problem, of course, is that there is no reliable way, as of yet, to understand their meaning. Still, between such abstract designs and pictorial mythograms, it seems highly probable that humans were creating a record of their ideas and observations long before the official beginning of written language and recorded history. Watson in fact suggests that Paleolithic art should be viewed as a form of writing. It is clear then that written language did not appear all at once around 4000 BC with the Sumerians, but has its antecedents in the designs and art of the late Stone Age. As of yet, we simply do not know how to read these messages from our deep past.

The Great Awakening marked a relatively abrupt change in the history of human evolution. Hominid history had already experienced several important earlier "evolutionary jumps". The first of these was when our ancestors moved out from the jungle and became erect; the second and third were the relatively quick and substantial increases in brain size, initially around two million years ago and more recently around 500 thousand years ago. Coincident with these anatomical and biological changes there were certainly significant cultural, behavioral, and psychological changes as well. The pattern of human evolutionary change that emerges is not so much a steady smooth advance, but rather relatively sudden evolutionary spurts followed by periods of relative stability.

In the 1970's, the biologists Niles Eldredge and Stephen Jay Gould proposed the theory of **"punctuated equilibria"** which described evolutionary change in terms of this idea of extended periods of stability or equilibrium followed or "punctuated" by relatively short, abrupt, and significant changes.[73] The theory of punctuated equilibria seems to apply to human evolution. Although there were various antecedents and building blocks being put into place prior to the Great Awakening, it is noteworthy how relatively recent, sudden, and dramatic was the appearance of the modern human mind and human culture.[74]

Based on this idea that historical change exhibits a pattern of sudden dramatic spurts, it is frequently argued that there have been three distinctive and fundamental cultural revolutions in the history of humanity – the Agricultural, the

Industrial, and the Informational.[75] Each of these revolutions was a jump forward that transformed all of human life. Yet, I would suggest that the Great Awakening should be included as a fourth fundamental cultural revolution – in fact, the founding revolution that truly created our modern species.

The Great Awakening appears to be primarily a mental, cultural, and technological jump rather than a biological jump. *Homo sapiens* seems to have had basically the same sized brain for at least hundred thousand years prior to the Great Awakening. What changed was what humans did with their brains.

It is a common belief that the emergence of culture represents an advance in the evolutionary process.[76] Cultural change can move much faster than biological change in that whatever is learned in a generation can be passed on to the next generation through education and the training of the young. Cultural change is purposeful and new ideas and technologies can be rapidly disseminated throughout a whole population. Genetic change only occurs once each generation at the time of conception. Genetic change appears to be based on random trial and error and requires many generations for new biological transformations to spread throughout a population. Ghiglieri argues that it was development of culture that gave humans a tremendous edge over other animals, both prey and competing predators. According to him, we became the most advanced and most dangerous animals with the evolution of culture.

Calvin sees culture as providing a new way to enhance the capabilities and the evolution of the mind.[77] Culture offers ideas, techniques, thinking principles, values, and conceptual schemes that the mind can learn to boost its abilities in dealing with the environment. Culture amplifies the powers of the mind.

Culture is not only a "tool" of the mind, but an environmental "space" in which the human mind must work. Human minds must be able to deal with the rules and values of culture. Culture has become a critical part of the human environment, at least as important as the natural environment.

I have already introduced the hypothesis that the human self and culture reciprocally evolved. Broadening this hypothesis, the total make-up of the human mind, which includes the self, as well as supporting cognitive, motivational, emotional, perceptual, and behavioral capacities, co-evolved with human culture. Humans born into the world of culture must learn its structure of rules and values in order to function and survive within it, but in turn, it is human minds that contribute new ideas, technologies, and values into the growing body of culture.

But it is not just mind and culture that reciprocally evolve; culture and genes intertwine, and co-evolve. For approximately the last 40,000 years, if not longer, the necessities and requirements of living in a cultured world have probably been a significant selecting mechanism on our genetic evolution. Humans that have been genetically selected are those that best survive, replicate, and flourish in a world of culture. Culture influences genetics, providing the environment in which continued genetic evolution occurs. Different genetic combinations compete with each other in the environment of culture.

Recent scientific evidence lends support to this idea that the growth of culture has had an influence on the genetic structure of humans. Although, as noted above, Homo sapiens during the Great Awakening appear basically the same anatomically as contemporary humans, detailed genetic research in two different studies has revealed some rather significant genetic differences between modern humans and *Homo sapiens* of 50,000 years ago. It may be that approximately seven percent of human genes have been altered over the last 50,000 years. The argument, presented by the geneticists who recently made this discovery, is that the cultural environment of humans has been selecting certain genetic types as most compatible with the special demands of civilization. If these experimental results are further validated, then although the gross anatomy of the human brain may not have changed much since the

Great Awakening, there are probably some important changes in the human brain that so far have gone unnoticed.[78]

There is the reverse argument from the discipline of sociobiology that human culture is a reflection and creation of our unique genetic make-up; that is, culture is in our genes.[79] Since culture provides a powerful mechanism for improving the capacity for humans to compete and survive within nature, hominids with genes that pre-disposed them to assimilate cultural principles were naturally selected for. Many of the basic features of culture, including cooperation, altruism, symbolization, and general principles of language appear built into us genetically. In fact, there may be hundreds of human "cultural universals" that are genetically inherited.[80].

Culture and genes therefore appear to form a reciprocity, each variable driving the further evolution of the other. It is not just simply that the individual psychological development of a child is an interaction effect of culture and genes – of nurture and nature – but nature and nurture intrinsically reflect the influence of each other. There is no pure nature or pure nurture – nature and nurture interpenetrate. Our nature (our genes) has been selected for and influenced by culture, and our ways of nurturing (our culture) is a reflection of our genes. Culture is in our genes and genes are in our culture. The interactive and interdependent evolution of genes and culture is a clear and highly significant example of reciprocal evolution in humans.

Another concept demonstrating how reciprocal evolution has operated within the history of humanity is **"The Red Queen Principle."** Based on an idea taken from Lewis Carroll's *Alice in Wonderland*, Calvin uses this principle to describe the dynamics of human evolution. In Carroll's story the Queen of Hearts explains the principle to Alice. Imagine being on a treadmill that keeps moving faster and faster. In order to just stay in the same place, a person would have to walk faster and faster. If a person walked at the same pace, they would go backwards since the treadmill is accelerating. Calvin suggests that our history shows clear examples of the Red Queen

Principle. As we became more adept hunters, the animals we hunted adapted to our predatory behaviors and became more elusive and quick in avoiding us. Hence, we were pushed into having to become even more adept at hunting just to stay even. Predator and prey reciprocally evolve, each advance in one causing the other to move forward as well. Although Ghiglieri does not use the expression "Red Queen," he provides another example of its operation in discussing intra and inter-group competition among males. Hominid males within a group competed against each other for females (sexual selection) and different hunting groups of males competed against each other for food. In both cases, males are continually forced into innovation and further development because their competitors - other males - are doing the same thing to get ahead as well. Ghiglieri thinks that the evolution of male intelligence and aggressiveness was fueled by males having to compete against each other for sex and food. In general, competition can lead to reciprocal evolution due to the Red Queen Principle.[81]

The science fiction writer Greg Bear provides a fascinating illustration of the Red Queen principle applied to genes and culture in his novels *Darwin's Radio* and *Darwin's Children*.[82] Bear speculates that the last big genetic jump in human evolution - that is the emergence of our species - was provoked by environmental stress, and that given our increasingly demanding and stressful present culture, a new genetic jump could soon occur in the human line. Genes, from a sociobiological perspective, generate culture, but then culture surges forward due to innovation and social learning eventually putting adaptive stress on the human population that created the culture.

In considering the significance of culture in human evolution, an important general trend in our history becomes very noticeable. Human evolution appears to be accelerating. In early hominid history, physical and behavioral changes (as for example evidenced in tool making) moved relatively slowly. It took approximately two million years for hominid brain capacity to significantly increase above the level of chimpanzees after our ancestors became erect. The earliest tools did not change much during the period of *Australopithecus* and *Homo habilis,* and then with the appearance of *Homo erectus* and new tools and behaviors there was not a significant amount of change for another million years. The next burst in evolution began around 500 thousand years ago, with increasing brain size and a variety of new innovations throughout Africa over the next few hundred thousand years. But the power of cultural evolution, which had been slowly building, eventually reached a critical threshold and around 50 thousand years ago accelerated further developments, producing a succession of new and distinct cultures which appeared approximately every ten thousand years. Subsequent changes, as we move from prehistory to agriculture, to the emergence of cities and empires, modernization, the Scientific Revolution, industrialization, globalization, and the Information Age, come increasingly more quickly. Human evolution may occur in bursts, but the bursts are getting closer and closer together.

As I noted in the opening chapter, it is a common view among contemporary writers that things are moving faster and faster.[83] This accelerative process extends back to the beginnings of human evolution. One way of explaining this accelerative trend is that the rate at which information is being created, stored, processed, and disseminated by humans is increasing.[84] Evolution is speeding up because information processing is speeding up. The emergence and growth of culture has greatly facilitated this acceleration of information and information processing. Tools, self-reflection, language, art, and increasing trade and exchange, all creations of human culture, speed up the evolutionary process and the growth of information. As Barbara Marx Hubbard states, these new components are "design innovations" in the evolutionary process.[85] These innovations are expressions and creations of evolution, but in turn enrich the evolutionary process and amplify the rate of evolution in a reciprocal loop.[86] The emergence of humans and subsequently human culture facilitated the "evolution of evolution."

One final important theme regarding early cultural evolution and the Great Awakening is humanity's realization of personal mortality. Explanations for the Great Awakening include competition with Neanderthals, the development of modern language, and the emergence of complex thought, but it has also been proposed by different writers that it was the discovery of death that lit the fire of cultural evolution. No other existing animal species indicates in their behavior any understanding that someday they individually will die.[87] Understanding personal death entails an extended view of one's future and a clear level of self-awareness - "I am going to die." Although the first undisputed examples of the burial of the dead extend back 100,000 years (for whatever reasons it was done), archeological evidence indicates that coincident with the Great Awakening there is unequivocal evidence that *Homo sapiens* and Neanderthals buried their dead with various artifacts placed in the graves, which would seem to reflect a belief in an afterlife.[88] A salient development in future consciousness - the realization of personal mortality combined with a belief in a hereafter - would then be at least partly responsible for the emergence of human culture.

Leonard Shlain believes that women first clearly realized the inevitability of personal death and that women accepted it better than men. According to Shlain, men seem to fear it more. Further, as others have also argued, Shlain thinks that the burying of the dead with beads, flowers, and various other artifacts implied that humans concluded that we didn't really die but somehow continued to exist. Hence, Shlain connects the burying of the dead with self-delusion based on fear and superstition. He states that men predominately invented mythical places after death to assuage their fear of death.

Shlain also thinks that humans developed art to create something to be remembered - to achieve some immortality - in the face of the conscious realization that life is transient and finite. Humans search for purpose and meaning, according to Shlain, because of the realization of death. He states that these early works of art were both self-satisfying and intended to be recognized by others. As early humans, we wanted to be remembered as unique individuals and from this we derived a sense of immortality. One meaning of the drawings on the walls of caves may simply be "Here I was – remember me."

Echoing similar themes, the philosopher of time J. T. Fraser states that humans find death "unacceptable" at a deep emotional level and all of the great creations of human culture, including art, religion, philosophy, and science, are attempts to combat the passage and end of personal time. All these high cultural achievements are efforts to discover or create permanence in a universe, which according to Fraser, is fundamentally one of unrest and flux. For Fraser, the "discovery of death" depended upon humans developing an extended sense of past and future. This expansion in temporal consciousness brought a great survival advantage to our ancestors, but at a price. Because we could see ahead, we had the capacity to plan, and thus emerged the realization of personal responsibility for our lives. But in realizing that we were responsible for our lives and that someday we would die, we lost our sense of peace.[89]

Hence, understanding personal death is not simply a cognitive insight about the future, it is a highly charged emotional experience regarding the personal future as well. We feel it – we are often terrified by it. The cultural anthropologist, Ernest Becker, in his Pulitzer Prize winning book, *The Denial of Death*, argues that personal death is humankind's most powerful fear and motivates a great deal of human activity and creation.[90] This psychological thesis obviously reinforces the views of Shlain and Fraser.

Fraser pays particular attention to myth and religion, the topic of the next chapter, as an early expression of denying death and the passage of time. The emergence of religion brought with it the promise of an after-life, a future beyond the death of the body. Almost all major world religions contain the idea of life after death, and numerous mythic tales describe the resurrection of both humans and various deities. Undoubtedly there are other contributing factors to the development of religion, but clearly one of the most important ones has been trying to find a palatable answer to our personal futures in the face of the incontrovertible insight that someday we will die. Perhaps this is a way to pacify the anguish and fear of the human soul, but following this line of reasoning, religion, one of the most powerful achievements of human culture, clearly emerges in the conscious realization of a fundamental fact about our personal futures.

Sex, Love, and Aggression – Women, Men, and Children

"Adam confronted a knotty problem no other male of any other species ever had to contend with – a female with a mind of her own"

Leonard Shlain

At this point in my survey of prehistory, I am shifting focus from the evolution of humans in general to the distinctive features of male and female evolution, in particular the unique qualities of male and female psychology and the reproductive challenges and strategies of the two sexes. It is also important to examine the evolving relationship between men and women. Our evolution is a co-evolution - a reciprocal evolution - of men and women - of two intertwining psychologies. The respective psychologies of the two sexes show up in our first myths and the two main theories of time that emerged in the prehistoric world. Our sense of the future is intimately connected with our dual sexuality as a species.

As argued throughout this chapter, it is important to look at the connection between biology, survival needs, and basic psychological and social activities to understand the evolution of future consciousness. Two of the most powerful biologically based human motives are sex and aggression, and both of these motives have played a significant role in the evolution of future consciousness. Leonard Shlain, in his book *Time, Sex, and Power*, presents an evolutionary explanation of the emergence of future consciousness based on a set of fundamental changes in reproduction, sex, male-female relationships, and hunting behavior which, according to him, took place in the last hundred and fifty thousand years.[91] Shlain argues that the Great Awakening was intimately connected to an evolutionary change in how we engaged in sex and bonded together, and the role hunting played in this process.

According to Shlain, the evolution of increasingly larger heads in our hominid line, coupled with our developing bipedalism and the resulting constriction of the birth canal, put great physiological stress on women in childbirth. The evolutionary "solution" to this problem was toward increasingly premature births, producing progressively more immature offspring. Because children were born increasingly premature, mothers needed to focus more on childcare for longer periods of time as our genetic line evolved.

On a related note, Diamond argues that as our food gathering culture evolved, which involved more sophisticated tools and behaviors, more time was needed to teach children the survival skills of human life. Children were less capable and more dependent and needed to learn more to be successful adults. Culture played a bigger role.[92] Again, the general point, now from a cultural and social learning perspective, is that, as humans evolved, children became more dependent and required more attention in their upbringing.

Shlain argues that as a result of increasing childcare demands, mothers needed more dependable, responsible, and committed male mates to supply security, food, and stability while they were busy tending to the children. Yet from a reproductive perspective, men are naturally motivated to engage in intercourse with as many different women as possible in order to maximize the number of potential offspring. Once a male impregnates a female, he can move on to another female and reproduce again. Further, male primates, in comparison with females, generally do not spend much time tending to childcare. In essence, although our female ancestors needed committed and attentive males, human males by nature and genetic heritage are polygamous and want to wander from the nest. Hence, how do you get the male to stay at home? The answer, according to Shlain, was sex.

One of the most unusual biological features of female humans is **cryptic** (or concealed) **ovulation**. Whereas other female primates and mammals exhibit estrus, a state of fertility with distinct and visible physical and behavioral symptoms, for human females there are no clear outward signs of fertility. Why would human females develop cryptic ovulation, since it does not seem to serve the function of maximizing the chances of reproduction during sex? According to Shlain, cryptic ovulation increases the time spent in sex and facilitates the development of committed males. The male must stick around and engage in sex more frequently with the same female since he doesn't know when the female is fertile.

Humans seem to be more invested mentally and behaviorally in sex than any other species. Testosterone levels are fairly steady and high in human adult males with a noticeable and dramatic peak during adolescence. Shlain argues, in fact, that adult human males are in a constant state of sexual arousal. Hence human males will have sex with a woman regardless of whether she is fertile or not. For her part, the human female is potentially receptive all the time, not only during her fertile period. Generally speaking, humans have sex anytime and anyplace, though not necessarily with anybody.

As Diamond notes, since most human sex does not directly connect to reproduction, it must serve some other important function. The answer that Shlain, Diamond, and others have presented is that it facilitates male – female bonding. For Shlain, not only does sex serve as a way to express and reinforce affection between the male and female, it provides a negotiation tool for solidifying long term bonding and commitment.

Although humans engage in sex on a frequent and continuous basis, with the female need for long term commitment from the male, sex moved beyond a simple impulsive act. The woman needed a mate who would not only impregnate her but stick by her. The male had to convince the woman that he would stay with her and provide protection, food, and parental care in exchange for regular sex. Consequently, according to Shlain, humans began to engage in sex with forethought; considerations regarding the future became an essential prelude to the act of sex. Sex became a negotiation between the female and the male and language became an important tool in this process. Where men used it as a tool of persuasion, women, after considering the implications of engaging in sex with a particular male and assessing his potential as a partner, used language to question, interrogate, and respond. The ancient Hebrew story of *Lilith*, the first wife of Adam, who would not bend to Adam's will is a reflection of this new relationship of male and female. The female was no longer automatically compliant.[93]

This negotiation holds special significance in regard to the evolution of future consciousness. Shlain argues that a basic exchange developed between women and men - the exchange of meat and iron for sex. Female humans lose a good deal of iron during their excessively heavy menstrual period and meat is one primary source of iron. Although it is traditionally the prehistoric male who is credited as the hunter and meat eater, according to Shlain, it was the female who drove the need for hunting meat. The woman needs meat to replenish her level of iron and the male provides it for the woman as part of the bargain for sex. As Shlain points out there is no other animal that trades meat for sex. As I noted earlier, there is evidence running back millions of years that hominids could delay immediate hunger gratification. Thus, it is not the appeasement of hunger men kill for but sex. Shlain's argument is a further elaboration on this significant point of human psychology and the evolution of future consciousness. Male hunters, instead of consuming meat from prey immediately, brought the meat back to the female in order to ensure (for the future) continued sex. Hunting, sex, and the eating of meat are interconnected and all highly future-oriented behaviors.

116

Shlain believes that males have had to adapt to a variety of changes that occurred in human females since the emergence of our species. He asserts that for the last 150,000 years men have been attempting to regain power lost to females. The male evolved more adept brains and higher forms of cognition to deal with the question of what he must do to convince the female to have sex with him. Males became what women wanted them to be. These changes in the male involved the capacity for enduring delayed gratification in both food and sex and achieving long term stability patterns in behavior, that is, monogamy. Yet this evolutionary change was a reciprocal trade-off. Monogamy gave the woman security and food, but made her dependent on the male. Further, through the process of sexual selection, females undoubtedly evolved a set of sexually desirable traits for the male. If the male is what the woman wants, the woman is what the male wants.

According to Shlain, the book of *Genesis* has it backwards. The pain of childbirth is the cause of self-consciousness and not the result. For Shlain, the unique qualities of human sexuality and male-female relationships are connected with the emergence of self-consciousness, self-control, and free will in humans. Sex and reproduction de-coupled in humans. Humans, both male and female, were able and potentially willing to have sex at any time. But the female also developed the capacity to withhold sex at any time; she gained control over this biological function. She could disengage from the "be-here-now" and turn sex into a negotiation concerning the future. Hence, Shlain states that she developed an ego and a self-consciousness that could stand back from the world and from her instincts and consider the consequences of her actions. She acquired "free will" and a decision making capacity regarding the future. The ultimate objective of this new capacity was the establishment of a better male-female connection in child rearing – a stable parenting situation. Shlain calls this the **"Original Choice."** Thus free will and future consciousness arose over the issues of sex, bonding, and child rearing – all psychological capacities connected with reproduction and the continuation of the species.

The capacity for choice and free will is often identified as one of the distinctive features of our species.[94] We are not ruled by instinct or set patterns of behavior. From this discussion of sex and bonding, we again see that free will is intimately connected with future consciousness. (Recall from the previous chapter the discussion on the interconnection between choice, free will, and future consciousness.) Only if we are aware of the future does it make sense to say we make choices. Making a choice involves thinking about different possible and potential actions pertaining to the future. In particular, when we make choices we consider the consequences of our actions. If a person doesn't consider the consequences of his or her actions, we say that the person is "thoughtless," rash, or impulsive, reacting to the moment. Freedom of choice arises in the opening of the mind to the future and the various possibilities of action and their resultant consequences. In the present discussion we see that the choices were over whether to engage in sex or not, as well as whether to commit and bond or not. These choices were made with an eye on the future.

Shlain also identifies the female as the first of the genders to conceptualize "deep time" and the future due to her understanding of the interconnection of the lunar cycle, pregnancy, and birth, as well as her grasp of the link between sex and reproduction. An apprehension of these processes all required an extended time perspective. While man had his gaze on the heavens, it was woman squatting in the dirt who noticed the synchronicity of the lunar and menstruation cycles, and figured out that the duration of pregnancy equated to a predictable number of lunar cycles.

As I have argued, there is a great evolutionary advantage in developing a more expansive sense and understanding of time – of learning from the past and anticipating and planning for the future. For Shlain, enhanced foresight develops in the female first – with the realization of the connection of sex and birth, but, as history unfolded, Shlain states that men would give themselves credit for the discovery of extended time. In early religion and myth, such as in ancient Mesopotamia, calendars and time originally fell under the sovereignty of female goddesses. But in later myths, male deities gained control of time and calendars. Also in Hinduism and Greek mythology, the sovereignty over time shifted from female to male deities.

For the female, it is self-evident that her children belong to her. For the male, it is not self-evident that he is the parent of his children. One of the main reasons behind the development of monogamy, from the male's point of view, was to ensure that the children he was protecting, and helping to raise, genetically belonged to him. He stayed around to protect his female mate from being impregnated by other males. But in order for a monogamous sexual relationship to have significance to him, the male needs to understand fatherhood - that males, through sex, are also responsible for the birth of children. Whether the male "discovered" the connection of sex and birth or was taught this basic fact of life by the female, the insight of fatherhood and commitment to the consequent time and emotional investment of child-rearing required of him was a significant step forward in the evolution of future consciousness in the male.

Shlain believes that for the father the child becomes both representative of the future and a way of conquering his own death. Through the child the father lives on. This evolution in male psychology brought with it the idea of honoring the father and further reinforced and enriched the mechanism of culture as a way to pass on the ideas and values of the father. This insight and all the consequent practices that resulted from it constitutes the beginnings of fraternal heritage, which on one hand implies that we should look to our male ancestors (the past) for guidance but is predicated on the desire of the father to pass on to his children (the future) his identity and values. The child looks to the past in reverence while the father looks with hope to the future.

Shlain states that males discovered paternity around 40,000 years ago. The male changed his attitude toward the child and the mother and drastically altered the future course of human society creating a patriarchal social system. Males increasingly "controlled" their own sexual behavior as well as that of their female mates. Males altered society so that their heirs could carry on their "names".

Let's turn now from love and sex to aggression and violence. In spite of the apparent oppositional qualities of these two motivational tendencies, we have already seen that, arguably, these motives are connected in our prehistory. According to Shlain and Ghiglieri, violent hunting behavior and aggressive competition among males were two primary means for achieving sex and bonding with females.

Human males are the most efficient hunters and killers in nature. In spite of the myth of the noble savage who lives in harmony and relative peace with nature, prehistoric human males are probably responsible for the extinction of numerous species across the entire face of the earth.[95] There has been an ongoing debate in the history of psychology whether aggression is learned or genetically inherited in humans, but our prehistory as well as contemporary physiological research seems to indicate a strong genetic component to male aggression.[96] As Howard Bloom illustrates with numerous examples, aggression, violence, and killing are ubiquitous throughout the animal kingdom and existed long before the emergence of human culture; aggressive and often violent human behavior is simply a manifestation of this fundamental propensity in nature.[97]

If anything, humans have taken violence and aggression to new heights. Shlain argues that *Homo sapiens* males have a highly developed aggressive nature. Only with *Homo sapiens* did organized kills develop to a high level. We became highly efficient and extremely dangerous social predators. For Shlain the biggest spike in increasing aggression occurred in the last 40,000 years. Diamond concurs on this point that "Man the Hunter" only emerged full-blown with our modern species. There is though disagreement over to what degree efficient hunting behavior was present earlier. Ghiglieri thinks that *Homo erectus* showed advanced hunting behaviors. At whatever pace male aggression and violence evolved, over the last three million years the hominid line transformed from being an animal of prey to being the most advanced and ferocious predator on the earth.

For Shlain, males had to develop the virtue of courage in order to hunt and kill animals, superior in size and ferocity. Courage is a future focused virtue in that the individual demonstrating this virtue exhibits determination to act in the face of anticipated danger. Although courage is a future focused virtue that is highly valued in many cultures, its evolution may be predicated on serving predatory behavior and human aggression. Throughout human history, it has been great warriors who most frequently are identified as courageous.

Predatory behavior is clearly future-focused since it serves the future end of obtaining food. Within human evolution, aggression also has a future focused dimension – it is not simply an impulsive instinctual reaction. Not only does aggression connect with predation, but following Ghiglieri, aggression within male competition helps the male in finding female mates and sexual partners. Reinforcing this point, Bloom argues that throughout history women tend to select as mates those males who are the most violent and aggressive.[98] Yet, aggression has mixed future benefits. As Diamond points out, modern humans have inherited two destructive traits from our ancestors. We kill each other in large numbers – our aggressive competitiveness – and we kill off other species. The general tendency to destroy and despoil our environment is an offshoot of our aggressive and highly effective hunting behavior.[99] In this case, the evolutionary heritage of future consciousness is a double-edged sword.

As we have seen, following the arguments of Shlain and other writers, sex and aggression are linked together and in humans both involve strong elements of future consciousness. As one final perspective on this topic, Harold Bloom, in his book *The Lucifer Principle*, presents some important and relevant ideas. As noted above, Bloom contends that violence is a fundamental and pervasive phenomenon throughout nature, and that, moreover, violence, killing, and destruction have a highly significant adaptive function or value. Nature is a competitive arena and violence is an expression of this competitive dimension; nature evolves through competition, and often this competition involves the aggressive or violent intimidation, if not elimination, of competitors. Of special note, as Bloom documents with countless examples, members within a species compete against each other, often through aggression and violence, for dominance in pecking order hierarchies; the strongest, most aggressive, and most intimidating make it to the top. For males, dominance within a pecking order brings with it the privilege to procreate with females; those males at the bottom of a pecking order have a difficult time finding receptive mates. Hence, competition and violence are connected with power in a social group and create opportunities for sex and procreation. Bloom also contends that violent competition between social groups has occurred throughout human history, again serving the function of securing female mates for procreation, as well as control over territory and resources. As documented throughout human history, conquering armies often kill the children and procreate with the females of the conquered group; males of various other species also show the same behavior after defeating the dominant male within a social group. Human males compete both individually and socially, often violently, for power and the opportunity to continue their genetic line.[100]

Hence, Bloom's analysis further reinforces the connection between sex, bonding, and procreation on one hand and aggression and violence on the other. Love and hate – the great polarities of human existence - often equated with the good and evil sides of our species – have evolved together in our history. Both human qualities reflect an inseparable mixture of genes and culture, of impulse and deliberation. As argued in the opening chapter, future consciousness is not simply a cognitive capacity, but one that also involves a powerful emotional and motivational dimension as well. Two of the key elements in the emotional-motivational dimension are sex and love and aggression and violence.

Agriculture, Reciprocity, Conquest, and Ecological Control

"The history of civilization is essentially a history of mankind's increasing ability to predict the future."

Anthony Reading

After the Great Awakening, the next great burst in human evolution was the development of agriculture and the consequent emergence of large urban settlements which coalesced into the earliest nations and empires. Agriculture, in fact, is frequently cited as the "greatest idea" humanity has ever created.[101] This developmental jump into what we would call "human civilization" occurred roughly between twelve and six thousand years ago, during a period when the climate on the earth warmed and stabilized, though the seeds (so to speak) of this monumental revolution in human existence go back to the Upper Paleolithic Age. The remains of settlements of a hundred people or more, containing relatively permanent large habitable structures, can be found in Eastern Europe and Western Russia dating back tens of thousands of years.[102] In general, humans probably began to move toward a more sedentary lifestyle before the appearance of large scale and systematic agriculture.[103] Also, based upon archeological evidence, it has been argued that the rudiments of civilization, which include art, symbolic notation, and large habitats, go back at least 20,000 years.[104] Trade and exchange, connecting settlements and different groups of people across long distances, also extend back tens of thousands of years. Also it seems clear that the first efforts at agriculture occurred in a relatively unsystematic, trial and error fashion, long before the appearance of the first major urban centers that depended primarily upon agriculture for food. Yet beginning in the "Fertile Crescent" in Mesopotamia and involving first the domestication of plants, followed approximately a thousand years later with the domestication of animals, humanity across much of the face of the globe moved from a predominantly foraging, hunter-gatherer, and nomadic lifestyle to a more sedentary, urbanized, and agriculturally based way of life. This occurred over roughly a five-thousand year period constituting the Agricultural Revolution.[105] Describing this transformation in more combative terms, Bloom depicts the process as a growing competition between burgeoning cities and pre-existing nomadic peoples, with cities eventually winning and progressively wiping out indigenous populations.[106]

The emergence of agriculture is a prime example of both reciprocal evolution and the expansion of future consciousness. As David Christian notes in his grand history of both physical and human evolution, *Maps of Time*, domestication, which is an essential element of agriculture, transforms both the life forms domesticated and those doing the domestication. Through selective breeding humans altered both animal and plant life, but in this process, humans became increasingly dependent on these altered life forms for their existence. Over the millennia our bodies have undoubtedly adjusted to the types of foodstuffs that we have nurtured and created through domestication.[107] And furthermore, we have progressively lost the abilities of our ancestors to hunt and forage for food in the natural environment. Paraphrasing the philosopher Hegel, the master became the slave of its own creation.

Agriculture is connected to future consciousness in a very simple and dramatic way. As agriculture spread across the globe, it represented the most pervasive and powerful intentional manipulation of the natural environment undertaken by humans up to that point in time. Agriculture is long-term, goal-directed planned behavior on a vast scale. Humans were not intentionally trying to change the biosphere of the earth when they began to domesticate and plant wheat, rice, and barley, but they were purposefully transforming the natural environment, bit by bit, and doing so in such increasingly great

124

numbers that the result was a global change of unparalleled proportions.

This purposeful manipulation of the environment included a significant reciprocal feedback loop. To begin with, agriculture was neither systematic nor large scale. But as humans became increasingly dependent upon whatever foodstuffs were planted and harvested, and learned about the process of cultivation, through trial and error larger scale and more systematic efforts emerged. Humankind, in interacting with nature, learned how to manipulate and control it. Although agriculture is a purposeful activity directed by humans, its growth was a co-evolutionary process.

Humans undoubtedly engaged in coordinated and long term planning prior to the Agricultural Revolution. The hunter-gatherer way of life also involved planning over an extended period of time. Hunter-gatherers had to learn the patterns of migration and the seasonal cycles of vegetation in the environment around them. Coordinating a hunt and executing it also required taking an extended view on the future. As noted earlier, Calvin has suggested that the general capacity for planning and foresight was dramatically enhanced through having to learn how to throw weapons and bring down moving animals in a hunt.[108] Still, as it evolved agriculture required a level of socially coordinated long term planning on a scale that far exceeded hunter - gatherer activities.

Agriculture involves a significant alteration of the environment. No longer were humans simply adapting in a passive or reactive way; our ancestors were altering the environment to serve our ends. This is active adaptation. Humans were undoubtedly purposefully manipulating their environment before agriculture, but agriculture is a significant jump forward in this capacity for active adaptation. The active and purposeful manipulation of the environment, and the degree to which we can accomplish this end, is one of the most distinctive features of our species.

Purposefully controlling the environment to serve human ends is one of the most fundamental expressions of future consciousness. It is a capacity that humans have increasingly improved throughout history. As noted earlier in this chapter, all life requires resources from the environment in order to survive, and all life has developed abilities for seeking out and procuring these resources. What humans have become more efficient at throughout evolution is extracting resources and finding ways to maximize the potential output of the environment.[109] Much of technology serves this end, from the stone tool for hunting and butchering animals to dams and massive electrical generators for providing energy. Manipulating the environment to serve needs and goals is a future oriented activity. Evolving this future oriented capacity serves human survival. Thus although we may criticize modern human society for using up natural resources and destroying the environment without any forethought regarding where it is all leading, such activities were built upon thinking about the future and purposefully acting to serve future ends. The world of agriculture, industry, cities, technology, and the global transformation of earth ecology is a manifestation of future consciousness. What we can say in response to present ecological concerns is that our goals and plans were probably too short sighted in the past – we did not sufficiently consider longer term consequences – in the future we need to think out further into time.

The growth of agriculture was connected with the emergence of increasingly larger urban settlements. Humans became more sedentary and clustered into progressively larger groups. Although Upper Paleolithic humans both created relatively permanent settlements and lived together in social groups (probably connected through kinship), Paleolithic humans in general probably lived in relatively small groups of ten to twenty individuals.[110] Although towns and cities are localized concentrations of population, these urban settlements became hubs of trading and exchange; the economic lines of interaction extended outward across large geographical areas. Again, although Upper Paleolithic humans traded with each other, often across long distances, with the emergence of cities trade and exchange increased dramatically. As Bloom describes it, with the growth of cities we see the emergence of *"Homo commercialis."* Through trade, cities provided a way in which humanity began to weave itself together in a more intricate, rich, and more extended fashion than ever before.[111]

As Bloom points out, every human society has some kind of principle of "give and take." In order "to get you have to give." Bloom refers to this principle as "reciprocity" and asserts that trade and exchange are built on the principle of reciprocity. In fact, for Bloom, reciprocity is the great power attractor in human affairs. It was one of the two major forces that weaved humanity together.[112] (I will come to Bloom's second major force momentarily.) As cities grew, with consequent increasing specialization, division of labor, and improved transportation, humans expanded the range and form of reciprocities of exchange among themselves. Interdependencies grew; the human social network progressively evolved.

Thus we see in Bloom's description another significant application of the idea of reciprocity to the structure and evolution of humanity. Not only is reciprocity the basic principle that underlies trade and exchange among humans, it is reciprocal trade and exchange that provides one of the most powerful integrative forces at work in human evolution. We come together and weave networks of interdependency through the creation of reciprocities.

In his book *Nonzero: The Logic of Human Destiny*, Robert Wright presents a related argument regarding the importance of reciprocity in cultural evolution.[113] Wright contends that there is a discernable and progressive general direction throughout all of human history. This direction is toward increasing social complexity based on the development of mutually beneficial relationships or transactions among people, individually and collectively. Although we all possess a basic need to serve our own individual interests, we repeatedly find that we can establish "win-win" transactions with others that benefit us individually. These arrangements, both supported and fueled by technological innovations, add to the complexity of our societies. In essence, social complexity grows through the evolution of reciprocities that mutually support the individual lives of those involved. Through the establishment of various types of exchange we evolve social complexity as well as serve ourselves. Note that this is another example of how the self and society form a reciprocity – each facilitates the evolution of the other.

Wright contends that if we look at human history we find that cultures do not remain static but, at different rates, invariably move in the direction of establishing more and more "win-win" relationships among the members, consequently moving in the direction of increasing complexity. There are, of course, many cases where individual societies collapse or disintegrate, e.g., the Roman Empire, the Egyptian Empire, and the civilizations of the Aztecs, Mayans, and Incas, but the overall direction across the entire globe has been increasing complexity and "win-win" relationships. Wright argues that this basic pattern of evolution or progress applies not only to all individual human cultures but to the total global scene of humanity. As a social species, our world is more complex and filled with more "win-win" reciprocities than in the past, and in fact, we are increasingly integrated and connected via these transactions and arrangements.

For Wright reciprocal exchanges do not just develop between urban settlements, as Bloom highlights, but as a general principle, such exchanges developed among humans at all levels of social organization. One important example of this would be the evolution of male – female relationships which I discussed in the last section. Wright identifies the formation of reciprocities as the essence of increasing social complexity or social evolution, whether it is among the members of a society or a family. Social evolution is the creation of new reciprocities. Wright notes that the creation of reciprocities benefits the individuals who establish the exchange. Social complexity supports individual survival or development.

This final point relates back to Bloom's contention that it was trade and exchange that increased the power and amplified the growth of cities. Cities as nodes in a network of exchange grew as the network got richer and stronger. The complexity of the whole benefits the parts. Thus, we come back to Bloom's argument that urbanized humanity progressively out-competed nomadic and indigenous groups of humans. Cities became much more complex, organized, and richer through trade and exchange than nomadic people. The web of influence and power of cities grew as urbanized humanity progressively assimilated the more ancient hunter-gatherer people of the world. There have been, of course, episodes where nomadic groups have conquered urban societies, but the overall trend through history has been in the opposite direction. We see this trend still continuing today.

We now come to Bloom's second force, one that has progressively integrated humanity. Bloom believes it is conquest. As I have already described, prehistoric human males developed highly efficient hunting behaviors as well as a strong disposition toward aggression and violence in competition among themselves that served them well in obtaining meat and mates. Still, there does not seem to be any clear evidence that prehistoric humans engaged in war. Yet when we come to the rise of cities and agriculture, war emerges among different cultures and groups of people. We show a history, since becoming "civilized," of engaging in almost incessant war and conquest. War as a means to conquer is a future oriented form of thinking and behavior; it involves goal setting, planning, flexibility, creativity, courage, and often complex strategic thinking. In fact, the capacity for military strategy emerged in our history as one of the most socially celebrated forms of future consciousness; many of the most famous people of the past are military strategists and leaders.

Military conquest leads to the assimilation and connecting together of different groups of people through domination. War and conquest leads to the formation of nations and empires. As Bloom notes, whereas reciprocity unites through exchange, conquest unites through domination.

Clearly, competition has been a major force in the evolution of life and humanity, although not all competition is aggressive or violent. Competition can occur along any dimension or skill that makes a difference in terms of survival. As noted earlier, competition fuels the Red Queen Principle. Competition leads to the extinction and elimination of species, or individual members within a species, in the interactive "struggle for survival." *Homo sapiens* probably out-competed and extinguished Neanderthals. Competition is a central principle in Darwin's theory of evolution (though not the only one).[114] War and conquest is one major manifestation of evolutionary competition in the history of humanity.

Bloom argues that there are two apparently opposite principles at work in the evolution of modern humanity, both of which produced integration - reciprocity and conquest. This duality of principles corresponds with a major theme in contemporary evolutionary thought. As Stewart argues, as does Lyn Margulis, who has achieved great notoriety and influence in her theory of evolution, cooperation is a powerful force in evolution.[115] Cooperation is often juxtaposed with competition as the two primary forces at work in evolution. Although competition is important, life also evolves through the development of cooperative relationships, e.g., symbiotic connections, multi-cellular aggregations, ecological interdependencies, and divisions of labor.

Reciprocity is a form of cooperation - in fact, it may be the essence and fuel of cooperation for without benefit to all those individuals involved, why would individuals establish cooperative relationships? Stewart argues that the development of cooperative relationships can be of benefit for all participants and provides an overall progressive direction to evolution.[116] On this point he sounds very similar to Wright. Thus reciprocity, in so far as it is built on cooperative exchanges, provides a complementary mechanism to competition in driving the evolution of life and humanity.

Reciprocity has not only served as a primary mechanism for the creation of biological and social complexity, it provides a universal principle upon which human values and ethics around the world have been developed and defined. Reciprocity is the foundation of the concepts of justice, equity, and perhaps even human care and kindness. To recall, as Bloom notes, all human cultures acknowledge and reinforce in their values and practices the importance of giving in order to get. If someone takes but does not give, humans in general find this objectionable. Exchange must be fair, or else someone feels cheated or robbed. Humans often give though not with any immediate return benefit, but with the hope and expectation that in the future their good deeds will be returned in kind. Social relationships and bonding are often cemented through gift giving and favors offered, with the expectation of future return and obligation. Sociobiologists have argued that apparently selfless and altruistic behaviors in humans are built upon the principle of "reciprocal altruism." We give of ourselves because in some way it benefits us (in particular, our genetic line) in the long run.[117]

The valuing of reciprocity can also serve as an instigator for retribution (retributive justice), violence, and even war. We enact punishments on those who do not follow this principle - who take without giving - and we go to war when we believe we are wronged or not given what we believe is our due. War is often started over perceived injustices or inequities. Thus although reciprocity and conquest at one level are opposites, the perception of not honoring the principle of reciprocity can lead to war and violence. Our history is filled with examples of this form of thinking and behavior.

Wright believes that internal social revolutions occur because leaders do not sufficiently abide by the principle of reciprocity. In Wright's mind, authoritarian systems of government invariably falter or fail at some point because they do not support "win-win" transactions among their members. They collapse or have to be revamped because of a lop-sided "win-lose" arrangement by which those at the top of the system accrue a disproportionate share of the benefits resulting from the social transactions among their members.

People rebel when they feel there is too much injustice and inequity in their social system.

Reciprocity therefore emerges in ancient times as an operative principle in human future consciousness. With an eye on the future, we exchanged and we gave in order to receive; we conceptualized early on that what goes round comes round; we developed mutual commitments and social bonding based on gift giving and "unselfish acts." We developed moral expectations that we must practice and honor just and equal exchanges. We made choices and guided our behavior in accordance with the value of reciprocity. Evidently, at the other end, often we took without asking or giving in return, but this frequently led to punishment, retribution, and even war. The history of humanity is of course filled with injustice, with selfish and inconsiderate actions, with conquest and violence without any concern for the other, but much of what we have purposefully developed and accomplished has been built on the creation and honoring of reciprocities.

As humanity was developing more complex social organizations, creating large urban settlements, and networking through trade and exchange, religion and myth were also evolving as important, if not central, dimensions in human thinking and behavior. As noted in the previous section, humans were burying their dead with artifacts, creating cave paintings with magical and shamanistic qualities, and carving fertility sculptures long before the emergence of agriculture and large urban settlements. As Shlain and Fraser argue, religion and the belief in an after-life emerged with the psychological realization of death, which may have occurred at least 50,000 years ago. Even if we discount such suggestive archeological evidence and arguments, there are strong indications that around 12,000 to 10,000 years ago religion and mythic belief, in a more modern recognizable form, emerged as a powerful and central force in human reality. This "religious revolution," as it has been called, may have occurred even prior to the full blossoming of agriculture. As Watson describes it, in the first large urban settlements, which have been uncovered in the Middle

East, people began to create "human-like" representations of deities. In particular, two figures predominate: a woman Goddess figure and her partner/offspring, a male Bull figure, presumably representing the female and male principles of life. The Goddess figure appears to be the supreme deity since she is depicted as giving birth to the male Bull deity. The woman figure seems to represent the power of fertility and the regeneration of life in the spring, whereas the male seems to represent, initially, virility and the "untameability of nature," and later, the domination over nature and animals. What is seen as especially significant in these representations and the worship of them is that first, humankind appears to be expressing a desire to control nature and animals, and second, the figures seem to reflect a belief in a higher level of reality above both humans and nature.[118]

Hence, at least three types of duality in thinking are expressed through these representations: The duality of male and female, with the female being the supreme deity; the duality of humans and nature, with the humans expressing a desire to control nature; and the duality of humans and higher beings, with humans presumably drawing their inspiration and power from these worshipped deities. What is particularly interesting about all of this is that as a prelude or stimulus to the growing conquest and control of nature through agriculture and the domestication of animals, there was a shift in mindset within humans that expresses this aspiration to gain control and dominion over nature and that this conscious desire was expressed through religious symbolism and worship.

Sexual power, religion, and the future make for an interesting combination of themes, and in our pre-history all three appear to be woven together. Our earliest religious thinking and myth making, while grounded in sexual symbolism, appears to also address concerns about the future. The Goddess and the Bull are personifications of the power of sex and reproduction, and also represent the conquest of nature and even death (it is the Goddess principle that recreates life in the spring). The female and male principles have also been connected in our history with the ideas of nurturance and conquest which leads us back to the ideas of cooperation and competition, the two fundamental forces behind the evolution of human society. The sexual duality and reciprocity of female and male makes the world go round and in many ways underlies humanity's conscious desire to mold and create the future.

References

[1] Fraser, J. T. *Time, the Familiar Stranger.* Redmond, Washington: Tempus, 1987; Russell, Peter *The White Hole in Time: Our Future Evolution and the Meaning of Now.* New York: HarperCollins, 1992; Stewart, John *Evolution's Arrow: The Direction of Evolution and the Future of Humanity.* Canberra, Australia: The Chapman Press, 2000; Shlain, Leonard *Sex, Time, and Power: How Women's Sexuality Shaped Human Evolution.* New York: Viking, 2003.

[2] Watson, James *The Double Helix.* New York: Mentor-New American Library, 1968.

[3] There is some degree of variability within new born members of a species that is due to prenatal inner environmental influences.

[4] Gould, Stephen Jay *Wonderful Life: The Burgess Shale and the Nature of History.* New York: W. W. Norton, 1989; Christian, David *Maps of Time: An Introduction to Big History.* Berkeley, CA: University of California Press, 2004, Chapter Five.

[5] Bloom, Howard *Global Brain: The Evolution of Mass Mind from the Big Bang to the 21st Century.* New York: John Wiley and Sons, Inc., 2000.

[6] Stewart, John, 2000.

[7] Fraser, J. T., 1987; Stewart, John, 2000; Reading, Anthony *Hope and Despair: How Perceptions of the Future Shape Human Behavior.* Baltimore, Maryland: The John Hopkins University Press, 2004, Page 118.

[8] Stewart, John, 2000, Page 78.

[9] Morowitz, Harold *The Emergence of Everything: How the World Became Complex.* Oxford: Oxford University Press, 2002, Pages 136 – 138.

[10] Bloom, Howard *The Lucifer Principle: A Scientific Expedition into the Forces of History.* New York: The Atlantic Monthly Press, 1995; Bloom, Howard, 2000.

[11] Bloom, Howard, 2000, Pages 29 – 30.

[12] Reading, Anthony, 2004, Chapter One.

[13] Miller, Kenneth *Finding Darwin's God: A Scientist's Search for Common Ground between God and Evolution.* New York: Perennial, 1999, Chapters One to Five; Morowitz, Harold, 2002, Chapters Twenty-three to Twenty-eight; Leakey, Richard, and Lewin, Roger *People of the Lake: Mankind and Its Beginnings.* Avon, 1978; Leakey, Richard *The Making of Mankind.* New York: E.P. Dutton, 1981; Johanson, Donald, and Edey, Maitland *Lucy: The Beginnings of Humankind.* New York: Warner, 1981; Putnam, John "The Search for Modern Humans" *National Geographic*, Vol. 174, No. 4, October, 1988; Sagan, Carl and Druyan, Ann *Shadows of Forgotten Ancestors: A Search for Who We Are.* New York: Random House, 1992; "New Look at Human Evolution", *Scientific American*, Special Edition, Vol.13, No.2, 2003.

[14] Morris, Desmond *The Naked Ape*. New York: Dell Publishing Co., 1967.

[15] Christian, David, 2004, Pages 125 – 127; Morowitz, Harold, 2002, Pages 140 – 145.

[16] Diamond, Jared, 1992, Chapter One; Calvin, William *A Brief History of the Mind: From Apes to Intellect and Beyond*. New York: Oxford University Press, 2004, Chapter One.

[17] Bloom, Howard, 1995, Pages 239 - 243.

[18] Ghiglieri, Michael *The Dark Side of Man: Tracing the Origins of Male Violence*. New York: Helix Books/Basic Books, 1999; Calvin, William, 2004, Chapter One.

[19] Ghiglieri, Michael, 1999.

[20] Shlain, Leonard, 2003.

[21] Wong, Kate "An Ancestor to Call Our Own" *Scientific American*, Vol. 13, No. 2, 2003; Leakey, Meave and Walker, Alan "Early Hominid Fossils from Africa", *Scientific American*, Vol. 13, No. 2, 2003.

[22] Tattersall, Ian "Once We Were Not Alone", *Scientific American*, Vol. 13, No. 2, 2003.

[23] Christian, David, 2004, Pages 154 – 155; Watson, Peter *Ideas: A History of Thought and Invention from Fire to Freud*. New York: HarperCollins Publishers, 2005, Page 23.

[24] There is some evidence that bipedalism extends back even further to 6 million years. See Watson, Peter, 2005, Page 22.

[25] Johanson, Donald, and Edey, Maitland, 1981; Ghiglieri, Michael, 1999.

[26] Morowitz, Harold, 2002, Page155; Shlain, Leonard, 2003.

[27] Whitehead, Charles "Evolution of the Human Brain" Paper presented at Toward a Science of Consciousness Conference, Tucson, AZ, 2004.

[28] Watson, Peter, 2005, Page 23.

[29] Christian, David, 2004, pages 159 – 163.

[30] Watson, Peter, 2005, Page 24.

[31] Watson, Peter, 2005, Page 39.

[32] Calvin, William, 2004, Chapter Three.

[33] Whitehead, Charles, 2004.

[34] Watson, Peter, 2005, Page 39.

[35] Tattersall, Ian "Out of Africa Again...and Again?" *Scientific American*, Vol. 13, No. 2, 2003 (b); Whitehead, Charles, 2004; Watson, Peter, 2005, Pages 25 - 27.

[36] Watson, Peter, 2005, Pages 4, 25 – 26.

[37] Watson, Peter, 2005, Page 45.

[38] Ghiglieri, Michael, 1999.

[39] Watson, Peter, 2005, Page 49; Reading, Anthony, 2004, Pages 31, 60 - 61.

[40] Bloom, Howard, 2000.

[41] Whitehead, Charles, 2004.

[42] Reading, Anthony, 2004, Pages 167 – 168.

[43] Reading, Anthony, 2004.

[44] Christian, David, 2004, pages 163 - 167.

[45] Calvin, William *A Brain for All Seasons: Human Evolution and Abrupt Climate Change*. Chicago: The University of Chicago Press, 2002.

[46] Calvin, William, 2004, Chapter Three.

[47] Watson, Peter, 2005, Pages 23 - 28.

[48] Whitehead, Charles, 2004.

[49] Whitehead, Charles, 2004; Donald, Merlin *Origins of the Modern Mind: Three Stages in the Evolution of Culture and Cognition*. Cambridge, Massachusetts: Harvard University Press, 1991, Pages 10, 137 - 141.

[50] Watson, Peter, 2005, Page 30.

[51] Whitehead, Charles, 2004.

[52] Bloom, Howard, 2000.

[53] Morowitz, Harold, 2002, pp. 141-147.

[54] Calvin, William *The Cerebral Symphony: Seashore Reflections on the Structure of Consciousness*. New York: Bantam, 1989.

[55] Dennett, Daniel C. *Consciousness Explained*. Boston: Little, Brown, and Co., 1991; Hawkins, Jeff *On Intelligence*. New York: Times Books, 2004; Gibson, James J. *The Senses Considered as Perceptual Systems*. Boston: Houghton Mifflin, 1966.

[56] Cann, Rebecca L., and Wilson, Allan C. "The Recent African Genesis of Humans" *Scientific American*, Vol. 13, No. 2, 2003; Shlain, Leonard, 2003, pp. 6-8.

[57] Wong, Kate "Who Were the Neanderthals?" *Scientific American*, Vol. 13, No. 2, 2003 (b); Sawyer, Robert J. *Hominids*. New York: Tom Doherty Associates, 2002.

[58] Tattersall, Ian, 2003.

[59] Ghiglieri, Michael, 1999, Chapter Three; White, Randall *Prehistoric Art: The Symbolic Journey of Humankind*. New York: Harry N. Abrams, 2003.

[60] Morowitz, Harold, 2002, p. 157. Also see Wong, Kate, 2003(b) for an alternative view that *Homo sapiens* and *Neanderthals* interbred.

[61] Diamond, Jared, 1992, Chapter Two.

[62] Diamond, Jared, 1992, Chapter Two; White, Randall, 2003, Chapters One and Four; Calvin, William, 2004, Chapters Seven and Nine.

[63] White, Randall, 2003, Chapters One and Four; Calvin, William, 2004, Chapters Seven and Nine.

[64] Christian, David, 2004, pages 178 - 182; Watson, Peter, 2005, Pages 30 - 32.

[65] Calvin, William, 2004, Chapter Eight; A thought - provoking speculative effort to describe the mind and experiences of earlier hominids, including *Australopithecus*, *Homo erectus*, and *Neanderthals* can be found in Steven Baxter's science fiction novel *Manifold Origin*. New York: Ballantine, 2002.

[66] Diamond, Jared, 1992.

[67] Reading, Anthony, 2004, Chapters 9 and 10.

[68] Watson, Peter, 2005, Page 38.

[69] Watson, Peter, 2005, Pages 34 - 35.

[70] White, Randall, 2003, Chapter Three.

[71] White, Randall, 2003, pages 86-88.

[72] Watson, Peter, 2005, Pages 35 – 36.

[73] Eldredge, Niles and Gould, Stephen "Punctuated Equilibria: An Alternative to Phyletic Gradualism" in Schopf, T. J. M. (Ed.) *Models in Paleobiology*. Freeman Cooper, 1972.

[74] Calvin, William, 2004, Preface.

[75] Toffler, Alvin *Future Shock*. New York: Bantam, 1971.

[76] Gell-Mann, Murray *The Quark and the Jaguar: Adventures in the Simple and the Complex*. New York: W.H. Freeman and Company, 1994, Chapter Two.

[77] Calvin, William, 2004, Chapters Eleven and Twelve.

[78] Holmes, Bob "Civilization Left its Mark on our Genes" *New Scientist*. December 24, 2005 – January 6, 2006.

[79] Hergenhahn, B.R. and Olson, Matthew *An Introduction to Theories of Personality*. 6th Edition. Upper Saddle River, NJ: Prentice Hall, 2003, Chapter Twelve.

[80] Pinker, Steven *The Blank Slate: The Modern Denial of Human Nature*. New York: Penguin Books, 2002; Brown, Donald *Human Universals*. New York: McGraw-Hill, 1991.

[81] Competition can also lead to the elimination or extinction of one of the competitors if one of the competitors can not keep up with other competitor.

[82] Bear, Greg *Darwin's Radio*. New York: Ballantine Books, 1999; Bear, Greg *Darwin's Children*. New York: Ballantine Books, 2003.

[83] Gleick, James *Faster: The Acceleration of Just About Everything*. New York: Pantheon Books, 1999.

[84] Russell, Peter *The White Hole in Time: Our Future Evolution and the Meaning of Now*. New York: HarperCollins, 1992, pages 33 – 41; Kurzweil, Ray *The Age of Spiritual Machines: When Computers Exceed Human Intelligence*. New York: Penguin Books, 1999, Chapters One and Six; Moravec, Hans *Robot: Mere Machine to Transcendent Mind*. Oxford: Oxford University Press, 1999, Chapter Three.

[85] Hubbard, Barbara Marx *Conscious Evolution: Awakening the Power of Our Social Potential*. Novato, CA: New World Library, 1998.

[86] Russell, Peter, 1992, pages 33 – 41.

[87] Shlain, Leonard, 2003, Part IV.

[88] Watson, Peter, 2005, Page 28.

[89] Fraser, J. T, 1987.

[90] Becker, Ernest *The Denial of Death*. New York: Free Press, 1973.

[91] Shlain, Leonard, 2003.

[92] Diamond, Jared, 1992.

[93] Shlain, Leonard, 2003, p.24.

[94] Reading, Anthony, 2004, Chapter 6.

[95] Diamond, Jared, 1992, Part V; Christian, David, 2004, pages 199 – 202.

[96] Ghiglieri, Michael, 1999, Chapters One and Two.

[97] Bloom, Howard, 1995, Pages 1 – 29.

[98] Bloom, Howard, 1995, Page 33.

[99] Diamond, Jared, 1992, Chapter Two and Epilogue.

[100] Bloom, Howard, 1995, Pages 37 – 39, 195 – 202.

[101] Watson, Peter, 2005, Pages 4 – 5, 53 – 56.

[102] Watson, Peter, 2005, Page 42.

[103] Watson, Peter, 2005, Page 58.

[104] Watson, Peter, 2005, Pages 50 – 52.

[105] Christian, David, 2004, Pages 185 – 202 and Chapter Eight; Watson, Peter, 2005, Page 56 – 58.

[106] Bloom, Howard, 2000, pages 109 – 117.

[107] Christian, David, 2004, page 216.

[108] Calvin, William, 1989. See also Robert Sawyer's *Hominids*, 2002, for a spirited, albeit speculative, account and defense of the intricacies of planning in the hunter-gatherer lifestyle of Neanderthals.

[109] Christian, David, 2004, pages 140 – 141.

[110] Christian, David, 2004, pages 185 – 190.

[111] Bloom, Howard, 2000, pages 109 – 117.

[112] Bloom, Howard, 2000, pages 109 – 120.

[113] Wright, Robert *Nonzero: The Logic of Human Destiny*. New York: Pantheon Books, 2000.

[114] Loye, David (Ed.) *The Great Adventure: Toward a Fully Human Theory of Evolution*. Albany, New York: State University of New York Press, 2004.

[115] Margulis, Lynn *Symbiosis in Cell Evolution*. 2nd Ed. New York: W. H. Freeman, 1993.

[116] Stewart, John, 2000, Chapters One and Two.

[117] Hergenhahn, B.R. and Olson, Matthew, 2003, Chapter Twelve; Shermer, Michael *The Science of Good and Evil*. New York: Times Books, 2004, Chapter Two; Pinker, Steven, 2002, Chapter Fourteen.

[118] Watson, Peter, 2005, Pages 59 – 67.

CHAPTER THREE

⊹⸺⊹

ANCIENT MYTH, RELIGION, AND PHILOSOPHY

"Progress, far from consisting in change,
depends on retentiveness...
when experience is not retained, as among
savages, infancy is perpetual.
Those who cannot remember the past
are condemned to repeat it...
this is the condition of children and barbarians, in whom
instinct has learned nothing from experience."

George Santayana

Thinking about the future has a rich and deep history.[1] There is much to be learned from past ideas and images of the future. In fact, contemporary views of the future, in many significant ways, are inspired and derived from earlier ideas and theories. Understanding our present views of the future requires looking at how different ideas and approaches to it have developed through the ages. The past puts the present into perspective; the present has been built upon the past. In biological evolution, many of the features of earlier life forms are carried over into later forms. Evolution in biology is to a great degree cumulative. The same is true for the history of future consciousness and, for that matter, the entire history of the human mind - ideas, insights, and discoveries build upon themselves. To borrow a metaphor from Isaac Newton, the futurists of the present can see outward as well as they can because they "stand on the shoulders of giants."

In this chapter, I examine the earliest recorded ideas, in printed word, on the future. These ideas build upon the prehistoric foundations described in the previous chapter and add new themes and concepts that have contributed to the ongoing evolution of future consciousness. Although prehistoric "mythograms" and other artistic representations may contain prophecies of the future, there is presently no accurate or reliable way to decipher the detailed meanings of these ancient images. According to J.T. Fraser, there is no clear evidence in prehistoric art that early humans thought in global or universal terms about either the past or the future.[2] Christian states that, as best as can be ascertained, prehistoric humans appeared to have thought in relatively concrete terms about local and specific concerns.[3] Questions about the origin or destiny of the universe, or even of humankind, do not seem to have occurred to them. Still, it was the basic themes and concerns of prehistoric life - of reproduction and death, of hunting and the kill - that led to the development of future consciousness and the first recorded views of the future.

The earliest written ideas about the future dating back around five thousand years are mythological and contain both descriptions of the past and prophecies of the future, including explanations of the origin and purpose of humanity and the cosmos. Within these ancient myths past and future are causally and thematically connected - the future flows out of the past. These ancient mythic views of the origin, history, and future of humanity and the cosmos invariably contain references to deities, gods, and goddesses. The past and future in ancient myth are personified. These deities are variously seen as supernatural - above or separate from nature - or as part of nature, actively involved in directing physical

142

events and human history – they are both transcendent and immanent. Often these deities are responsible for the creation of the universe and humankind and often they significantly influence or determine the future and the ultimate purpose of the cosmos. The future is often seen as controlled by destiny, fate, and the will of the gods. Mythic views are usually expressed in narrative form, involving personalities and personal challenges, interpersonal conflicts, adventure, and drama. The life of the universe and the saga of humankind are conceptualized as stories.

Ancient myths with their gods and goddesses would provide one primary source of inspiration for the development of traditional religions around the world. Socially organized religions incorporated into their belief systems earlier mythic stories and prophecies, as well as rituals and moral systems of behavior that provided direction for how to live. Myth would also impact the development of ancient philosophical views regarding reality, time, morals, and the future.

In this chapter I cover the history of myth and religion from around 3000 BC (or BCE – Before the Common Era) to the rise and flourishing of Christianity and Islam around 1000 AD. I examine both Eastern and Western religion and myth, including Egyptian, Mesopotamian/Babylonian, Zoroastrian, Hindu, Buddhist, Taoist, Greco-Roman, Judaic, Christian, and Islamic ideas on the past and future. As a prelude to the next chapter, where I describe the rise of rational-scientific approaches to the future in modern times, I also describe in this chapter the beginnings of Western philosophy in ancient Greece (600 to 300 BC). Although Greek philosophy, in so far as it approached reality from a rational and abstract point of view, was in many ways at odds with religious-mythic thinking, Greek philosophy did influence the development of Christianity in the first Millennium and any complete explanation of the Christian vision of the future needs to discuss the influence of Greek thinking.

The Power of Mythic Narrative

*"...myths are archetypal patterns in human consciousness
and where there is consciousness there will be myth. ...
In the moments when eternity breaks into
time, there we will find myth."*

Rollo May

Myths provided the first systematic explanations of history and the first prophecies of the future. For most of recorded history, the primary mode of understanding both the past and the future has been the myth – stories and sagas describing the challenges, meaning, and purpose of life.

The first meaning listed for the term "myth" in *The Oxford American College Dictionary* is "a traditional story." The term "myth" can also mean a superstitious or fanciful tale without factual support. Although early myths do indeed contain references to spirits, fantastical creatures, and supernatural beings, and describe cosmic or earthly events that are scientifically implausible, ancient myths were invariably grounded in important facts about nature and the meaning and psychology of human life.

The fact that our earliest written explanations of the past and future were mythic in form can be placed in an evolutionary context. According to Merlin Donald, in his *Origins of the Modern Mind*, the third fundamental stage in the cognitive evolution of hominids, after the "episodic" and the "mimetic," was the "mythic," which was associated with the emergence of modern *Homo sapiens*. (For Donald, *Homo erectus* behaved and thought in a "mimetic" fashion, an advance over *Australopithecus*, who functioned at an "episodic" level.) Probably coincident with the emergence of cave art and mythograms, humans began to develop organized explanations of nature and human reality in the form of narrative myths. When the earliest modern *Homo sapiens* (circa 100,000 to 50,000 years ago) thought or spoke about their understanding of nature it was in the form of myths or stories. This new integrative cognitive capacity was intimately connected with the emergence of language - language provided the tool to produce and communicate integrative explanations. According to Donald, in fact, the initial primary function of complex spoken language was narration and myth creation.[4] Hence, when humanity began to record in the written word (around 3000 BCE) the explanations of nature handed down from previous generations, the form these explanations took represented a certain way of thinking characteristic of the early history of our species - a narrative mythic way of thinking. Myth and narration represent a way of understanding the world that was an early stage in our cognitive evolution.

According to Leonard Shlain in his book *The Alphabet and the Goddess,* the two central myths in prehistoric times seem to be the stories of the "**Goddess**" and the "**Hunter.**" These myths provided two different interpretations of the saga and meaning of human existence.[5] The goddess myth highlighted the eternal cycle of life and death, whereas the hunter myth emphasized the necessity of killing in order to survive. The goddess myth was connected with feminine qualities such as nurturance, the giving of life, and the importance of community with both nature and fellow humans. The hunter myth was connected with masculine qualities such as dominance over nature, conquest, and physical violence.

We should recall from the previous chapter that the two primary deistic figures found in early urban settlements in the Middle East around 10,000 years ago were the goddess and the bull. (Shlain's description of the hunter myth closely corresponds with those qualities associated with the bull.) The Goddess figure seems to occupy a central or supreme position in these early representations and Shlain, in fact, does acknowledge that the goddess was the central deity early in our history.

But as a general trend, Shlain sees a movement away from worship of the goddess toward an elevation of the male with his hunter traits and values in the period roughly from 2000 to 600 BCE. As one example, Shlain states that the earliest Sumerian and Babylonian myths in the Middle East dating back to the beginnings of recorded history (circa 3000 BCE) identified the goddess – the giver of life – as the central deity, but later Babylonian mythology displaced the goddess as the supreme deity and "life myth" with a supreme male deity *Marduk* and a central "death myth".[6] Marduk achieved dominance and control of the world, as the story goes, through the slaying of the more ancient and primordial goddess *Tiamat*. In Shlain's mind, this mythical tale symbolizes the social and religious transformation that occurred in ancient Babylonia as it moved from a goddess centered culture to a culture dominated by men and masculine deities.

Taking the opposing point of view and based on his review of Upper Paleolithic art and artifacts, Bloom contends that more homage was given to the hunter than to the goddess even in prehistoric times. In particular, the bull as a male mythic symbol of fertility was revered as a great source of power and the giver of life.[7] But as Bloom also acknowledges, prior to 2000 BCE, Mediterranean trading cultures practiced Mother Goddess centered religions. This is clearly the case in the highly advanced Minoan civilization of ancient Crete.[8] Overall the bulk of scholarship supports the view that the goddess was the central deity in prehistoric religion and myth throughout much of the world.

According to one popular theory, early goddess centered cultures in the Middle East and the Mediterranean were overrun and replaced by waves of Indo-European invaders that came out of the north starting around 2000 BCE. These nomadic horsemen had a patriarchal social order and they valued war and conquest more than trade. A similar shift, one that had an equally negative impact on goddess worship, took place in India around the same time when it was also overrun by Indo-European people coming from the west. The feminist writer Riane Eisler describes in great detail and passion, and with much psychological insight, the demise of earlier goddess centered cultures during this period of nomadic invasions in two of her books *The Chalice and the Blade* and *Sacred Pleasure*. According to Eisler, goddess centered cultures valued social partnership, cooperation, balance and equality of the sexes, whereas the male god centered cultures that subsequently emerged valued social hierarchies, sexual and philosophical dualism, and the superiority and dominance of males over females.[9]

An opposing theory is that the shift from goddess centered to male centered myth and religion was due to the growth of large cities and urban civilization. As cities grew, and along with them standing armies to protect these cities, males achieved greater control and leadership in human life. Military leaders - all men - became increasingly powerful. Of particular note, early cities often had large central areas that were occupied by religious temples and structures, and the individuals who controlled these urban religious centers were male priests. It was these male priests who often "conferred godlike status" on the male political rulers of early cities. Political, religious, and military power in such cities resided in the males and quite naturally, the cosmologies and theologies came to reflect this male dominance in urban social affairs. It wasn't the nomads who destroyed the goddess, but city life and the consequent growing social power of men within cities.[10]

Undoubtedly both hunter and goddess mythologies had a great influence on ancient human cultures. In fact, we should recall from the previous chapter that the respective roles of the hunter and mother were intertwined in our ancestral biology, psychology, and behavior. The hunter served the mother by providing food and protection and the mother provided the hunter with sex and offspring. Love, bonding, and commitment united the male and female - the hunter/father and mother/nurturer. It also seems to be the case that the respective power and influence of the feminine and the masculine in both myth and human society has oscillated throughout recorded history, and has varied among different regions of the world. Whatever the specific details of the relative power of male and female deities across time and ancient cultures, and I more fully examine this topic in later sections of this chapter, it seems clear that our earliest myths were connected with fundamental themes of human survival and reproduction (which are future oriented themes) and highlighted the central contributions and values associated with each of the two sexes. To restate and expand upon the conclusions of the previous chapter, sex and the contribution of the two sexes, religion, and the future were intimately tied together in the minds and the myths of prehistoric humans.

Aside from the primary female and male deities, Watson lists the following additional "core elements" of pre-historic myth and religion: Sky gods associated with the sun and the moon, sacred stones (such as the megaliths of Stonehenge), and the beliefs in the power of sacrifice, in an afterlife, and in a soul which survives death.[11] It should be noted that all of these other core elements are also connected in one way or another with understanding, predicting, or controlling the future.

Although myths are often seen, especially from a scientific viewpoint, as forms of superstition without any rational or factual support, they are an expression of humanity's desire to understand the world in a coherent and meaningful way. As Donald argues, mythic thinking is an evolutionary step in humanity's attempt to make sense out of reality. Even the highly regarded scientist Murray Gell-Mann acknowledges the positive values associated with myths. According to Gell-Mann, myths give order to reality, provide inspiration to individuals and cultures, and give a society a distinctive identity.[12] The existential psychologist, Rollo May, concurs listing four primary functions to myth: Myths provide a sense of personal identity, a sense of community, support moral values, and deal with the mysteries of creation.[13] Fraser and others would add that religious myths provide a sense of stability within the flux of time and address the anxiety provoking fact of human death.[14] All of these functions of myths are of essential importance to the psychological and social well-being of humans. In that myths gave ancient humanity purpose they gave humanity its first verbally articulated sense of the future.

What is especially important about myth is that it embodies a distinctive mode of experience and way of thinking about life, history, and the future. Fundamentally, a myth is a story - a narrative - involving a sequence of connected events often containing a dramatic plot with both a resolution and some intended moral or meaning. Both history and the future can be described in narrative form. The narrative is a dynamical and temporal description of reality - change occurs - something happens. But also, as many writers emphasize, the narrative provides a mode of understanding that gives life coherence and value.[15] The events of life are causally connected together within a story and some overall meaning or point to the story unifies all the events within it. Conceptualizing life as a story is a form of temporal consciousness and gives meaning and order to change and the procession of time.

Further, myths affect people at an emotional and personal level. People identify with mythic stories because the stories contain human or human-like characters that encounter various life challenges and experiences. Mythic characters exhibit the whole plethora of psychological traits and moral qualities, both good and bad. The Greek gods and goddesses, for example, were each connected with distinctive personality types and traits. In most religions, the gods and goddesses, as personifications of characteristic attributes -are variously wise, playful, adventurous, and terrifying or frightening. Again, with the Greeks, each deity embodied a particular skill or ability - an area of excellence - be it warrior like, as in the case of *Ares*, or erotic, as in the case of *Aphrodite*. People found meaning, inspiration, and wisdom in these mythic characters. For Joseph Campbell, mythic characters and their exploits provide a form of "music" for experiencing life.[16]

Mythic characters, as **"archetypes"** or prototypes, often symbolize essential qualities of life or human psychology. The term "archetype" signifies a fundamental idea, theme, or motif usually represented through some image, persona, or symbol. Examples of mythic archetypes include the goddess, an archetype that represents love, procreation, and nurturance; the hunter who represents courage; and as in many early religions, the sun or sun god, who is the giver of light and life and often the ruler of time. Various gods and goddesses in ancient myths stood for justice, war, wisdom, fertility, renewal, and the forces and patterns of nature. All ancient cultures created and worshipped their own characteristic set of deities and mythic beings, whose exploits, adventures, and achievements were recounted in the myths of the culture and represented the central units of meaning or archetypes for the society. Depending on the important challenges and features of different environments and ways of life, different central archetypal deities were created. All ancient cultures though conceptualized the fundamentals of their distinctive reality in terms of some set of archetypes.

Modern scholars, including the psychologist Carl Jung, the anthropologist Claude Lévi-Strauss, and the preeminent modern spokesperson for the "Power of Myth," Joseph Campbell, have extensively studied myths and the archetypes embodied in myths.[17] One general conclusion all these scholars have reached is that in spite of some differences there are common human themes that run through all mythologies and relatively common symbols and archetypes across different cultures. Carl Jung attributes these universal archetypes to common historical events and common thematic structures within the human mind.[18] All cultures talk about love and strife; birth, life and death; men, women, and children; and morals and virtues.[19] All cultures seem to have myths about the past and the future. As I argue in this chapter, reflecting such common mythic archetypes, the different religions around the world show a great deal of overlap in terms of important themes and issues.

One essential question concerning the meaning and order of things that many myths attempted to address was the creation and origin of humankind and the cosmos. As Morowitz notes, speculation about the origins of the earth and the universe seems to be part of the human condition.[20] We look for answers to the big questions. According to Fraser the earliest creation stories occur in the 3rd and 2nd millennia BCE in Egypt, Mesopotamia, China, and India.[21] These earliest creation myths were an important step forward in the evolution of temporal consciousness for such myths provided a way to conceptually organize the entire grand panorama of time – they provided a cosmic perspective on past, present, and future. The myths connected the deep past with the present and often identified key themes, such as the primordial struggle between order and chaos, which provided a way to understand the flow of events through time.

These earliest cosmic stories frequently personified the process of creation, seeing various deities as intimately involved in the process. An essential feature of narrative myths is that ancient people conceptualized reality as a personalized drama and saga, filled with archetypal characters defining the procession and meaning of time. Ancient humanity personified or anthropomorphized the forces and fundamental patterns of nature, and in particular, the reality of time. In ancient Egypt, for example, the journey through the sky of the sun god *Ra* is represented as a procession of different subordinate gods with different personal attributes and different meanings associated with each deity. In the Taoist *Yin-Yang*, which represented the basic rhythm of time, each polarity – *Yin* and *Yang* - is personified as female and male respectively, and associated with different qualities, such as darkness and the earth for the *Yin* female and light and the heavens for the male *Yang*. The ancient Greeks personified time as the god *Cronos*, who later became *Saturn* in Roman times, and "Father Time" in the Middle Ages. In Indian mythology, *Shiva* was seen as the god of creation and destruction – of becoming and passing away – of the giving of life and the inevitability of death. The ancient Zodiac, with its twelve signs or characters, represents the passage of each year, as well as successive ages, in terms of personified qualities, traits, and meanings.[22]

In general, for ancient humankind, nature was filled with and animated by spirits (**"animate naturalism"**). Spirits and deities populated the earth and powered and directed natural events. As Karen Armstrong notes in *History of God*, in ancient times the "world was full of gods."[23] Further, within many mythic narratives, humans and deities interacted and communicated. Such an encounter with a deity is referred to as an "epiphany." For example, in the epochal stories of Homer's *Iliad* and *Odyssey*, the Greek and Trojan characters felt the presence of their gods and goddesses, interacted with them, and spoke with them. The Judaic God communicated with Adam and Eve, Noah, Abraham, Job, Jacob, and Moses, among others. In fact, ancient people believed that they frequently communicated and communed with gods and goddesses. The supernatural and archetypal personalities of ancient myths not only provided an explanation of life's origin, meaning, and purpose, they also personally and intimately inspired and guided people (so they believed) through the struggles and adventures of life. Hence, future consciousness in pre-historic times often took the form of "internal dialogues" or voices guiding or directing individuals in their actions.[24]

Ancient rites and rituals are connected with myths and mythic characters. Gods and goddesses presumably possessed various powers, such as fertility, wisdom, creativity, and even the capacity to resurrect the dead. As noted above, ancient deities pervaded, inspirited, and animated the processes of nature. Myths conveyed stories of the actions of such deities, and if one attempted to imitate these behaviors, one could hopefully share, so the ancients believed, in the natural and spiritual powers of the gods. Humans could participate in the reality of the gods and even become god-like.[25] Religious rituals are often efforts to emulate the archetypal behavior of deities as described in myths and partake in their god-like powers. Also, as noted earlier, one of the core elements of early religions was the ritual of sacrifice where the killing of animals, or even humans, presumably influenced the gods and increased the chances of good fortune in the future.[26]

On a related note, myths also provided moral guidelines. Although not all deities or humans described in myths embodied admirable ethical ideals in their behavior, often they did, showing justice, compassion, loyalty, and other positive traits. Attempting to emulate the behavior and ideals of gods and goddesses gave ancient people moral standards and a sense of ethical direction. Myths provided role models for making ethical choices. There were, of course, mythic characters that represented negative, immoral, and evil aspects of life. They were archetypes of the "dark side." Figures, such as Satan in Christianity, still served a moral purpose, for they illustrated what to avoid, fear, or fight against in life.

The dark or evil side of reality, often projected onto the "other," is, in fact, a common belief in many cultures. This creation of a moral enemy is connected with a basic feature of human psychology and social organization: the "us versus them" dichotomy.[27] The beliefs and values of a culture, which include their myths, not only provide a core set of ideas that unite a culture, but these beliefs and values (what Bloom calls an "ideology") frequently create an oppositional adversary of other cultures. Those outside a group – the "other" - are often seen as a morally inferior enemy to which a double standard of values is commonly applied. Cooperation (and other associated virtues such as love and harmony) is encouraged and reinforced within a cultural group, whereas war, violence, and conquest are morally sanctioned toward other people and cultures outside of the group. (It is interesting that these two sets of standards closely align with the two fundamental forces of competition and cooperation discussed in the previous chapter.) Thus the ideology of a people not only gives them a sense of identity and distinguishes the culture but also sets it in opposition to other cultures and creates an "us versus them" mindset.[28]

Myths contain drama, challenge, and frequently conflict. The "us versus them" dichotomy, as an essential element in these conflicts and challenges, is a feature in many myths. Judaic mythology, for example, describes an ongoing conflict between the chosen people of *Yahweh* or *Jehovah* and all other people – in this case the "other" persecutes the followers of the one true God. Zoroastrian myth sets up an opposition between the forces of good and evil, and Islam creates a dichotomy between the followers of *Allah* and all others. In all these cases, the journey into the future is portrayed as a moral struggle between believers and non-believers the desired end of which is the defeat of non-believers through war and violent actions.

The stories that describe the lives, character traits, and beliefs of great religious figures of the past reflect many of the qualities of the myths described above. Understanding the meaning of life and finding inspiration and direction through the traditional stories of religious figures is a form of mythic thinking and consciousness. Buddha, Jesus, Moses, Abraham, and Mohammed, all probably real historical figures, led mythic lives. They all embody archetypes that address the meaning, purpose, and value of human life. They encountered challenges in their lives and successively overcame or transcended these difficulties through character virtues such as faith, compassion, love, courage, and commitment. All these figures presumably made contact with some deep or ultimate reality – something spiritual or divine – that provided enlightenment and a sense of direction. They all saw "the good" and attempted to live their lives in response to such ethical revelations – often in opposition to what they saw as evil or ungodly. Billions of people worldwide still read the stories of their lives and attempt to follow, through ritual and general behavior, the ideals and practices expressed by these religious figures. Mythic thinking about the future is still very much alive.

Ethical ideals expressed through myth clearly address the question of the future. Ideals are prescriptions for how to think, what to feel, and how to behave. The ideals in myths are usually conveyed concretely through the particular actions of the characters in the stories. Their behavior and choices provide ethical direction for the future. Although myths often describe events and characters of the distant past, they also point to the future and provide guidance for how to live.

Also of great relevance, myths frequently contain prophecies and plans of action for the future. Polak describes prophets as projectors of the future who draw upon the heritage of a particular culture.[29] The stories of the Old Testament are especially noteworthy in this regard. *Yahweh* spoke with many of the early Judaic prophets and religious leaders, telling them what was to come in the future and what they must do to realize the prophesized future. *Yahweh* made promises and gave directions to "His people" and early Jews attempted to have faith in such promises for the future and follow their god's directions.[30]

As we will see through a variety of examples in this chapter, mythic stories of creation and the deep past are often connected with prophecies of the future. As one example, events of the distant past and the prophesized future in the stories of the *Bible* are tied together giving the total history of time an overall pattern and direction. The drama begins with the fall of Adam and Eve. Humankind's eventual redemption and salvation are foretold in the coming of the Messiah, and a final battle pitting the forces of good against evil is foreseen bringing resolution to the conflict of good and evil. Temporal consciousness is integrated into a whole, connecting past, present, and future.

As can be seen, myths have great psychological power. Myths address all the main components of the human mind, including cognition, emotion, motivation, and the self. As narratives, myths organize time in a way that people can readily understand. These ancient stories provide meaning and a sense of purpose to the passage of time. Myths represent reality in a concrete, personalized, and dramatic way. Myths depict fundamental challenges and conflicts, giving life drama and, frequently, a fundamental enemy against which we must do battle. Following Campbell, myths provide a framework - a "music" and "song" - within which to experience "being alive."[31] Further, myths have great instructive value; myths contain ethical guidelines and values to live by. Myths both prescribe and explain. Mythic consciousness has been and still is a fundamental mode of future consciousness.

The Mythic-Religious Quest for God

"Fear was the first mother of the gods.
Fear, above all, of death."

Lucretius

"In the beginning, human beings created a God
who was the First Cause of all things and Ruler of
heaven and earth…That, at least, is one theory…

Karen Armstrong

Religion grows out of myth. In fact, one blurs into the other. Myth is narrative, whereas religion integrates myth with organized practices of worship and ethical behavior, and often general forms of social organization with religious leaders and figureheads. Religion includes both explanations of reality and prescriptions for behavior, institutionalized and codified. Karen Armstrong, in *A History of God*, emphasizes the practical dimension of religion; according to Armstrong, religion provides a way of life.[32] Similarly, Michael Shermer, in *The Science of Good and Evil*, acknowledges the explanatory function of religion while highlighting its ethical and social dimensions. Religion evolved as a social system to articulate and enforce rules of ethical behavior.[33]

As with myth, religion usually includes ideas and beliefs about deities - of gods and goddesses. Armstrong contends that humans have been worshipping gods since we were first recognizably human. Armstrong contends that gods find their origin in prehistoric times, a view supported by the archeological evidence reviewed in the last chapter. Yet, in contrast to the view, that religion emerged as a response to the fear of death and the transitory quality of life, Armstrong states that the worship of gods expresses the wonder and mystery humans feel in response to the world.[34] Both explanations, positive and negative, certainly are true. The two views embody the two polarities of human motivation - to approach and to avoid. Life is something to fear and something to love and embrace, and so is God.

According to Armstrong, there are numerous and often contradictory meanings associated with the nature of God, and ideas about God keep changing and getting redefined with each new generation. Reciprocally, the meaning of atheism keeps changing as well, for what is being denied in atheism depends upon what is being asserted in theism.[35] Yet, Armstrong does acknowledge that God is frequently identified with transcendence and ultimate reality. In this sense, the quest for understanding God (or the gods and goddesses if one is a polytheist) is a quest to understand what is fundamental and absolute within the world. Although religion and myth are often seen as archaic and primitive modes of thinking, many of the ideas first articulated in the mythic and religious quest for knowledge of God still influence contemporary thinking on time, the nature of existence, ethics, and the meaning of life.

Babylonia and Egypt: Order, Chaos, Life, Death, and Sexual Creation

One of the cradles of human civilization was ancient Mesopotamia surrounding the Tigris and Euphrates rivers. Some of the earliest indications of agriculture and large urban settlements can be found in this region. The Sumerians who lived in this area early on provide the first evidence of written language - cuneiform writing on stone tablets. The Sumerians also invented the wheel, the chariot, schools, libraries, clocks, and created the first recorded written history and proverbs.[36] They produced some of the earliest religious texts, and, as Polak argues, along with the ancient Egyptians were "intensely preoccupied" with the future. Many predictions and omens on the future emerged from the ancient cities of this region.[37]

157

According to J.T. Fraser, since these early people in ancient Mesopotamia lived in a turbulent and unstable environment with violent weather conditions, they imagined the world beginning in conflict. Their creation myths described an interplay and struggle between darkness and chaos and light and order.[38] The biologist Elisabet Sahtouris describes this ancient cosmology of order and chaos more in terms of rhythmic balance – the two forces seen as alternating in dominance.[39] Armstrong states that the Babylonians, who lived in this region after the Sumerians, believed that the gods and creation emerged out of chaos. Chaos preceded the emergence and evolution of order – it preceded the gods. This, in fact, was a common belief in the ancient world. (In one early Sumerian story of creation, chaos took the form of the primordial sea connected with the goddess *Nammu*.[40]) According to Babylonian myth, the earliest gods were relatively formless and primitive, but through a procreative process among themselves evolved into more advanced and defined beings. Echoing Fraser's view, Armstrong argues that for the Babylonians creation and the emergence of human civilization was an ongoing struggle against the primordial and the forces of chaos.[41] Babylonian rituals, laws, and ethical principles were ways to maintain order and direction amidst the presumed destructive and dissipative forces of chaos. As another common theme within ancient myth and religion, people often described their deities in a perpetual battle with the forces of chaos and death, and such deities were routinely invoked through prayer and ritual to assist humans in their personal struggles with these forces.

In early Sumerian myth, Ishtar was the "universal goddess" – the "queen of fertility" and heaven; she was extremely popular and revered and was responsible for the return of life in the spring.[42] But as was mentioned earlier, in later Babylonia there was a transformation from a goddess-centered religious culture to a male-centered one. According to Shlain, this transformation took place around 1800 BCE. It was at this time that Hammurabi, the famous king of Babylonia, codified and instituted a set of basic laws to govern all the people in his kingdom.[43] For Shlain, the rise of patriarchal society and religion is invariably connected with a top-down hierarchical system of social control and abstract rules of order, as clearly represented in the social and legal system created by Hammurabi.

Central to this new male-centered religion was the story of *Marduk*. *Marduk* was the central god of Hammarubi's new capital city of Babylon. *Marduk*, the Sun God, who was one of the younger, more evolved gods, defeated and killed in a great struggle the female goddess, *Tiamat*, who was an older, more primordial deity representing chaos and the abyss. After destroying *Tiamat*, so the story goes, *Marduk* brought increasing order to the world and created humanity.[44] Based on historical records, it appears, in fact, that as Marduk rose to pre-eminence in Babylonia, he appropriated many of the powers previously associated with other earlier gods and goddesses. Thus, in this one mythic story (and its historical backdrop) we see the evolutionary struggle of order versus chaos, of light versus darkness, and male versus female all tied together – with the male usurping the powers of the female. The male god is the deity representing order and light, while the female goddess is associated with chaos and darkness. Evolution or progress involves conflict, violence, and the conquest of the female by the male.

This general theory of progress involving an ongoing struggle and the triumph of order over chaos has been a highly influential idea throughout history. It provides a general description of the direction of time, connecting events of the past, present, and future. The original formulation of the theory though was evidently sexist and, as I will note later, this sexist slant on the struggle of order and chaos would continue into later historical periods.

The ancient Babylonians also thought about time in cyclical terms. They are well known for developing the discipline of astrology which is based on the cyclic theory of the **Zodiac**. Through astrology, they attempted to foretell their future. They believed that events in the heavens affected events on the earth. The year was divided into twelve repeating astrological periods or signs of the Zodiac which were connected with twelve major constellations in the sky. The sign of the Zodiac under which one was born presumably indicated one's personality and pathway in life. The Babylonians also charted and recorded the seasonal movements of the "planets" (wanderers of the sky) which, together with Zodiac signs, provided the basis for creating horoscopes that predicted the future.[45]

As one final point regarding Mesopotamia and Babylonian ideas on the future, we come to the great literary epic *Gilgamesh*. Gilgamesh was probably a real person and ruler of Uruk around 2900 BCE. The tale of Gilgamesh revolves around his quest for the secret of immortality – an obvious future related theme. Filled with great adventure, including an account of a great flood upon which the Biblical account of the Great Flood is probably based, Gilgamesh is ultimately disappointed in his quest; immortality is not to be gained for humans. Polak argues that in general Mesopotamia was rather pessimistic about the future; it is with the ancient Egyptians that he finds a more optimistic view including the belief that humans survive death in an afterlife.[46]

Egypt was a second major cradle of civilization in the ancient world. As with Sumerian and Babylonian culture, Egyptian religion is populated with a great variety and assortment of gods and goddesses who often engage in conflicts and struggles with each other for dominance and control. As with the Babylonians, the gods and goddesses of Egypt were an integral part of human life. Ancient Egyptians routinely turned to their gods and goddesses for assistance and protection in all aspects of life. As Armstrong notes, in early pagan religions the separation between the world of humans and the divine was not totally distinct. Instead the pagan vision was holistic, where humans and deities were all part of the same interwoven and interconnected reality. As one important and illustrative example, according to myth the Egyptian pharaohs were direct descendents of the gods.

Throughout most of its history, ancient Egyptian religion was predominately polytheistic and its gods and goddesses were personifications (often in animal or partially animal form) of the diverse forces of nature. Yet among the pantheon of Egyptian gods and goddesses, the sun god (who had several different names) generally occupied a position of supremacy.[47] In fact, for a short period of time this supreme deity was seen by the famous Egyptian pharaoh Ikhnaton (ca. 1392 - 1362 BC) as the only real and true God, foreshadowing the subsequent rise of monotheism in Judaism.[48]

Egyptian accounts of creation (and, as with the Mesopotamians, there were various versions) are similar to Babylonian views in that the world of order presumably rose out of a "sea of chaos." Having observed how mounds of sand arise out of the Nile after the river's flood waters recede each year, the Egyptians, in an analogous fashion, described creation as the emergence of a mound of earth - the "Risen Land" - out of the primordial waters of chaos. Egyptian myth is also similar to Babylonian in that the gods and goddesses then arose out of this fundamental creation through a process of procreation. In fact, it is especially evident in Egyptian mythology that sex among the deities is generally the means by which the great assortment of gods and goddesses came into being.[49] This central theme of sexual creation is undoubtedly an expression of the primordial human association of sex with the creation of life as evidenced in mother goddess mythology.

Two interesting exceptions to this rule concerned the "creator gods" *Atum* and *Ptah*. In one story of creation *Atum* is the first god to emerge out of the primordial mound but he does so in an act of self-creation; *Atum* is described as "he who came into being of himself." *Atum* is self-caused - a view of the nature of God we will see more fully expressed later in Judeo-Christian-Islamic religions. The story of *Atum* anticipates the more modern theological and philosophical idea of a supreme being who is the cause of its own existence. In the story of *Ptah* on the other hand, the civilized world and all the gods and goddesses of ancient Egypt came into being through *Ptah* thinking and naming things into existence. The thoughts of *Ptah* create the world.[50] This more intellectual account of creation anticipates later Judaic thinking that God created the world through simply willing or thinking it into existence. The story of *Ptah* also anticipates the Judeo-Christian-Islamic rejection of coupling and sexuality as integral to creation, substituting a single masculine God who creates through a spiritual-mental act without the need for sex.

One central, highly influential myth in ancient Egyptian religion that combines a variety of important archetypal themes is the story of *Isis*, *Osiris*, and their son *Horus*. The goddess *Isis* and the god *Osiris* are sister and brother and grandchildren of *Atum*. Though brother and sister, *Isis* and *Osiris* marry and become rulers over Egypt bringing great "abundance and prosperity" to the land. Yet, a problem arises in the form of a rivalry. *Osiris*, the god of order, had a brother *Seth* who is the god of disorder. Again, as an expression of the idea that order and chaos (disorder) are fundamental conflicting forces in reality, a bitter rivalry develops between *Osiris* and *Seth* over who should rule the world. *Seth* kills *Osiris* and scatters his remains across the face of the earth, sending his soul to the underworld. Unwilling to accept the death of her husband, *Isis* with the help of the god *Anubis*, resurrects *Osiris* and they procreate, producing a son *Horus*. With the protection and assistance of his mother, *Horus* engages in battle with *Seth* to gain rule over the earth – his presumed birthright. *Horus* finally defeats *Seth* and becomes the ruler of Egypt. According to Egyptian mythology, all the pharaohs were descendents of *Horus*.[51]

Isis was probably the most popular of many fertility goddesses in ancient Egypt. She was the "Great Mother" as well as the goddess of magic and healing. As the above story illustrates, she also had the power of life over death. She not only resurrects *Osiris* but then gives birth to a "son of God" in the form of *Horus*. As Armstrong notes, the "death of God" and the subsequent resurrection by the Goddess, who possesses the power of fertility, is a common story in ancient times.[52]

Shlain argues, in particular, that the procreation and loving care of *Horus* by *Isis* predates and anticipates the Christian story of the mother of God and the love of Mary for Jesus. For Shlain, as it is represented in the story of *Isis*, (and one can also include the myth of Ishtar) it is the woman and the goddess who originally had the power of life over death, but with the subsequent rise of male dominated religions across the world, this power was taken away from the female and relegated to the supreme male god instead.[53]

The Egyptians clearly believed in life after death. One central ritual performed each year revolved around the renewal of life through the goddess. As with the Babylonians, the function of rituals was to partake in the power of the gods and goddesses – in this case, the power to create new life. The resurrection of life after death was seen by Egyptians in the return of life in the spring and in the return of the day and the sun after the darkness and the night. The Egyptian practice of mummification to preserve the body of the dead was another expression of their belief in life after death. *Anubis*, mentioned earlier in the story of *Isis* and *Osiris*, was the god of mummification. The Egyptians believed in the eternal cycle of time and the ascension of soul and body (as a whole) after death to a higher eternal realm.[54] Such was their vision of the future.

Although *Isis* as a female was clearly associated with the power of the resurrection of the dead, the sun god was also emblematic of this power. Because the sun symbolized the re-emergence of life after death, as well as being the "giver of life" through the light that it shines down upon the earth, the sun god became the central and most powerful deity throughout much of Egyptian history. The sun god had many names and was connected to and synthesized with the original creator god *Atum*. The name most frequently used for the sun god was *Ra* (or *Re*). *Ra* in fact was frequently identified as the father or grandfather of all other gods and goddesses, including *Isis* and *Osiris*. *Ra* is the god who presumably gave

163

Osiris, and later *Horus,* dominion over the earth and human civilization.

Aside from being the giver of life, *Ra* was also the god of time. The movement of the sun through the sky during the day was seen as the journey of *Ra* on his boat across the heavens. As *Ra* traveled across the sky, he assumed the persona of a different animal god (for example, the scarab and the falcon) for each successive hour in his journey.[55] Each of these personae was archetypal representing different qualities associated with the animal gods. Time was therefore not some abstract and general quantity, but rather personified and connected with different aspects of reality. To recall, the Babylonian Zodiac also represented the passage of time through the year as a succession of archetypal figures associated with major stellar constellations.

When the pharaoh Ikhnaton (reigned 1379 - 1362 BC) came to power, he initiated a full scale revolution and transformation in Egyptian religion. He rejected all the popular gods and goddesses and replaced them with a single god - *Aten* - represented as a "sun disc" with neither a face nor any human feature. Ikhnaton destroyed many of the statues and images of the various deities in polytheistic Egypt and attempted to force all the priests and the general population, as well, into worshipping this single, faceless god. He attributed the total power of creation and dominion over all existence to *Aten.* As Durant argues, Ikhnaton is the first historical figure to clearly formulate and attempt to practice a monotheistic religion.[56] Although Shlain disputes the view that Ikhnaton believed in a pure, unadulterated monotheism, it is noteworthy that Ikhnaton represented his supreme deity without a human face or human characteristics.[57] This predates the idea that first developed in Judaism and early Greek philosophy that the supreme God (or the primary creative force of nature) can not be reduced to human or concrete terms. Ikhnaton also elevates *Aten* above all existence, again anticipating the Judaic idea that God is somehow separate and "above" the world. To quote from one of Ikhnaton's poems dedicated to *Aten,*

"Thy dawning is beautiful in the horizon of the sky,
O living *Aten*, Beginning of life.
When thou risest in the eastern horizon,
Thou fillest every land with thy beauty.

Thou art beautiful, great, glittering, high above every
land,
Thy rays, they encompass the land, even all that thou hast
made.
Thou art *Ra*, and thou carriest them all away captive;
Thou bindest them by thy love.
Though thou art far away, thy rays are upon earth;
Though thou art on high, thy footprints are the day." [58]

However we judge the intellectual and theological insights
of Ikhnaton, his revolution ultimately failed and, after his
death at a young age, the worship of the entire polytheistic
and personified array of gods and goddesses was restored to
ancient Egypt. Monotheism would later rise up again in the
Middle East, in a stronger form, in the writings of the Judaic
prophets. But it is also important to note that at least as early
as Ikhnaton and the first Judaic prophets, monotheistic ideas
were being formulated in ancient India as well.

Hinduism and Buddhism:
The Eternal One and Cosmic Consciousness

A third cradle of civilization in the ancient world emerged
around the Indus River in India. Archeological evidence seems
to indicate that the early cities in this area worshipped both
a female goddess of fertility and a male "horned" god. [59] But
just as nomadic Indo-European invasions from the north swept
down into the Mediterranean world around 2000 BCE, the
Indus valley was also invaded and conquered by nomadic Indo-
European or Aryan people from the Middle East during the
same period. [60] These invaders brought a warrior and caste
system to Indian civilization and a set of religious ideas that

eventually coalesced into a polytheistic ten-book anthology of religious poetry and hymns, the *Rig-Veda ("Songs of Knowledge or Wisdom")*. In this new social and religious-spiritual order, the chief gods are male, including, for example, *Indra*, the creator and warrior god who overcomes evil, and *Varuna*, the sky god and, prior to the popular ascendancy of *Indra*, the ruler of all other gods.[61] Though predominately polytheistic, the *Rig-Veda* did contain intimations of monotheism as well.[62] To quote a few selected lines from the *Hymn of Creation*:

> "By its inherent force the One breathed windless:
> No other thing than that beyond existed...
>
> The One by force of heat came into being.
> Desire entered the One in the beginning:
> It was the earliest seed, of thought the product."[63]

Those who created and followed the religious ideas contained in the *Rig-Veda* emphasized the importance of sacrifice, the universe and all the gods presumably having been created in such an act. (Recall Armstrong's point that rituals were often attempts to imitate the actions of gods.) According to Watson, the composition of the *Rig-Veda* occurred through revelation in drug induced trance-like states around sacrificial ceremonial fires.[64] As we will see, visions of both the divine and the future as contained in other religious traditions were often attributed to revelations – presumably communicated from spirits and gods.

Beginning around 700 BCE a new set of religious writings emerged in ancient India, the *Vedanta*, literally "appended to or after the *Vedas*." The *Vedanta* includes the famous set of writings known as the *Upanishads*. The writers of the *Upanishads* had become dissatisfied with Vedic beliefs and practices and wished to create a new spirituality that downplayed the importance of sacrifice and focused more on inner development. The concept of "Atman" – the eternal soul and most innermost self – first appears in the *Upanishads*.[65]

The first clear expression of the idea of **karma** can be found in the *Upanishads*. Karma is destiny (or the life force), but a destiny that is determined by the individual. In order to understand the meaning of karma, the notion of reincarnation first needs to be explained.

Within the *Upanishads* is the theory of **reincarnation** (or *samsara* meaning rebirth), an idea that is common to many different ancient cultures. According to the theory of reincarnation, when a living being dies, its soul migrates to another living being, be it animal or human. Death is not final, but just the end of one cycle of life to be followed by the beginning of another cycle. In reincarnation, again we see the idea of life arising after death.

The *Upanishads* connects the idea of karma to reincarnation. The total set of deeds, good and bad, within a person's life creates a person's karma and determines the quality of his or her next life. A person may rise to a higher level of existence if his karma is positive or sink to a lower level if it is negative.[66] A soul can realize *moksa* and total salvation, becoming one with the absolute spirit of *Brahman* (discussed below), if all negative karma is eliminated and the soul overcomes the *maya* - the mistaken belief that the phenomenal world is real and that the self is a separate being.[67]

The connected ideas of karma and reincarnation contain a clear conception and vision of the future, as well as of the connection between the future and the past. As a general principle, it is ethics that determines the quality of the future. After death, a soul moves to a higher or lower level of existence depending on his or her ethical choices and actions in the previous life. Each life brings with it a destiny created in the past life, but provides the opportunity for the person to improve his or her karma for the next life.

The idea of karma expresses a cosmological principle of justice. Unethical behavior must be paid for in this life or the next. Ethical behavior is rewarded through the building up of positive karma and the reduction of negative karma. In essence, karma is the Hindu version of the principle of "what goes round comes round." Life is a great balancing act of positive and negative; justice is always served in the long run. In several important respects Hindu thinking reflects the fundamental principles of cycles and balance in understanding time and the nature of reality. Reincarnation and karma is one prime example.

The future of the soul, as noted above, is also a consequence of achieving knowledge or enlightenment - in particular, realizing the "oneness" of the self with the cosmos and the illusory nature of the physical or phenomenal world. Both of these insights are central to Hindu thinking and, in fact, as we shall see, the renunciation and transcendence of the physical world is a common idea in many religious and philosophical belief systems. The ultimate future and absolute reality exist beyond the world of matter and time.

As noted, a soul, through enlightenment and the elimination of all negative karma, can be released from the cycle of reincarnation and achieve unity with the eternal oneness, transcending the wheel of time. A soul can become one with *Brahman*. The term "Brahman" which in the *Vedas* meant the sacred power of prayer came to mean in the *Upanishads* the sacred power that pervades, sustains, and animates everything. (Again we see the idea that nature is empowered and in-spirited by deities.) Everything is a manifestation of *Brahman* - *Brahman* is the absolute, all encompassing One. Although within the *Upanishads* there is the idea of *Brahman* made manifest, where personified qualities are associated with *Brahman's* reality, there is a deeper sense of *Brahman* as the "un-manifest" One that is the ultimate source of all being - that is, in fact, "Being" itself.[68]

With the emergence of Hindu monotheism in the idea of *Brahman*, God acquires a de-personalized quality. Whereas polytheistic religions personified the forces of nature, *Brahman*, as the ultimate spirit that animates all reality, is not in the strict sense a "person" at all. We will see this idea of an abstract and depersonalized ultimate reality become increasingly influential in the centuries ahead. Shlain associates this view of reality with the masculine mindset that grew to dominate both religious and secular thinking in the ancient world.

This distinction between the manifestations of God and the intrinsic real nature of God is also a common theme throughout many world religions. A related distinction also common to many religions is "transcendence" and "immanence." In Babylonian, Egyptian, and Hindu thinking we have found the idea that deities pervade natural reality - they are "in the world" - and it is there that they manifest themselves - they have an immanent presence. Yet, beginning with Ikhnaton and Hinduism, we also see the idea that there is a supreme God who is beyond the world that is transcendent. The Judeo-Christian-Islamic tradition clearly identifies a transcendent God that is beyond human comprehension, but yet one who also directs events on earth from "on high." This idea that the flow of time and specifically the unfolding of the future are being directed by some transcendent being has been very influential throughout the history of humankind. It is part of the legacy of our religious heritage and traditions around the world.

Yet as Armstrong repeatedly notes, humans thirst for divine immanence. God must be part of our lives. The futurist Barbara Marx Hubbard believes that beginning around 3000 BCE a "spiritual impulse" arose in humanity, where at least some people attempted to make contact with the "Oneness" of all existence. Through subsequent ages humans have repeatedly tried to achieve a "cosmic consciousness" with God and ultimate reality.[69]

Throughout the ages Hindu mystics have sought this oneness with *Brahman* through ritual and meditation, but so have practitioners of many other religions who have sought contact and immersion with the ultimate transcendent One. According to Armstrong, what we find in Hindu descriptions of this state of "cosmic consciousness," as well as in descriptions from other religions and mystical traditions, is that the experience of enlightenment - of becoming conscious of the ultimate Oneness - is beyond language and beyond the powers of reason to grasp.[70] There is a common human experience of the One and this experience transcends human categories of understanding. This mystical – religious view of enlightenment has been ubiquitous throughout human history, and conflicts with the rationalist view (to be discussed later) that through reason and language humans can grasp, understand, and describe ultimate reality.

The Hindu belief in the unity of all existence extends to the relationship of individual souls and the universal spirit of *Brahman*. Reflecting the pagan belief that the divine is within everything, Hinduism contains the idea that the inner self of each person embodies the "spark" of *Brahman*, which is referred to as **"Atman."** This most inner self is one with *Brahman* – enlightenment involves discovering this unity. Whereas generally Western religions clearly separate God and individual human souls, in the Hindu doctrine of *Brahman-Atman*, God and all souls are one.

Hinduism is a complex and multifaceted religion and contained within its teachings are also many polytheistic elements. There are numerous Hindu deities and **avatars** (incarnations of deities) who have been worshipped through the ages; avatars provide a way to personalize the impersonal quality of Brahman – the ultimate One.

Contained within Hindu doctrine is one well-known effort to synthesize the complex assortment of figures and deities into a trinity ("Trimurti") of primary deities who are both a one and yet three distinctive supreme personae. This trinity consists of *Brahmā* (the male persona of Brahman), *Shiva*, the god of creation and destruction, and *Vishnu*, the preserver of the world. All the varied gods, goddesses, and avatars of Hinduism are offspring or manifestations of one of these three personae.[71]

Within Hindu cosmology, the world is created as a dream in the mind of *Vishnu*. The world of time, though lasting trillions of years, is eventually destroyed with great violence by *Shiva*. Yet after the world of time comes to an end, *Vishnu* dreams the world anew, and the cycle of creation and destruction begins again. In a somewhat different version of this tale, *Shiva* is the creator of the world as well as the destroyer, and the cyclic and epochal quality of time and creation/destruction is seen as "the dance of *Shiva*."[72]

There is, in fact, a famous statue of Indian art representing the dance of *Shiva*, with *Shiva* manifesting himself as the god *Nataraja*; with four arms and surrounded by a ring of fire, he gracefully balances on top of the demon of ignorance. Dance is highly significant in Hindu thinking for it is associated with creation. The ring of fire represents both destruction (through fire) and the light of truth. In contrast to the serene stillness and abstract quality associated with Brahman, Shiva is a god of fire and energy connected with the ideas of beauty and sexuality. In numerous Hindu temples in the cities of Orissa and Tanjore, considered among the greatest works of architectural and sculptural art in the world, can be found highly explicit erotic figures and scenes. Often these temples were dedicated to the worship of Shiva where the depictions of sex presumably represented a higher, more ethereal beauty awaiting souls in heaven. Even more strongly connected with sex was the development of Tantric Hinduism (and Tantric Buddhism). It is noteworthy that in these religious movements, female goddesses and the female principle again achieved a central divine power; in fact, true worship of the female divinity is realized in sexual intercourse.[73]

Given the above examples, it is clear that not all features or expressions of Hinduism involved a rejection of the physical. Still, according to J. T. Fraser, in general Hindu cosmology is a prime example of how major world religions deny the finality of death and the ultimate transitory nature of reality. Within Hindu thinking what is absolutely real is eternity, rather than the passage of time. First, recall that *Brahman* – the One – the all pervading spirit - is eternal. Second, note that though the world is periodically destroyed by *Shiva*, eventually it begins again, making the ending of things only temporary. Existence (within the world of time) is fundamentally endless – an ever repeating cycle of beginnings and ends.[74] Since the world of time is ultimately an endless cycle, the future is fundamentally just a repetition of the past. Also, to recall, given the Hindu belief in reincarnation, personal death is only apparent for souls are perpetually re-cycled through multiple incarnations until all negative karma is worked off and the soul unites with the eternal *Brahman*. Finally, within Hindu doctrine, the changing physical world is illusory and enlightenment is achieved through seeing beyond this false reality. Fraser also includes Buddhism - the next tradition I will describe - as another belief system that denies the ultimate reality of time and personal death.

During the time of approximately 700 to 400 BCE, a period generally referred to as the **Axial Age**, (for human history seems to pivot on this age), a number of influential religious and philosophical figures emerged across Asia and Europe and produced a rich variety of new ideas and teachings on reality, knowledge, God, and the meaning of life. They were responsible for the creation of many new religions, spiritual practices, and theories of philosophy. They emphasized self-responsibility, abstraction, literacy, and a rejection of the authority of royalty. They expressed a movement away from polytheism toward inner development, morality, and an enhanced sense of individuality. This group of religious and philosophical leaders and figureheads included Isaiah, Socrates, Zoroaster, Lao-tzu, Confucius, and Siddhartha Gautama, or as he is more popularly known, the Buddha.[75]

Buddha (563 to 460 BCE) was born a wealthy prince in India but observing the poverty and suffering of much of humanity, he renounced his wealth and noble position and went in search of knowledge, self-enlightenment, and a solution to the apparent misery and difficulties of life. Buddha was critical of the materialist and mercantile way of life he saw emerging around him and turned inward to find a better way. The answers that Buddha found formed the starting point for one of the major world spiritual traditions, Buddhism, which eventually spread over much of Asia in the centuries that followed his life.[76]

Certain aspects of Buddhism reflect the influence of Hinduism, the predominant religious tradition in the region where Buddha lived. Buddha believed in reincarnation and the illusory quality of time, and he believed that the ultimate goal of life was to achieve a unity or oneness with ultimate reality. But Buddha rejected, or more correctly, he transcended, all aspects of personification in his thinking, including the most powerful form of all - the fact that we personify ourselves.

For Buddha, the ego or self is the source of all craving, suffering, and misery. The ego separates us from ultimate reality in that we distinguish ourselves in opposition or contrast to the world through this psychological construct. We conceptualize reality as "me and the world." Because we separate ourselves from the world, we desire what we perceive as not part of us; where there is desire, there is frustration, disappointment, and suffering.

According to his teachings, the solution to life's miseries is "right living" and "right thinking" (*dharma*) achieved through inner awareness and discipline. The truth lies within. We need to see that all worldly existence is flux and impermanent, and that the self is simply an idea that we use to conceptualize or understand our reality and not some intrinsic or absolute essence that defines who and what we are. We must see through the veil of illusion - including the illusion of the individual self.[77] For Buddha, it is egoism and the karma of the ego that binds humans to the Wheel of Sangsara - the cycle of birth and death.[78] Buddhist teaching puts forth the alternative idea of the Wheel of Salvation, which leads to the transcendence of the ego and time.[79]

Once we achieve this insight - this state of cosmic consciousness - and break free of the wheel or cycle of unending reincarnations, we transcend time and achieve **Nirvana**.[80] Nirvana is the ultimate reality, even beyond the gods, and the state of complete "liberation."[81] Nirvana can not be adequately described with words. Reason can not grasp or understand its reality. In ways Nirvana sounds like the ultimate oneness of *Brahman*, for all differences and distinctions (all separation and structure) evaporate within Nirvana. But even the term "oneness" carries with it a conceptual meaning and Nirvana transcends all linguistic or conceptual categories.

As noted, the influence of Buddhism spread across much of Asia, including China and Japan. In China, a form of Buddhism called *Mayahana* Buddhism arose and became extremely popular in the first few centuries after Christ. Of particular note, in *Mayahana* Buddhism Buddha achieved the status of a deity and was believed to be the creator of the world. Furthermore, as a fundamental prophecy for the future, followers thought that incarnations of the Buddha would time and again return to the earth to rescue it from the evil which, presumably, arose at regular intervals.[82] Buddha, in essence, becomes the savior, with the world as a stage on which good and evil vie for dominance.

Thus, for Buddhism, as for Hinduism, time has a cyclic, derivative, and ultimately illusory reality. What is fundamentally real is beyond time - in Hinduism it is *Brahman*, in Buddhism it is Nirvana which is beyond earthly description. Hence, in both Buddhism and Hinduism, there is a metaphysical dualism encompassing the changing and differentiated world of appearance and the changeless, undifferentiated world of ultimate reality. And the ultimate goal of life - the defined trajectory into the future - is to achieve immersion or unity with a realm that is beyond time.

Taoism and Confucianism: The *Yin-yang*, Reciprocity, and Balance

A fourth major cradle of human civilization is ancient China. As with India, two great philosophical systems arose in China as well. Since neither of the philosophical systems, Taoism or Confucianism, is associated with any primary gods, goddesses, or deities, it would be somewhat inaccurate to describe these belief systems as religions. In fact, since Buddhism (discounting *Mayahana* Buddhism) did not depend upon any important deity in its principles or practices, it would be inappropriate to label it as a religion as well. Yet, Taoism, Confucianism, and Buddhism, are usually identified as spiritual or religious traditions in general histories of world religions.[83]

There are a variety of different readings and interpretations of Taoist cosmology. Piecing together ideas from several sources, the *Tao* (literally "the way") is eternal – having no beginning or end - and is the cause, principle, and reason behind of all existence. The *Tao* animates and harmonizes all motion in the universe. Because the *Tao* is unbounded (some would say infinite) it cannot be restricted, abstracted, or captured in words. (Note the similarity with the idea of *Brahman*.) The *Tao* literally encompasses both being and non-being. As stated in the first stanza of the *Tao Te Ching* ("The Way of Life") – the central philosophical work of Taoism,

> "The Tao that can be described
> is not the eternal Tao.
> The name that can be spoken
> is not the eternal Name."

In describing the creation of the universe, in the beginning was *Wu Chi* – the nonexistent and without limit – the "ultimate state of nothingness." Within this undifferentiated state of pure potentiality stirred the first motion, and out of this motion emerged the germ of the universe identified as the "Pearl of the Beginning." From this arose the *"Tai Chi"* – the ridge or oneness that creates the fundamental duality of *Yin* and *Yang* – the two primary forces within the universe. Or as stated more poetically from the *Tao Te Ching* in Chapter 42,

> "The Tao begets One;
> One begets Two;
> Two begets Three;
> Three begets the myriad things"

In the diagram below, *Yin* is the dark form, *Yang* the light form, and the *Tai Chi* is the sine wave defining the interface of the two forms. The expression *"Tai Chi "*also refers to the process of balanced interaction and interplay between *Yin* and *Yang*. Since the balanced interplay between *Yin* and *Yang* is also thought of as the *Tao* manifested within the world, the sine wave interfacing *Yin* and *Yang* can likewise be thought of as representing the *Tao*.[84] The *Tao* is the oneness that creates the two-ness of the *Yin* and *Yang*.

The Taoist Yin-yang

The flow of time and the organization of reality is described and symbolized in Taoism in terms of its fundamental archetype, the *Yin-yang*. In general, *Yin* and *Yang* refer to the basic polarities of existence, such as darkness and light, passive and active, and sky (heaven) and earth. *Yin* is the earth and the feminine principle of reality, whereas *Yang* is heaven and the masculine principle. *Yin* is often associated concretely with a bird and abstractly with matter and space; *Yang* is associated with the dragon, and spirit and time. It is noteworthy that the West traditionally had roughly the same set of oppositional associations around the male and female – the male being associated with heaven, light, and the active, and the female with the earth, darkness, and the passive.

In Taoism though, these two principles of the feminine *Yin* and masculine *Yang* are mirror images of each other and are united in their complementarity, interdependence, and balance with each other. Note in the diagram above that at the center of *Yin* is *Yang* and at the center of *Yang* is *Yin*, further emphasizing their interdependency. The forces of *Yin* and *Yang* are intertwined in everything and rhythmically oscillate within all processes of nature. The natural flow of time has a cyclic rhythm involving the alternating dominance of *Yin* and *Yang* – time is a balancing, a circling of complementary forces. Hence, Taoism emphasizes the inherent unity, harmony, cooperation, and balance within the universe. The two are a one. In fact, the *Tao* – the way – is this unity, harmony, cooperation, and balance. As I stated above, the interface and interplay of *Yin* and *Yang* is the *Tao*. The rhythmic oscillation of the "Two" is the "One."[85]

Time then for the Taoists is an orderly process. There is a general pattern to the ebb and flow of events – a waxing and waning of *Yin* and *Yang*. Further, the Chinese, like their Middle Eastern counterparts, developed a cyclic and repeating Zodiac to describe the orderly procession of time. While the natural flow of the *Tao* can be momentarily unsettled by chaos, demonic influences, or human willfulness, in the long run, the *Tao* reigns supreme. In essence, it is a form of cosmic justice – an ultimate balancing that always finds its way to fulfillment and realization. Hence, it is important for humans to attempt to move "with the *Tao*" – not to resist or attempt to counteract the natural flow of events. The practice of Taoism at one level is simply trying to live in harmony with nature – with the forces of *Yin* and *Yang*.[86]

Followers of Taoist philosophy did attempt to predict the future however. In fact, if there is a natural order to time, then it makes perfectly good sense to think one can predict what is to come. Yet the popular method that was employed is based on magical thinking and incorporates an element of chance. The Chinese believed they could predict the future through **divination.**

The *Tao Te Ching*, a philosophical book describing in poetic terms many basic principles of Taoism, was presumably written by Lao-tzu though the actual authorship of this book is debatable.[87] A second major book of Taoism is the *I Ching* (The Book of Changes), which is the book often used as a means for foretelling the future. The *I Ching* is divided into 64 sections, each section corresponding to one of the fundamental Taoist hexagrams. Each Taoist hexagram is a sequence of six *Yin* and *Yang* in different orders and combinations, and each hexagram has a specific meaning. For example, the first hexagram are six successive *Yang* that symbolize "The Creative"; the sixth hexagram is three *Yang*, a *Yin*, a *Yang*, and a *Yin*, representing "Contention"; and the sixty-fourth and final hexagram is *Yang-Yin-Yang-Yin-Yang-Yin*, standing for "Unfinished." Through some type of quasi-random activity, such as the tossing of coins, the throwing of sticks, or the selection of plant stalks, a particular hexagram is identified and the hexagram is read, presumably providing knowledge about what is to come in the future.[88]

Even if the Chinese believed they could predict the future, this did not mean that they believed they should attempt to control or direct it. In fact, knowing what was to come was a way to prepare oneself to stay in harmony with the flowing *Tao*. This general attitude of passivity toward life and the future is actually expressive of the feminine side – the *Yin* – according to Taoist philosophy. As the religious historian David Noss states, this feminine attitude in Taoist thinking and practice is due to the perception that human society is too *Yang* and needs to be counter-balanced with *Yin*. Philosophically the Taoists believed in the balance and equality of masculine and feminine principles, and through their passivity attempted to bring balance into Chinese society.

According to Shlain, although the *Yin-yang* implied an egalitarian way of life between men and women, ancient Chinese culture was decidedly patriarchal.[89] As noted above, the Taoists themselves would have agreed with this assessment of their society. This quality of male dominance within their culture, according to the Taoists, was reflective of Chinese society being too "Confucian." Confucianism as a philosophy stood for male domination over women, as well as for strong social hierarchies, written laws, literacy, reason, urban civilization, control, and abstraction – all values that were almost the antithesis of Taoism. Taoism valued intuition, the inexpressible, fluidity, nature, the concrete, and non-resistance.

This almost contradictory mix of philosophies in ancient China is itself a *Yin-yang*, but as the teachings of Confucius (551 – 479 BCE) gained influence, Confucianism increasingly superseded the earlier teachings of Taoism. Accordingly, women lost power. As Shlain states, even Taoism was perverted in this process, turning from the principle of "make no dams" to "making dams and stopping the flow." China became increasingly more *Yang*.

Taoism and Confucianism were not completely at odds though. Confucius studied the writings of Taoism and frequently spoke of the value of Taoist principles. Confucius, in fact, identified the Tao as one of the key principles in his philosophical thinking.[90] If we look at the principle of the **"Golden Mean"** however, we can see how each philosophy, while supporting this idea of balance in life, interprets it differently. The "Golden Mean" teaches that one should never do anything in excess; rather, one should follow the middle road. The *Tao* and the *Yin-yang*, of course, symbolize the ultimate supremacy of balance in the universe, and presumably in following the *Tao* one would lead a life of balance. Yet in Confucianism, balance seems to turn into an authoritarian rule of action and thinking – something to be achieved through mental and behavioral effort and control. For Confucius, the Chinese world in which he lived was corrupted and in turmoil, and principles of order needed to be developed and implemented to bring happiness to his land. Thus, in seeing the chaos of things, he pushed for order, and if the ultimate order was balance, then balance was something that needed to be consciously and rationally imposed. Hence, whereas for the Taoists balance was something that would come naturally if one didn't fight against the *Tao*, in Confucianism balance required self-effort and conscious direction.

Another important connection between Taoism and Confucianism, again having to do with the ideal of balance, concerns the principle of reciprocity. As noted in the previous chapter, early human cultures acknowledged the importance of reciprocity and practiced it in human interaction and economic exchange. Reciprocity served as a foundation for justice and equal and fair treatment of each other. The Taoist theory of reality is built upon the idea of reciprocity, in the sense of complementarity and interdependency. *Yin* and *Yang* require each other for their existence. Confucius, in turn, raises the idea of reciprocity to a central ethical principle. This is illustrated in the following story:

"Tzu-kung asked, 'Is there one word which may serve as a rule of practice for all one's life?' The Master (Confucius) said, 'Is not Reciprocity (mutual consideration) such a word? What you do not want done to yourself, do not do to others.'"[91]

As can be seen, this statement sounds very much like the Christian **Golden Rule** – "Do unto others as you would have them do unto you." In the mind of Confucius, this general ethical principle of behavior was based upon the idea of reciprocity. Yet as the religious historian David Noss points out, Confucius did not extend this principle to those people who commit evil acts, hence his sense of benevolence was limited in a way not found in the Christian version of the rule. For Confucius, if one commits an evil act, it should be repaid in kind; (actually he uses the word "justice" in the sense of retributive justice). His attitude is more in line with the Judaic principle, "An eye for an eye - a tooth for a tooth."

In general, ancient Chinese philosophy and culture valued the principles of order and harmony, and both Taoism and Confucianism emphasized this fundamental perspective on life, each in its own way. If Taoism is more passive and accepting in its particular approach, Confucianism is more active and controlling. Further, Taoism highlights intuition and oneness with nature, whereas Confucianism highlights rationality and a strong concern for social community and social order.

As a clear expression of the importance of order in both Chinese society and Confucianism, during the highly organized Han dynasty (from approximately 200 BCE to 200 AD) Confucianism was an integral part of its dominant philosophy and system of government. The cosmos was believed to possess a natural order and harmony and this structure to things was reflected in Chinese social order as centrally controlled by the emperor. The Han dynasty, reinforced by its Confucian philosophy, was a hierarchical and authoritarian system that emphasized rules and duties. There was a continued concern with "right conduct" presumably to maintain resonance with the natural cosmic order of things.[92]

If we turn to Neo-Confucianism, as it developed during the Song dynasty (960 - 1279 AD), we again see an explicit connection made between ethical principles for living and the cosmic order of things. For Neo-Confucian thinkers, such as the highly revered Zhu Xi (1130 - 1200), the principle of *li* explained the order and development of matter and natural reality, and if understood, provided direction for how to live ethically and achieve wisdom and happiness. Similar to Confucius, Zhu Xi was a rationalist and did not highlight supernatural forces or deities in his thinking; in fact, Zhu Xi believed that natural forces could explain the structure and pattern of nature.[93]

As two final notes on Chinese philosophy as expressed in both Taoism and Confucianism, we find that with the emphasis on order and harmony, there is a relatively static conception of time; everything has its place, balance is important, and the *Yin-yang* cycles through time, forever changing, forever the same. Also, the orderly make-up of nature and the cosmos provides a template and guidance system for achieving happiness, wisdom, and the ethical good life; the microcosm should mirror the macrocosm. This second theme provides a bridge to the next religious tradition to be discussed - Zoroastrianism.

Zoroastrianism:
The War of Good and Evil

Whereas Taoism conceptualized time and reality as a circle and an ultimate harmony of all forces in nature, Zoroastrianism, which emerged in ancient Persia, viewed time and reality as a battle between opposing forces that eventually would lead to a conclusion and victory of one force over the other. In this cosmic struggle the force of good conquers the force of evil. If Taoism is built on reciprocity, one of Bloom's two primary forces that shaped human history, Zoroastrianism is inspired by the other fundamental force, conquest.

Although Zoroastrianism is not a very popular religion in the present day, it is a highly influential belief system in the evolution of future consciousness and Western religion; in fact, its impact on Western religion as well as on futurist thinking is immense. The ancient Israelites, who were exiled in Babylon during the 6th Century BCE, encountered and appear to have taken many of their key ideas from Zoroastrianism which was then the dominant religion in Persia and the Middle East.[94] To begin with, as one example, the Judaic and Christian view of time as a necessary sequence of events leading to some culminating moral resolution, an idea that would also be taken up by Islam, probably derives from Zoroastrianism and the prophecies of its founder, Zoroaster (660 – 583 BC). According to Zoroaster, the supreme God *Ahura Mazda* ("Wise Lord") has been engaged in a struggle of "good and evil" with *Angra Mainyu* ("the Bad Spirit") throughout history and this struggle will eventually lead to a final apocalyptic battle culminating in the defeat of evil and the salvation of all good souls.[95] As another significant anticipation of Judaic-Christian thinking, Zoroastrianism contains the idea of a messiah, who will lead the forces of good against evil at the end of time.[96]

Zoroaster framed time, both individually and cosmically, in moral or ethical terms. The flow of events through time is due to an ongoing conflict of good and evil. This conflict occurs at a cosmic level, but also occurs in the souls of all human beings. Through each of our lives we are engaged in an ethical struggle, attempting to pursue what is good, but continually being tempted and led astray by the forces of evil. The soul is the micro-cosmic battlefield and reflection of the macro-cosmic war of good and evil.

Zoroaster also saw time in violent and combative terms. We have already encountered the idea in Babylonian mythology that the flow of events in reality is due to an ongoing struggle between order and chaos - a conflict of opposites. For Zoroaster, time is a war of good and evil that will eventually lead to an ultimate and final battle. Good will triumph and the spirit of evil will be conquered and destroyed at the end of time but only through destruction and obliteration.

Zoroaster personifies the fundamental forces and dynamics of the universe. There are two ultimate spirits that represent the essential qualities of cosmic and human existence. There are also many lesser spirits who are expressions or servants of these two major cosmic personae. Although there is debate over who first invented the idea of a supreme God, it is fairly clear that it was Zoroaster who invented the idea of the Devil. For Zoroaster, life is a drama - in this case a personified struggle between God and the Devil - with a concluding chapter that resolves the underlying conflict played out in the story.

Within Zoroastrianism, and in contrast to Taoism and all Eastern thought, time is a line rather than a circle. Further, time is not eternal but finite. Time eventually comes to an end. This linear and finite view of time would become a central tenet in the Christian conception of time. Also within Zoroastrianism, time is a progression rather than a repetition, as it is within Taoism. The triumph of good over evil signifies a positive resolution to the flow of time. A similar notion of progression can be found in Christianity.

Although Zoroaster believed that humans possessed free will - to choose between what is good and right and what is evil - the universe as a whole is destined to follow a particular direction. *Ahura Mazda* will triumph over *Angra Mainyu* at the end of time. So the progressive direction of time in the universe is pre-determined. This view of time is **teleological** (from the Greek "*telos*" meaning end). What happens in the world is determined by some ultimate goal to be realized in the future. In essence, the foreordained future determines what comes before. Christianity adopted a similar set of ideas regarding free will and the destiny of the universe. Still it is important to also highlight the significance of free will or choice in Zoroastrianism. Each individual can choose to follow either what is good or what is evil. This idea also anticipates Judeo-Christian thinking; each individual has the power to select his or her own destiny.

In his theory of reality and time, Zoroaster connected **eschatology**, which deals with the "end times" of humankind and the universe, and the idea of the **"apocalyptic"** which means the revelation and perception of the ultimate truth. At the end of time *Ahura Mazda* reveals the final and complete truth about reality. Again this anticipates Christian thinking. Zoroaster also introduces the idea that God (*Ahura Mazda*) will judge all human souls at the end of time according to whether they followed a life of good or a life of evil ("Judgment Day"). Hence one's ultimate individual future is determined by one's ethical behavior or lack of it, as seen through the judgment of a supreme being. Those who are judged good and worthy, having followed *Ahura Mazda*, will be rewarded with an eternal afterlife in Heaven or paradise; those who are judged evil, having followed the spirit of *Angra Mainyu*, will be punished and damned in Hell. For Zoroaster justice is ultimately served at the end of time.[97]

In several respects, Zoroastrianism is strongly dualistic. **Dualism** is the theory that reality consists of two distinctive - often opposing - sets of qualities, components, realms, or forces. We have already encountered the philosophical doctrine of dualism in Hinduism and Buddhism, where the eternal or timeless realm is distinguished from the temporal realm of change. Zoroastrianism supports an absolute dualism of good and evil, truth and falsity, and body and soul.

The metaphysics and ethics of Zoroaster is though an interesting mix of dualist, polytheistic, and monotheistic elements. Although Zoroaster sets up a fundamental dualism in his theory of good and evil and truth and falsity - *Ahura Mazda* is good and the source of truth - *Angra Mainyu* is evil and the teller of lies - Zoroaster sees one side of this dualism as superior to the other side. There is a single God - *Ahura Mazda* - who is all powerful, all good, the creator of the universe, and who orchestrates the direction and resolution of events in the world. This is similar to later monotheistic ideas in Judaism and Christianity. Zoroaster also carries this asymmetrical or lop-sided dualism into his theory of humans and his theory of ultimate value. He believes that humans possess a non-material soul connected with their material body, but it is the disembodied soul that survives physical death and is superior to the physical body. We are both matter and spirit but the spiritual side is on a higher level. In resonance with this lop-sided dualism of body and spirit, Zoroaster places more importance on the "other worldly" over the physical world. Paradise lies beyond time and the physical world. (This last idea is similar to Hinduism and Buddhism.) All these ideas on reality and the future anticipate similar notions in Christianity, and in many respects, Islamic religion as well.

Greco-Roman Myth and Philosophy: The Apollonian, the Dionysian, and the Theory of Progress

Ancient Greece is often identified as the fountainhead of Western Civilization. Yet, just as in China where we find the opposing philosophies of Taoism and Confucianism, we find in Ancient Greece a combination of opposites – of chaos and order – of love and hate - of madness and reason – of mysticism and rationalism – of myth and abstraction - this is the heritage of the West. Greece had its myths and personified deities, such as *Zeus, Aphrodite, Athena*, and *Hermes,* who engaged in all manner of melodrama, conflict, and mayhem, frequently involving earthly humans in their machinations. The Greeks conceptualized reality and creation in narrative form. Also, Ancient Greeks participated in numerous rituals and "mysteries" in which they believed they mystically shared in the powers of their deities. In particular, Ancient Greece had the cult and "mystery" worship of the resurrected, dark deity of *Dionysius* – the god of wine, dance, sex, emotion, and reverie. *Dionysius* is in many respects the Greek god of chaos. But ancient Greece is also the birthplace of Western philosophy – of the abstract and rational systems of Parmenides, Plato, and Aristotle that came to challenge the validity of the whole edifice of mythic thinking around the world. If we follow Merlin Donald's theory of the developmental stages of human cognitive evolution, it is in ancient Greece that we see the blossoming of the "theoretic" mindset layered on top of and juxtaposed with mythic thinking.[98] The theoretic mode of cognition involves de-personalized, abstract, analytic, and often dialectical thinking about reality. Yet, interestingly, the Greeks also had a god associated with reason and order – the god *Apollo*.

Aside from the contrast of mythic and theoretic thinking in ancient Greece, it is important to also highlight the contrast of reason and thinking with passion and emotion. Emotion and reason are frequently viewed as two different and opposing modes of consciousness (though to recall from my earlier discussion of the psychology of future consciousness these two psychological processes are interconnected). In ancient Greek mythology the rational and passionate approaches to life were personified in the gods *Apollo* and *Dionysus*. The "**Apollonian**" perspective emphasized reason and order while the "**Dionysian**" perspective highlighted passion and disorder. Ancient Greek culture acknowledged both dimensions of consciousness, and in fact, valued a balance between passion and reason.[99]

Although religion is often characterized as more emotional than rational, especially when it is contrasted with science, both reason and passion can be found in all religions. There are strong rationalist traditions in Judaism, Christianity, and Islam that emphasize the value of reason in the search for truth and enlightenment. This rationalist side of Western religion owes much to the Greeks. Still it is important to note that various religious practices often highlight the passionate dimension of human experience. The sense of personal abandonment in rites, ceremonies, and rituals – of great collective expressions of reverie, ecstasy, and often music and dance – are Dionysian rather than Apollonian in character.

Hence, although emotion and reason are undoubtedly intertwined in human consciousness, we have two traditions in the West, going back at least as far as the Greeks, that have respectively emphasized either the passionate/emotional or the rational side of humanity. These two traditions, the Dionysian and the Apollonian, not only have influenced religious thinking but secular thinking as well. In modern times, the contrast emerges as a fundamental ideological conflict between the rational philosophers of the Enlightenment and the expressive art and literature of the Romanticists. The Greeks valued both sides of the human mind, but Western human history has witnessed conflict and oscillation between these two dimensions of human consciousness. Approaches to the future and how to guide and direct human life have been significantly impacted by which dimension of human consciousness has been emphasized.

Two of the most influential books ever written in the West are the *Iliad* and the *Odyssey* by Homer (ca. 800 BCE). These are classic works of literature that many ancient Greeks read and revered. The *Iliad* tells part of the story of the siege of Troy – the saga of Helen, Paris, Agamemnon, Achilles, Hector, and a host of other Greek and Trojan characters. The *Odyssey* describes the journey and return of Odysseus to his home in Greece after the battle of Troy. In both tales the Greek gods and goddesses frequently interact with the humans. In fact, most of the major events in both tales are orchestrated and manipulated by the Greek deities. For example, the attack on Troy by the Greeks was presumably instigated by a personal conflict among the Greek goddesses *Hera, Athena*, and *Aphrodite*.

In his novel *Ilium*, a science fiction retelling of the siege of Troy, the contemporary writer Dan Simmons conveys, from a contemporary perspective, the vivid and psychologically compelling sense of the ongoing presence of deities in the experiences of the ancient Greeks and Trojans. The characters in *Ilium* behave and talk as if they are perpetually on stage

before the gods and are being watched and judged; the characters in *Ilium* have great theatrical egos.[100]

The Greeks believed that the gods and goddesses were a living and active reality in their lives. The gods and goddesses moved about and through the world influencing everyday events and controlling the forces and patterns of nature. "The gods ...were in the streets and houses of the people."[101] In fact, the gods may have also been in their minds. Not only did the gods and goddesses determine the events of nature, they were also seen as guiding or determining the thoughts, feelings, and actions of humans. As conveyed in the *Iliad* and *Odyssey*, and recreated in Simmons' *Ilium*, the ancient Greeks frequently attributed their behavior to the will and thoughts of the gods. They heard the voices of gods and obeyed their commands.

Using the writings of Homer as one primary source of evidence, the psychologist Julian Jaynes has argued that ancient people actually did hear or experience the voices of deities in their minds. Prior to the development of our modern rational mode of consciousness, where we experience our inner self as the source, cause, and instigator of our actions, ancient people did not have such a clear and singular conscious sense of self-determination. They felt the presence and heard the voices of other selves – which they identified with their ancestors, gods and goddesses, and various spirits. Their minds were more a multiplicity of wills and personalities than a singular voice. According to Jaynes, ancient people did not have a clear modern sense of self-responsibility.[102] Whether one agrees with Jaynes' theory or not, ancient people spoke and acted as if gods and goddesses appeared to them and gave directions for how to live. There are indications that during the Axial Age, which immediately followed the time of Homer, the human mind did go through a fundamental change in thinking and consciousness associated with an increased emphasis on linear rationality, literacy, and abstraction, and most importantly perhaps, a highly enhanced sense of self-responsibility. Records from the ancient world leading up to the Axial Age are populated with innumerable prophets, soothsayers, and visionaries who saw and felt the presence of deities and spirits. Even after the Axial Age, we still find some significant individuals, such as Paul and Mohammed, who experience voices and visions from God. But there seems to have been an overall shift sometime during the Axial Age toward an increasing sense of self-determination regarding the future.

Both Watson and Polak take a different view of *The Iliad* and *The Odyssey*; they believe that the beginnings of modern human consciousness are evident in these books and that, in fact, these books express the sense of the struggle of the human mind attempting to break free of subservience to the gods and achieve self-determination. This theme of struggle against the forces of the gods is one of the key developments in Greek thinking that would significantly influence later Western views on the future – in particular, the idea that humankind can determine its own future. For Watson, *The Iliad* and *The Odyssey* are not histories but rather, in some respects, the first modern "narratives" populated as they are by heroic figures who are fully developed characters with both human strengths and weaknesses. The gods and goddesses in the stories are not unknowable or elevated above the world of humans but involved and present in the story, and these deities clearly have their own failings as well – the wisdom of the gods is questioned. Of particular note, Watson argues that Odysseus achieves a sense of rational self-determination and independence from the gods by the time we come to the conclusion of the *Odyssey*. Polak states that these books embody both the tragic and the heroic, as the human characters struggle, not always successfully, against the will of the gods. There is a growing consciousness of free will expressed in the stories and a hopeful sense that the future can be positively directed through the efforts of the human characters. As both Watson and Polak argue, the ancient Greeks saw life as a struggle to realize one's potential and to determine one's destiny in the face of the gods, the forces of fate, and the inherent weaknesses of human character. As embodied in the heroic and yet tragic hero, the human character in Greek literature and thinking had come to the realization that fate was in one's own hands. This insight also brought with it the understanding that directing one's own life was filled with obstacles and challenges, some of which came from within. Still, Polak contends that the Greeks, beginning with *The Iliad* and *The Odyssey*, developed an optimistic vision of their own capacity to create a positive future.[103] In an interesting parallel, toward the end of Simmons' *Ilium* and continuing into its sequel *Olympos*, the human characters rise up in defiance against the manipulation of their lives by the gods.[104]

Thus it could be argued that *The Iliad* and *The Odyssey* capture the inherent psychological struggle occurring in the human mind as we evolved from a species that attempted to follow the will of the gods to a species that saw its destiny as a product of its own will and self-determination. As many have argued, especially for those who attempt to follow the will of God as they understand it within their particular religious tradition, this change in thinking brought with it human arrogance and hubris and a false sense of independence. It is a great debate within human history whether this shift in thinking represents an act of courage and maturity or one of arrogance and naiveté.

Turning to the early mythic elements in Greek thinking, approximately at the same time as Homer was creating his epic tales of humanity and the gods, the poet Hesiod was writing his *Theogony* in which he described the creation of the world and the origin of the gods. According to Hesiod, in the beginning was Chaos, and from Chaos came *Gaia* (the earth), *Tartarus* (the abyss), and *Eros* (love). Chaos also produced Night and Darkness, which mated with the help of *Eros* to produce the Day. Earth brought forth the Ocean, the Mountains, and Heaven (*Ouranos*). Earth and sky – *Gaia* and *Ouranos* – then mated producing the first gods and goddesses of ancient Greece. *Cronos* – a son of *Ouranos* and the ruler of the Titans who would eventually become the God of Time – usurped the power of his father, but in turn was overthrown by his son, *Zeus* who became the supreme ruler of all the Greek gods and goddesses. Most of the gods and goddesses in Homer's tales are either children of *Cronos* (brothers and sisters of *Zeus*), or children of *Zeus* and one of his goddess mates. Beginning in chaos, the Greek pantheon of gods and goddesses, consumed by internal rivalry, tumultuous romances, jealousies, and conflicts, emerged as an all-encompassing patriarchy with *Zeus* reigning on high from his thrown on mythical Mount Olympus.[105] It is important to see in this mythic tale of creation that the two fundamental forces at work are sexual reproduction and violent war and conquest; in the final analysis a dominance hierarchy of power is achieved by a male deity through conquest. As Polak states it, in their myths the ancient Greeks described creation as a result of battle, war, and chaos.[106]

Greek myth and religion is an amalgamation of many traditions, waves of immigration and invasion, and local customs. Prior to the time of Homer, ancient Crete to the south of Greece, whose culture would influence the Greeks, appears to have practiced a strong earth Goddess centered religion. But Crete was eventually destroyed by the Indo-European invasions from the north that brought with them a patriarchal belief system. The god *Zeus* probably reflected the mythic beliefs of these invaders from the north. *Zeus*, a highly assertive patriarch and impulsive deity, raped innumerable goddesses and mortal women alike, and had a great sexual appetite.[107]

One highly popular view of creation and time in ancient Greece is the myth of *Oceanos*, the great river that flows around and encircles the world. *Oceanos*, at times identified with *Cronos*, is eternal and represents the infinite and unending cycle of time. *Oceanos* is the world soul and the source of all creation, including the gods and goddesses. *Oceanos* was also connected with the mythical animistic creature, the snake or worm *Ouroboros*. *Ouroboros* was pictured swallowing its own tale – symbolic of the circular and endless nature of time. *Ouroboros* carried on its back the signs of the Zodiac, representing the necessary progression of events in time.[108]

As can be seen in the above tales and myths of ancient Greece, the Greeks personified the creation of the universe and the forces of nature. The world was animated by various deities and spirits. Elisabet Sahtouris contends that up to the approximate time of 500 BCE, prehistoric and ancient humans saw nature as fundamentally alive, what she calls the "organic view" of reality.[109] Yet, according to Sahtouris, beginning in ancient Greece and eventually spreading around much of the Western world, a second world view arose, a "mechanistic" and "rationalist" view that transformed the human mind and human society. She describes this second, newer world view as proposing that there exists a fundamental unchanging order and single God underneath the flux of nature.

This transformation in thinking, as Sahtouris describes it, occurred during the period I have referred to as the Axial Age. There are different descriptions and explanations of this psycho-social transformation. I have already identified a variety of explanations in this chapter – the emergence of theoretic thinking, the blossoming of a belief in self-determination, and a turning inward to find truth and direction. One thing seems clear – there was a significant advance during this period in the powers of abstraction and reason in humans. In ancient Greece, this change is associated with the emergence of abstract philosophy (circa 600 – 400 BCE). A somewhat similar change in thinking occurred in Judaism in the Middle East around the same time, connected with the ascendancy of an absolute monotheism and transcendent God. Similar changes in thinking also took place in the Far East, in the abstractions of Taoism and Hinduism and the introspective philosophies of Buddhism and the *Upanishads*. Although Sahtouris identifies the Greeks as instigating this change in thinking, the change seems to have occurred in concert across various areas of the world. Also, it should be re-stated that the Greeks valued both reason and order, and passion and chaos; even with the development of abstract rational philosophy, they continued to pay homage to their gods and goddesses and practice mystical rituals, such as the rites of *Dionysius*, and struggled with the issue of self-determination, as evinced in their great works of literature and dramatic tragedy. Still, the emergence of abstract Greek philosophy clearly epitomizes a new way of thinking and mode of consciousness that appeared during the Axial Age.

A good place to begin a review of ancient Greek philosophy is with a fundamental dispute over the nature of reality and time that emerged between the pre-Socratic philosophers, Parmenides (ca. 515 - 440 BCE) and Heraclitus (ca. 535 - 470 BCE). Parmenides saw ultimate reality as ordered and eternally permanent, as "being" rather than "becoming and passing away." For Parmenides, time and motion were illusion and mere appearance. He espoused the concept of an eternal oneness, primary, absolute, and all enveloping. For Parmenides what is ultimately real is an all pervasive eternal oneness and unity. We have already encountered a similar view in Hinduism and the idea of *Brahman* (and in fact, Parmenides may have been familiar with the Hindu philosophy of his time). Yet for Parmenides, the basis for his monistic philosophy was logical reasoning rather than insight, revelation, myth, or mystical intuition. Parmenides reasoned that change must be unreal because it contradicted the "**Law of Identity**." The "Law of Identity" states that what is, is, and what is not, is not; that a thing either is or is not - it cannot be both. Change involves becoming and passing away; what is not becomes what is and what is becomes what is not. This is logically impossible, according to Parmenides, hence time, change, and everything associated with the world of flux, matter, and distinct particulars must be unreal and mere illusion. Again, in parallel, Hinduism also saw the world of time and change as unreal.

Heraclitus, in almost complete opposition, saw reality as flux and change. Heraclitus is reputed to have said, "The father of all things is war" (conflict creates everything); "You can't step into the same river twice" (although we treat things as constant or the same over time, everything keeps transforming); "The only thing that stays the same is that nothing stays the same" (what is ultimately stable is change); and "That which is in opposition is in concert, and from things that differ comes the most beautiful harmony" (an apparent contradiction of the Law of Identity, a well as similar in meaning to the Chinese *Yin-yang*.)[110]

There are a variety of significant aspects to this dispute between Heraclitus and Parmenides. For one thing, the philosophical disagreement is framed in an abstract form without reference to deities, spirits, or other personifications of reality. Heraclitus does use the concrete metaphor of fire to describe the world, but basically the argument is abstract. Heraclitus speaks of a "Law" of the universe to explain change (see below), rather than deities or spirits, and Parmenides bases his argument on the "Law of Identity" – another abstraction. Second, Parmenides supports his philosophical view through logical deduction and reasoning. There is no reference to religious authority or mythological inspiration. Third, the disagreement between Heraclitus and Parmenides reflects a fundamental difference in attitude regarding the nature of existence. Is reality fluid and changing, or is reality constant and permanent? Do we believe that time – becoming and passing away – is basic, or do we think that the eternal is what is real? How we view the past, as well as the future, reflects whether we put an emphasis on permanence or change in our understanding of reality. Heraclitus did believe that there was an underlying order to the world of time – the "*Logos*" – but the *Logos* is a fundamental pattern to change, whereas Parmenides basically denied the reality of change and saw the eternal realm of stable, unchanging order as the only true reality.

Plato (427 – 347 BCE) is generally regarded as the most influential philosopher in Western civilization. Not only did he articulate most of the key philosophical issues that would be discussed and debated in later centuries, he also had a powerful impact on the subsequent development of Western religion, political theory, and science. His ideas, both in positive and negative ways, have significantly affected the evolution of thinking on the future in the West.

Plato states in one of his *Dialogues*, the *Timaeus*,

"We must make a distinction and ask, what is that which always is and has no becoming; and what is that which is always becoming and never is? That which is apprehended by intelligence and reason is always in the same state; but that which is conceived by opinion with the help of sensation and without reason, is always in a process of becoming and perishing and never really is."

Plato attempted to synthesize the philosophies of Heraclitus and Parmenides by proposing a metaphysical dualism of eternity and time. Plato, though, elevated permanence above change, in arguing that what was ultimately real was an eternal order, and that time and change were derivative and mere appearance. (In this sense, he was closer in spirit to Parmenides than Heraclitus.) Plato separates reality into two different realms and elevates one realm above the other. Eternity was the realm of abstract or ideal forms, which could be known through reason; time was the realm of appearance and opinion revealed through perception. Eternity was the realm of order; time was the realm of chaos. Eternity could be understood. Time was confusion. Plato's dualism of eternity and time was connected with his dualism of matter and spirit, for it is matter, which is temporal, and spirit, which is eternal. Further, Plato is usually seen as the primary inspirational source of Western philosophical rationalism – that ultimate reality can be understood through reason.[III]

Plato's philosophy of reality also reflected the influence of another ancient Greek, Pythagoras (ca. 581 – 507), who believed that the universe was fundamentally mathematical. Pythagoras first coined the term *"philosophia"*, meaning the "love of wisdom." According to Pythagoras, mathematical truths are both abstract and eternal, and underlie the order of nature. Pythagoras is associated with the expression **"harmony of the spheres"** which refers to the order and coordinated orchestration observed in stars and planets in the heavens. Pythagoras is also well known for arguing that music is mathematical in form and that reality is a form of music. Plato adopted the Pythagorean metaphor of reality as music. He believed that eternity had a mathematical harmony, beauty, and order, and he associated the eternal with the heavens above, whereas time and corruption he associated with the earth. Further, mathematical calculation is a prime example of rational thinking and for Plato, reality is something that fundamentally can be understood through reason.

Plato believed that the eternal realm was populated by abstract and ideal forms – in essence, the pure and presumably changeless forms of ideas, such as truth, beauty, and the good. These ideal forms were arranged in a hierarchy with the supreme "Good" at the top. Temporal reality consists of mere shadows or approximations of these forms, but as noted above, the ideal forms could be known through the mind and reason. Armstrong proposes that Plato's forms are a rational (or rationalized) version of mythic archetypes. Reality is no longer personified and understood in terms of concrete metaphors; reality is organized in terms of a set of fundamental principles, ideas, and forms – reality is abstraction.[112]

As Armstrong and many other historians, philosophers, and theologians have noted, Plato would have an immense impact on later religious thinking and, in particular, on Christianity. Following Plato, many of the ancient Greeks believed that ultimate reality was eternal. The classical and medieval periods of Western thinking, both philosophically and religiously, tended to be Platonic in their metaphysics. Through Plato's influence on St. Augustine, in particular, Christianity emerged as Platonic with its emphasis on the eternal and spiritual over the temporal and physical. Following Plato, the Judeo-Christian-Islamic tradition saw the divine as changeless and eternal.[113] The Platonic trend in Western thought clearly identified the eternal and spiritual realm as more important than the temporal and physical realm. For Plato, and his numerous intellectual descendents, spirit and reason were elevated above matter and bodily desires. Consequently, Western thinking for a long time was generally not very concerned about the temporal or natural future, since it was not viewed as an important issue regarding the basic meaning or purpose of life. What is the promise or significance of the future on this earth if what is supreme or of a higher level of reality lies outside of time?[114]

Yet as we have seen, dualist thinking concerning eternity and time is not unique to Western philosophy and religion. Philosophies and mythologies in Eastern thinking also developed a dualism of a higher eternal realm and a lower temporal realm. Within Hinduism and Buddhism, the soul ascends through reincarnation and the cycle of life and death to Nirvana and eternal oneness. The world of time - of birth, life, and death - of individual desires and flux - is a journey and way station to a higher eternal realm.

Plato's general theory of creation also connects with both Eastern and Western religious thinking. In religions such as Judaism, Christianity, Islam, and Hinduism, a Supreme Being or God is postulated as the source of all order and the ultimate foundation of all reality; time is relegated to a derivative status, presumably having been created by the eternal being. Hence, what is eternal not only exists above the world of time but creates the world of time. Similarly Plato believed that the realm of eternal order was the creative source of the world of time. Order precedes chaos and not the other way around, as had been the common view in ancient mythologies such as those in Babylonia, Egypt, China, and pre-Platonic Greece.

Although Plato is usually seen as the major starting point in Western philosophy for rationalism, there is another side to Plato - the mystical - that would also influence later thinking in the West.[115] Reasoning is a linear cognitive process. When we reason, we move through a process of sequential inference, moving from one thought to the next one. Further, reason is connected with analysis and articulation - ideas are sharply defined and delineated. Yet, when people think, they often use intuition or insight as well as reason. Intuition (and insight) is an all-at-once process - a person "sees" or understands some fact, principle, or truth as a whole, in a flash. When Plato spoke about the acquisition and contemplation of knowledge he often described the process in "intuitive" terminology. The truth was seen or grasped by the human mind. According to Watson, what Plato meant by reason was the intuitive grasp of eternal abstractions, though as we will momentarily see below, Plato clearly also embraced and taught an analytical and logical form of reasoning as well.[116] Western mystics in later years, who were influenced by Plato, would frequently describe the contemplation of eternal and higher truth as an intuitive process. Further, Plato described the contemplation of ideas as an aesthetic experience. There existed a sense of rapture – of love – of beauty – in the mental contemplation of the eternal forms. Truly, philosophy was the "love of wisdom."[117]

Two of the most important influences on Plato's thinking were Pythagoras, and Socrates (469 – 399 BCE). Socrates was Plato's teacher, and Plato used Socrates as his invariable spokesperson in his *Dialogues* to present his own philosophical views. In the *Dialogues* Plato portrayed Socrates as a cunning and determined investigator who would question and debate others on innumerable philosophical topics. Socrates would attempt through cross examination, incessant clarification, and tenacious reasoning to arrive at the truth. Howard Bloom describes Socrates as a left-brain extrovert. Left cerebral hemispheric functioning is usually associated with an emphasis on linear logic, analysis, and language. Shlain supports this point contending that Socrates was one of the key figures of the Axial Age, which brought an increased emphasis on literacy and linear thinking to human society around the world.[118] In the *Dialogues*, through the debates of Socrates with other individuals, Plato illustrates the dialectical mode of thinking where an idea is proposed and then criticized, with rebuttal and counter-rebuttal; the truth is eventually arrived at through this process of back and forth discussion and debate – hence the title of *Dialogues*. In this process of dialectical reasoning, analysis, the questioning of assumptions and logical deduction are clearly seen as critical in discovering the truth. In contrast, Bloom describes Pythagoras as a right-brain introvert. Right hemispheric thinking is usually described as holistic, visual, and intuitive.[119] To recall, Pythagoras saw the universe in terms of music and harmony – intuitive and holistic metaphors, and Plato was influenced by this vision of the cosmos as well. Hence, just as Plato had brought together Heraclitus and Parmenides, one can also see Plato as synthesizing in his thinking both Socrates and his method of rational inquiry and Pythagoras with his emphasis on intuitive insight. Plato's concept of reason involves both these modes of thinking.

Plato would not only influence philosophy and religion but Western political thinking as well. His most famous dialogue, *The Republic*, which contains the famous "Myth of the Cave," where he metaphorically distinguishes between shadowy appearances and the light of eternal reality, is also a treatise on an ideal society; it is probably the first fully developed example of Utopian thought in Western civilization as well.[120] Plato lived in a time of upheaval and change within Greek civilization, and his quest for stability and certainty in his metaphysical philosophy reflects a desire to find order amidst the world of disorder around him. His political philosophy, as developed in *The Republic*, also expresses an aspiration toward order in the midst of chaos.

According to Bloom, Plato did not approve of the liberal and democratic practices of Athens, where he lived. Instead he was attracted to the more authoritarian system of Sparta. Athens had become in Plato's time a web of commercial exchange, with many sub-cultures and a definite international flavor. Sparta, on the other hand, was more isolated, less materialistic, and based on a rigid system of conformity and control. Sparta was order – Athens was chaos. In the *Republic*, Plato argues against democracy as a viable form of government and instead supports the idea of a "Philosopher King" who would rule with wisdom, benevolence, and a sense of justice – much of the Republic deals with the idea of justice, approached dialectically and conceptualized as an ideal form or abstraction to be understood through reason. Not just anyone or everyone can rule in Plato's ideal society and Philosopher Kings must be educated and trained from youth to rule wisely and competently. Philosopher Kings must gain an understanding of the eternal principles of truth, beauty, the good, and justice, and not be overpowered by the flux and corruption of popular opinion, materialism, and time. In essence, for Plato, the determination of society and its operations can not be left up to the uneducated masses – it must be ruled by rational thinkers from above.[121] Though Plato rejects democracy as a form of government, it is important to note that he does place the responsibility for governance and social control in the hands of humans rather than gods; in this regard, Plato expresses the growing Greek ideal that humankind, rather than the gods, is the master of its own fate.[122]

As one final theme to consider regarding Plato, quite relevant to the topic of the future, Plato believed (as also did Pythagoras before him) in reincarnation, as well as an afterlife. At least some Greeks, including Plato, believed in reincarnation – the idea that the soul could return to the earth in a new human form, or even animal form. (The Hindus believed this as well.) Plato thought that the rational part of the human soul was immortal, that upon the death of the body it survived in a higher plane of existence, and that the soul could return from this higher realm. In fact, Plato also thought that the rational soul existed prior to its incarnation in a physical body and was of divine origin. As Watson puts it, Plato saw humans as "fallen angels." The idea of a divine origin of the soul was associated with the mystical Orphic tradition or cult of ancient Greece, which influenced Pythagoras in his thinking and, in later times, would form an integral part of Neo-Platonic thinking as well. Also of relevance, connected with the Dionysian mystery cult was the idea of resurrection for, according to legend, Dionysius, after having been killed and torn to pieces by the Titans, was resurrected in a new physical body. (Orpheus, the central figure behind the Orphism, was reputedly also torn to pieces, interestingly by members of the Dionysian cult.) In addition, contributing to this set of connected ideas was the Greek idea of an afterlife. Though many Greeks did not believe in an afterlife, we do find the idea in Homer, where souls after physical death journey to Hades, in the underworld; later we find the idea of the Elysian Fields, a more appealing land – a paradise in fact – that human souls go to after death. Plato was aware of all these ideas and, as an expression of his mystical and Pythagorean side, he thought that souls did journey to a higher realm after death – the divine realm of eternity where they originally came from – and that souls could return to the earth in a new physical form.[123]

In closing this review of Plato, a quotation from the contemporary cosmologist Lee Smolin is quite appropriate. Smolin does not mention Plato in this quote, but Plato is the most important architect of the philosophy that Smolin describes. Smolin states, "...we can see how Western cosmology and political theory arose together from the opposition of the spirit and the body, the eternal and the decaying, the externally imposed order and the internally generated decay." This dualist contrast is part of the legacy of Plato, but so is Plato's combination of the rational and the mystical-intuitive.

Plato's most famous student was Aristotle (384 – 322 BCE). In some ways Aristotle carries on and further develops the ideas of his teacher, for example in his study of reasoning and logic. In other ways though, Aristotle goes off in a different, if not diametrically opposed, direction. In particular, Aristotle advocates in his writings for an empirical and naturalistic approach to the development of knowledge, and in this regard is much closer in spirit to modern science than Plato.

Along with the increasing emphasis on logic and reasoning found in Greek philosophy, a second important emphasis was the attempt to explain nature in terms of natural causes and principles, rather than in terms of spirits and deities. This philosophy of naturalism is evident in many of the pre-Socratic philosophers, including Heraclitus, but also Thales (640 – 546 BCE), who believed that everything was reducible to "water," Anaximenes (585 – 528 BCE), who argued that everything was composed of air, and Democritus (460 – 371 BCE), who thought that all of reality was made up out of extremely small physical "atoms." Such theories moved away from supernatural explanations of reality and rejected the dualistic systems of thought that proposed two realms of spirit and matter (though Thales is reputed to have said that "the world is full of gods"). Instead these naturalistic views explained the world in terms of primary or fundamental physical substances, entities, or laws.[124] This shift in thinking is highly significant for it represents the beginnings of science - of naturalistic explanations of reality and time.

For Watson, the emergence and development of Greek science, as well as Greek philosophy and mathematics, was a pivotal event in the evolution of the human mind. According to Watson, a new way of thinking came into existence: the world could be known without the aid of the gods. Science in ancient Greece was free, individualistic, and argumentative, rather than ruled by doctrine and authority. For Polak, philosophy, science, and natural law, as expressed by thoughtful and argumentative individuals, replaced myth and the gods as a way to understand and control the world.[125]

Aristotle further develops the pre-Socratics' naturalistic and scientific approach to the universe.[126] He identifies four basic causes behind any given natural phenomenon - the material, efficient, formal, and final causes - and through his investigations into physics, biology, psychology, and other areas of nature, attempts to describe natural processes in terms of these causes. Further, he follows Heraclitus in viewing reality as fundamentally change; Aristotle's causes are, in effect, factors that explain how and why things change. Also, Aristotle initiates the empiricist tradition in Western philosophy and science, attempting to understand nature by observing it rather than consulting sacred texts and myths. All told, in many respects he rejects Plato's dualism of two realms, contending instead that the form (formal cause) and the matter (material cause) of natural objects co-exist in the object. There is not for Aristotle a separate realm of ideal, abstract non-material forms as there is in Plato's philosophy.

One especially noteworthy and central theme in Aristotle is his emphasis on the process of natural growth. Because Aristotle studied biological life so extensively, he tended to see all of reality in terms of growth and self-actualization. Things in nature move toward natural ends (final causes) which are simply the realization of their inherent potential (formal cause). Aristotle was a teleologist, believing that change is directed toward specific ends ("telos") but his teleologism was naturalistic rather than supernatural or other-worldly. Events move toward ends which are determined by their own inner potentials. Although Aristotle saw time as basically cyclical, he highlighted in his naturalistic philosophy the concepts of growth and directionality in nature. This emphasis on naturalistic growth would contribute to the early Greco-Roman sense of progress in nature (see below).

If Plato was the inspirational source for Western rationalism, Aristotle was the most famous and influential teacher of the principles and practice of reason. Aristotle formulated and codified the various syllogisms of deductive reasoning and identified many of the basic types of logical fallacies. When logic was taught in the West in the coming centuries, it was Aristotle's writings that served as the foundation.

Aristotle's logic contains the well-known "**Law of the Excluded Middle.**" This law basically states that a thing can not both be A and not-A, or a statement can not be both true and not true. The Law of Identity is a version of this principle – what is, is, and what is not, is not; it is, or it isn't – it can't be both. Another way to state the principle is that logical contradictions can not be true. If a person is married, the person can not also be single. If an object is hot, it can not also be cold. This principle is the foundation of Aristotle's logic, and in fact, has been the basis of Western logic for the last two thousand years. Through the eyes of Western logic, the world is to be understood in terms of "either – or", "black and white", and "right and wrong." One could say that in the West, following the logic of Aristotle, the clear and absolute distinction is central to all thinking and inquiry.[127]

Yet, if one considers the logic of the Taoist *Yin-yang*, it appears that the Law of the Excluded Middle is not only rejected, but its opposite is embraced as fundamental to the nature of reality. Features of reality that in the West we would describe as "opposites" and mutually exclusive, in Taoism are seen as co-implicative, co-existing, and mutually interdependent. You can't have A without non-A. The darkness and the light, male and female, hot and cold, and hard and soft, among other opposite pairs, co-exist in reality and actually interpenetrate each other. If Aristotle and Western logic describes a world of black and white, Taoism describes a world where "truth is the color of gray."[128]

Although subsequent Western philosophy was strongly influenced by Aristotle's logic and Plato's theory of an eternal and unified order, two pre-Socratic philosophers expressed views that, contrariwise, had strong affinities with Taoist philosophy. These two philosophers are Heraclitus, already introduced above, and Empedocles. Though not as influential as Plato and Aristotle, Heraclitus and Empedocles embody a line of thinking that would be highly significant in later thought on the nature of reality and time.

Heraclitus seems to have rejected the Law of the Excluded Middle. In fact, on reviewing statements and ideas attributed to Heraclitus, he sounds very much in resonance with Taoism. The historians of philosophy, G.S. Kirk and J.E. Raven, summarizing his views, state that according to Heraclitus there exists an "essential unity of opposites," that "each pair of opposites ...forms both a unity and a plurality", that the "unity of things...depends upon a balanced reaction between opposites", and that "the total balance in the cosmos can only be maintained if change in one direction eventually leads to change in the other, that is, if there is unending 'strife" between opposites."[129] Heraclitus refers to this underlying unity and perpetual balancing of opposites as the "*Logos*," or the logic of change. The *Logos* of Heraclitus sounds very much like the *Tao*.

In these ideas on the nature of reality, Heraclitus juxtaposes conflict and plurality with harmony and unity. Although Heraclitus is remembered for his conflict theory of time (a view we have already seen expressed in Babylonian myth and Zoroastrianism), the contemporary psychiatrist Hector Sabelli, in his book *Union of Opposites*, argues that Heraclitus not only believed that "War was the father of all things" but that "Harmony was the mother."[130] It should be noted that this "union of opposites" is the abstract analogue to the "Hunter" and "Goddess" archetypes and of the masculine and the feminine modes of thinking – of conflict and togetherness.

Empedocles (493 – 433 BCE) also emphasizes the themes of opposition and balance in his theory of reality and time. Starting from a naturalistic perspective, Empedocles argues that there are four fundamental elements – fire, water, air, and earth – of which everything is composed. These four elements are constantly being rearranged by two fundamental forces – love and strife. Love, which he identifies with the goddess *Aphrodite*, brings things together, while strife, which he connects with the god of war, *Ares*, pulls things apart. Following Parmenides, there is no real becoming or passing away, only rearrangement of the primary elements. Love and strife exist at both a cosmic level and a personal level and respectively represent the "good" and "bad" sides of reality. Hence, discord is bad and togetherness is good, and the soul, as in Zoroastrianism, is the micro-cosmic mirror of the universal battle of good and evil. Further, love and strife oscillate in dominance in life, creating a cyclic nature to time. Balance is achieved, as is also the case in Taoism, through each force repeatedly oscillating in dominance with the other force. The philosophy of Empedocles is a clear expression of the idea that history involves a back and forth swinging between the forces of unity and plurality, peace and war, and love and hate. Yet, because he sees this oscillatory process as cyclic and eternal, there is no resolution or final conquest of one force over the other. Life is forever love and strife, inextricably bound together.[131]

Yet, Empedocles does include in his theory of the origin of the universe and humankind a set of ideas that sounds similar in some ways to contemporary evolutionary theory. According to Empedocles, life begins in a haphazard assortment of body parts which randomly combine together in all possible configurations. Some of these configurations are viable, whereas many other configurations are not. Those viable configurations survive, leading to our present array of life forms that function and are adapted to the environment.[132]

Pulling together several lines of thinking in ancient philosophy and religion, it appears that two different views of reality emerged in ancient times, one which emphasized duality and one which emphasized reciprocity. Plato clearly separated reality into two distinct realms, a philosophy that would be taken up later by both Christianity and Islam. Also, we saw that Zoroaster, as a religious precursor to all later Western religions, divided the world into the supreme forces of good and evil set in opposition to each other, and, similar to Plato, distinguished higher and lower realms of existence, as well as body and spirit. Although Aristotle rejected Plato's metaphysical dualism, he articulated in his principles of logic, an "either - or" system of thinking. On the other hand, both Taoism in the East and Heraclitus and Empedocles in the West formulated a philosophy that emphasized the interdependency and complementarity of presumed "opposites." Whereas the Zoroastrian - Platonic line of thinking separated reality, the Taoist - Heraclitian - Empedoclean line treated reality as an interdependent whole. The holistic vision is realized by treating the forces of unity and plurality, harmony and conflict, and love and strife as reciprocities. The archetypes of the Hunter and the Mother, for example, are conceptualized as equal and co-dependent in the latter framework. This sexual or gender interpretation of theories of reality is supported by the historical fact that dualist views of reality tended to be sexist, with the male side clearly dominating the feminine side. The rationalist Greeks and the dualist religions of Judaism, Christianity, and Islam all suppressed the female, in both their societies and their ideologies. The philosophy of Taoism, on the other hand, treated the feminine and the masculine as equal.

These different theories of reality, whether philosophically or religiously inspired, were connected with a variety of ideas regarding the nature of change and time. Parmenides rejected time; Plato gave it a secondary status, and though in the *Timaeus* he refers to time as "the moving image of eternity," in general he sees time as corruption and chaos, at best only approximating the perfection of eternity. Zoroaster sees time as linear and progressive, leading to the triumph of good over evil, but still, treating the reality of time as below the higher reality of spirit and eternity. Heraclitus sees a *Logos* underlying time, but like the Taoists, sees time as basically rhythmic and cyclic. A common theme among many of these views is that conflict and discord is an essential feature of time, often contrasted with the idea that eternity is harmonious, unified, and peaceful.

Love and war, unity and plurality, and order and chaos are themes that run through ancient mythology and religion. Especially connected with the theme of order and chaos, is the issue of necessity and chance in the flow of time. As noted above, ancient mythologies, in personified and archetypal form, described the passage of time, giving time an ordered and structured reality. As Fraser notes, they provided a sense of stability in a world of change.[133] The circle or cycle - a fundamental archetypal form in both Eastern Taoism and the Western Zodiac - depicted time as possessing a necessarily ordered pattern that repeated itself over and over again. The ordered and necessary progression of the circle of time also shows up in Greek mythology in the image of the world river of *Oceanus*, which later transformed into the Greek god *Cronos* and the god of time. Yet the element of chance also appears in both ancient Greek thinking (*Kairos* or lucky coincidence) and later Christian thought (*Fortuna* the blind goddess). And in particular, although the Taoist *Yin-Yang* represents time as an ordered sequence, the *I Ching* acknowledges an irreducible element of chance in understanding the significance of events in life.[134]

The orderly and repeating pattern of the cycle was not the only way in which necessity was conceptualized in ancient religious thinking. **Fate** or **destiny** plays a significant role in most ancient religions, from Egypt and early Hinduism (The Law of Karma) to Greek mythology (*Nemesis* the goddess of necessity), Judaism, and Christianity. The Greek gods, including Zeus, were at times powerless over fate.[135] As we have seen, the Greek mind wrestled with the idea of fate in its growing awareness and aspiration toward self-determination. Judaism and Christianity derived their concept of necessity from the Judaic idea that God had a plan for the world, and that the events of the world were guided by this plan of God. There was a historical necessity to the events in the world, following from the fall of Adam to the trials of Job and Abraham and the coming of the Messiah.

Hence, long before the emergence of science and modern philosophical thinking, in both religion and mythology, the question was asked, pondered, and debated: Is the future certain or uncertain? Different cultures and religious traditions have seen the future as filled with luck and chance, and conversely, as having a set purpose and direction. As a consequence of these different cultural and religious views there has been, on one hand, a philosophical attitude throughout history that emphasizes individual control and responsibility over one's future, or on the other hand, another attitude that emphasizes acceptance of one's destiny or even fatalistic resignation.

The most popular historical interpretation concerning the ancient Greeks' view of time is that generally they saw time as either cyclic, or a corruption and decay of something higher, either in the distant past (a Golden Age) or a higher realm (eternity). Yet, there are indications that the Greeks also believed in a progressive theory of time. One central insight on their part that led them to this alternative viewpoint was their discovery of history.

As we saw, the writings of Hesiod contain a mythic account of the origin of the world. Yet, also in his writings we find the "myth of the ages," a historical description of the various ages of humankind preceding his own time. Hesiod clearly describes both advances and regressions in this history, but overall there is a sense of progression from very primitive beginnings to the present. Hesiod also recounts the **"myth of Prometheus"** who stole the secret of fire from the gods. Presumably, prior to this event, humanity existed in a less advanced state. The step forward represented by man's achieving control over fire is, however, the consequence of robbing the gods, rather than a singularly human achievement. Zeus punishes Prometheus for his act, symbolically reflecting the ambivalence and guilt humanity often feels about accomplishment and advancement. Perhaps we are filled with hubris and vanity, believing that we can ascend the ladder of progress and become equal with the gods – this theme we have noted is central to Greek tragedy and their striving for a sense of self-determination. The myth of Prometheus is not the only story in antiquity in which humans are punished for attempting to move upward, become masters of their own fate, and hence perhaps threaten the gods. The story of Eve in *Genesis* is another famous example of this theme – the theme of human hubris. Yet Hesiod also states in his writings that social progress and the growth of civilization can be accomplished through the efforts of humans and the implementation of principles of justice. Thus Hesiod clearly believes that humans have power over their own fate.[136]

There are other ancient Greeks who, in considering the question of humanity's history, see a sense of progress across time. The great historian Thucydides (ca. 455 - 400 BCE), who described the military conflict between Sparta and Athens in his famous *History of the Peloponnesian War*, saw Greek history as involving an advance from a more primitive and barbarous state. Even Plato, in *The Laws* and *The Statesman*, describes humankind as first existing in a state of moral innocence with no art or organized society; as he states it, "Men lacked all tools and all crafts in the early years." Further, describing the process of social and political development from these early beginnings, Plato states that "doubtless the change was not made all in a moment, but little by little, during a long period of time." Plato invokes the metaphor of growth from a seed to describe the progressive advancement of civilization, a concept of growth we have already seen within Aristotle. The seed represents the idea of potential in describing the developmental processes of nature.[137]

As I argued in the opening chapter, historical consciousness is intimately tied to future consciousness. Though prior to the writings of Hesiod and Thucydides there were various mythic historical accounts of the development of humankind and the universe as a whole, according to Watson, modern history begins in the work of these two Greek writers. What Watson emphasizes is that beginning with Hesiod and Thucydides there is an effort to research history, to collect data and evidence, and consider different points of view, rather than simply passing on the ideas of a singular tradition and authority; that is, there is an effort to be empirical and thoughtful in the histories of Hesiod and Thucydides. Further, Watson notes Thucydides in particular as doing away with gods, spirits, and supernatural explanations in his recounting of the past; history becomes naturalistic and secular.[138] This shift in understanding the past coincides with a similar change in understanding reality as a whole (the naturalistic and logical methods of Greek philosophers and scientists), and opens the door to further discovery unshackled by the dominating influence of tradition and religious authority.

According to the contemporary historian Robert Nisbet, the clearest example of the idea of historical progress to be found in the Classical period is within the book *On the Nature of Things* by the Roman philosopher and poet Lucretius (99 - 55 BCE). Lucretius has been seen as anticipating the modern concept of evolution, for he describes the cosmos as beginning in chance and physical forms coming together through collision and "conformation of atoms" (the order out of chaos theme) from which eventually comes forth life. For Lucretius, different living forms emerged in the primitive beginnings of nature; some survived and some became extinct, depending on their capacity to secure food and protect themselves. Those forms that survived reproduced and passed on their traits to their offspring, through a process that sounds similar to Darwin's notion of natural selection. When Lucretius comes to the development of humanity, he describes early humans as existing in a hunter - gatherer state without clothes, weapons, fire, huts, or communities, and in general, not possessing any social constraints on their behavior. Slowly - through ingenuity and natural intelligence - humans develop all the different aspects of organized society, technology, and the crafts. For Lucretius, progress is not something stolen from the gods - it is a creation of humanity. This history provided by Lucretius is, of course, speculative, but it is quite striking how much of it comes close to the truth as we understand it today. Also, it is naturalistic.

In the tradition of the Greek naturalist philosophers, Lucretius saw change as due to natural rather than supernatural forces and the will of the gods. What Lucretius adds to this naturalist viewpoint is that progressive change is due to forces in nature. The line of religious thinking running from Zoroastrianism to Judaism and Christianity also articulates a vision of progressive change, but one that is orchestrated by God. Lucretius sees historical progress as due to inherent forces and principles in physical nature. In fact, he turns the question of the relationship of God, nature, and humanity on its head. Instead of gods having created the world and humankind, the physical world is the origin of humankind and it is humanity that, in fear of nature and attempting to comprehend the causes of things, invents gods as an explanation.[139] Thus, in Lucretius we see the psychological emancipation and the triumph of the Greek ideal that humans have been and can be, even more so in the future, the masters of their own fate and not simply pawns of the gods.

As Nisbet argues, in the Greeks and those Roman philosophers who were influenced by them, we find the beginnings of the

211

insight that "civilization has advanced, is advancing, and will continue to advance" - which in essence is the modern theory of progress. Although many Greeks saw time as cyclic and filled with either conflict or decay, the idea of progress also can be found in their writings. And given the rise of rationalist, empiricist, and naturalist thinking within ancient Greek science and philosophy, this progressive vision of past, present, and future was not tied to supernatural or mythic thinking. It would though take another two thousand years before this secular progressive mindset would really take hold in the West. In the interim, the West was generally dominated by mythic and religious thought, and a spiritual sense of progress. The beginnings of this religious mindset are usually traced back to the development of Judaism in the ancient Middle East.

Judaism:
Prophecy and Monotheism

Greek philosophy and Judaic religion are frequently identified as the two major systems of thought that would influence the subsequent development of Western civilization. Although Judaism and Greek philosophy are often contrasted, as expressing two different modes of thinking about reality, in at least one important respect the two mindsets are in agreement.[140] Both Judaism and Platonic philosophy elevated order above chaos and, in fact, saw a supreme order as the source of all creation. We have already seen this idea expressed in Plato's philosophy of the abstract eternal forms as the source of all order. In Judaism, the supreme order behind the world was a singular and all powerful God.

Just as the Jews personified their metaphysics of reality in the form of God, they also personified and dramatized their whole system of belief in the form of a collection of stories - a threaded narrative - about their people, their history, their sacred principles, and their relationship with God. (In this sense, the Jews exemplify in their tradition and mode of thinking "mythic consciousness.") This narrative, written, compiled, and edited by various historical individuals, is the Judaic *Bible* or *Holy Scriptures*, and as it is referred to in Christianity, the "Old Testament" of the *Bible*. The Judaic *Bible* contains stories associated with various famous historical figures, many who probably actually existed and others who were probably fictitious. Among these historical and mythic figures were Adam and Eve, Cain and Abel, Abraham, Noah, Job, Moses, Daniel, Isaiah, and Joshua. These characters in the *Bible* endure hardship, challenge, disappointment, and at times defeat and conquest by their enemies. They are tempted, punished, and at times beaten into the ground. But throughout the narrative, a sense of hope, faith, and determination repeatedly rises up again and is expressed in response, if not defiance, to the difficulties and apparent chaos of life. Although the Judaic *Bible* incorporates numerous elements and ideas from ancient Babylonia, Zoroastrianism, Greek culture, and other Middle Eastern influences, an overall distinctive philosophy and sense of direction emerges, emphasizing faith in the future founded upon an ongoing, living covenant with God.[141]

The traditional starting point for the saga of the Jewish people is the story of Abraham who is instructed by God to sacrifice his son as a demonstration of his faith and obedience. At the last moment before Abraham carries out this act, God speaks to him again, telling him that he does not have to go through with the sacrifice, as he has shown, by his actions and his resolve, his faith and obedience. The drama of such stories in the Judaic *Bible* is to illustrate morals and lessons of life – in this case it is critical to have faith in God. After this test of obedience, God tells Abraham that he will be the father of a new nation of people that eventually will achieve greatness and power. God explains that He and the children of Abraham, as His "chosen people," will form a covenant. God promises to guide and protect His chosen people if they worship and obey Him. Abraham, once again showing his faith and obedience, agrees to the covenant. Thus the Judaic God makes a promise for the future to Abraham; Abraham demonstrates his faith in his belief that God will make good on his promise, and the subsequent saga of the Judaic people begins.[142]

Polak argues that the idea of a covenant between God and humankind, as a foundation for the future, is the key element in the unique image of the future created within Judaism. If one believes in God, then God will give His blessings and salvation. For Polak, the idea of a covenant places control and responsibility for the future in the hands of the individual. If a person follows the commandments of God, a positive future is secured; arbitrary fate is replaced with human control over destiny. Yet, as Watson points out, although the concept of a covenant with God is a central feature of Judaism, the idea may have been taken from Zoroastrianism where individuals, in choosing between following the good spirit *Ahura Mazda* or the evil spirit *Angra Mainyu* incur, depending on their choice, the consequences of either eternal reward or punishment.[143]

The next major historical figure in Jewish history is Moses (ca. 1300 - 1200 BCE). In the book of *Genesis*, it is recounted that Moses is given the Ten Commandments from God and told to journey to Egypt and lead the Jewish people there out of bondage. (As a recurrent theme in the Jewish historical drama, the Jews are frequently held captive, conquered, or exiled from their homeland by other nations or people.) According to Armstrong, the God of Moses is *Yahweh* - a deity who evokes fear and terror - and in fact, inflicts great destruction and catastrophe on the Egyptians when the pharaoh resists Moses's request to release the Jewish people. He is a lofty God who stands distant and above humankind and hands down his commandments from on high. But again there is a promise made between God and His people - Moses will lead them out of bondage and they will journey under God's direction to the "promised land" and find happiness and fulfillment. Again, God demands loyalty and uncompromising commitment in exchange for a positive future.[144]

Another important theme in Judaism we see emerging in the story of Moses is utopianism. An ideal land and nation is envisioned, defined in terms of the ethics and values of

Judaism, which will be realized in the future. As Polak notes, this "promised land" will involve the remaking of the earth; further, the idealized Jewish nation of the future will occupy an exalted and central position in the human world.[145] The realization of this utopian future is not only conceptualized as a reward for following the word of God but also a victory of the Jewish people over its numerous worldly oppressors and enemies. Thus, although Jewish utopianism has a spiritual and ethical dimension, it is also has a retaliatory quality. As Watson argues, throughout their early history the Jewish people felt trapped and buffeted about by other nations and empires and their evolving utopian vision of the future expressed a fundamental desire to defeat their enemies and achieve a sense of greatness and recognition.[146]

Armstrong argues that even by the time of Moses, the Judaic religion had not completely articulated a thorough-going monotheism. The early Jews acknowledged other gods and goddesses besides *Yahweh*. In fact, the early Jews were probably polytheistic and *Yahweh* only gradually achieved a central or dominant position over time.[147] During the period of Moses, as well as Abraham before him, the Middle East was home to numerous deities and different religious practices. (Recall the various deities of Mesopotamia and Babylon.) In one of His commandments the God of Moses states, as Armstrong translates it, that there should be "no strange gods for you before my face." This commandment can be interpreted as *Yahweh* demanding allegiance from among the many other deities. According to Armstrong, the "One God" of Judaism did not start off as an all-enveloping deity in the minds of His believers, but rather was in competition with other deities for allegiance. *Yahweh* was warlike, in part as a symbolic expression of His struggle to conquer and defeat the other gods and goddesses of the Middle East. (Thus the early Jewish God embodies the militant and competitive psychology that the Jewish people felt in relation to neighboring cultures.) Watson relates that *Yahweh* was probably originally a god of fertility and fire represented by the bull – which to recall was a popular animistic icon and archetype of aggression associated with the male.[148] Armstrong argues that the early history of the Jewish people shows an ongoing struggle to achieve complete loyalty to *Yahweh* amidst the temptations of other gods and goddesses. According to Armstrong, the early Jews wanted the sense of immanence and holism that derives from polytheistic religion, but *Yahweh* emerged as a transcendent and terrifying God separate from nature and humankind – thus there was an ongoing struggle among the Jewish people to feel a sense of complete allegiance to this one rather aloof God.[149]

Shlain has a somewhat different view of Moses and the development of Judaic monotheism. Following traditional thinking, Shlain believes that Moses is, in fact, the starting point of monotheism in Judaism and that the Ten Commandments is a significant first expression of this religious doctrine. It is worth describing in detail Shlain's views on this topic for his ideas dovetail and connect with several other themes in this chapter.

According to Shlain, a new dimension in religious consciousness became increasingly dominant during the Axial Age which, to recall, is identified as the historical period of approximately 700 to 400 BCE. Shlain emphasizes that during this time intellectual abstractions became progressively more important in religious thinking. He attributes this shift in religious consciousness to the rise of literacy and the creation of the alphabet. For Shlain, during the Axial Age there occurred a change in dominance from the visual image to symbolic language as the preferred way to represent and understand reality. In general, humans began to see and understand the world more through abstract concepts symbolized through written language than through concrete and personified images and visions. In the previous section, I described how Greek philosophy and early science, as revealed in the thinking of Socrates, Plato, Aristotle, and the Pre-Socratics, clearly demonstrates a shift in consciousness toward the abstract and what Donald refers to as the "theoretic." Shlain argues that the change in mentality during this time occurred in religious thinking as well as in philosophy, that it was connected with a shift in emphasis from imagery to abstract symbolism, and finally, that it was this transformation that further solidified the decline of the goddess in favor of male centered religions. In Shlain's view, during the Axial Age, any remaining central female deities or goddess cultures - which revered the image - lost power, giving way to law and text-centered male dominated cultures. (Shlain, in fact, traces this ongoing transformation as far back as Hammurabi (died 1750 BCE), king of Babylon and worshipper of the god *Marduk*, who is credited with creating one of the first extensive codes of civil law.)

According to Shlain, the first of the "word-centered" cultures, ancient Judaism, developed a supreme and absolute male God, a monotheistic belief system as opposed to the prevalent polytheisms of the ancient world. Moses presumably lived long before the Axial Age, but for Shlain, Moses anticipates the mode of religious consciousness that would spread around the world in the centuries ahead. The Jews, by the time of Moses, had developed an abstract alphabetic script and, starting with the Ten Commandments, they began to emphasize the printed word and the reading of a sacred text as the primary vehicle for religious understanding.

Shlain interprets God's first commandment to Moses ("I am the Lord thy God...Thou shalt have no other gods before me") as a clear expression of a singular God and as a rejection of the need for a female Goddess as a counterpart to *Yahweh*. Shlain interprets the second commandment ("Thou shalt not make unto thee any graven image...") as a rejection of imagery in representing God. Unlike the earlier gods and goddesses with their visual and concrete embodiments, this new God, according to Shlain, had no face. The Judaic God was beyond the image - in fact, He explicitly forbids the making of any image of "any thing" as a way to represent Him. This new God had revealed Himself through symbolic language and had laid down His moral commandments to Moses as abstract rules. This supreme God of Judaism - a male - becomes an abstract *"Logos"* - the "Word" - the logic of the world. And interestingly, as Shlain notes, nowhere in the Ten Commandments is there a directive or rule concerning loving fellow human beings. Love had been associated with the goddess and the emotional, as opposed to the rational side of life.[150]

Reading and writing in alphabetic script are left-brain functions and support and reinforce linear abstract thought. The idea of a single God that transcends all concrete imagery and stands apart from physical reality is a supreme abstraction. The idea of a single unifying God is analogous to the philosophical belief that all of reality could be explained and subsumed under some absolute and unifying principle or law. The idea of a single God connects with the idea of a single absolute truth and a single set of moral principles and, according to Shlain, supported the authoritarian mindset of male dominant religions that arose in the West.[151]

Shlain places particular importance on the system of representation that humans use in thinking, understanding, and communicating. He believes that the human capacity for abstraction was given a significant boost with the development of a pure alphabetic system of writing. Previously, the earliest forms of writing, found in Mesopotamia, Egypt, and China, were all **pictographic** to degrees – representing ideas in stylized visual signs or symbols that resembled in form the objects signified. Shlain argues that, on the other hand, the Jews may have been the first to develop a completely symbolic alphabet, even prior to the Greeks. The Judaic God of *Yahweh* had no face because of the emergence of alphabetic writing and abstract ideas in Judaic thinking where the idea did not require a visual representation of its meaning. Symbolic abstraction led to the codification of laws and the centrality of the text (in this case the earliest books of the Judaic *Bible*) in defining the nature of reality, of morals, and of humanity's meaning and purpose in life.[152]

One essential element in the rise of modern religions was the emergence of central texts or sets of writings associated with each religion. For Judaism there was the *Bible*, for Hinduism there were the early *Vedas*, the *Upanishads*, and later the

Bhagavad-Gita, and for Taoism there was the *Tao Te Ching* and the *I Ching.* Later still, Christianity incorporated the Hebrew *Scriptures* with the New Testament to form the Christian *Bible* and Islam produced the sacred *Koran,* based upon the teachings of Muhammad. The functions of these texts were to provide authoritative statements on the nature of reality and morality, the existence of specific deities, humanity's relationship with these deities, and the origin, history, and future of humanity and the cosmos. These texts generally were comprehensive in scope, as best as their creators understood the world around them, and definitive and authoritative in tone. Although all these texts have generated extensive commentary by both followers and critics down through the ages, in every case the effort has been made to establish a standardized and official version of the text.

Shlain's general historical point is that a new mode of thinking emerged during the Axial Age – the capacity to form abstract ideas without the need for visual representation – and this new mindset significantly influenced and changed how people explained and described reality. Shlain connects this abstract capacity with linear, analytic, and linguistic thought as opposed to the holistic, insightful, and visual mode of thinking associated with images. Where the image supports the psychological capacity of intuition, the word and abstraction support rationality and logic. According to Shlain, this new abstract and rational mode of consciousness would dramatically affect the future evolution of religion. Religions began to acquire an abstract dimension, above and beyond the narrative, visual, and personified mode of consciousness. This new dimension of thinking asserted itself as the alphabet spread across many parts of the ancient world.

Shlain argues that the two different modes of consciousness have produced significant differences in terms of social behavior. Although one of the main features of religion has been to provide a way to "connect with the whole" – to identify a cosmic meaning and purpose to everyday life and commune with the forces and beings that direct and determine nature and reality – religious belief systems do not necessarily produce peace and togetherness within humanity. Shlain notes, in fact, that image and goddess centered cultures have been much less war-like than abstract, male centered cultures. In the latter type of culture, one who did not accept the belief system adopted by that culture was vilified, killed, or conquered; to refer back to an earlier discussion, there was a strong "us versus them" psychology. On the other hand, pictographic and goddess-centered cultures have been nowhere near so militant and aggressive. Thus, according to Shlain, one of the most often cited weaknesses or flaws in organized religion – its intolerance toward non-believers – is a consequence of a particular culture being too male-centered and abstract in its mindset.

Hence by 500 BCE, two different modes of consciousness influenced human belief systems: the personified, concrete, and visual versus the abstract, logical, and textual. According to Shlain, the first mode of consciousness was mystical and passionate yet more peaceful and tolerant; the second mode was more rational yet paradoxically more militant and intolerant. As modern religions evolved, in both the East and West, these two different mindsets showed up in different cultures with varying degrees of influence. Further, the abstract and logical mode, though, according to Shlain, first appearing in religious thinking, began to separate from and eventually oppose religious belief systems.

To pause for the moment, and reflect and summarize, we see in Shlain's historical analysis, along with the previously cited nomadic invasions and increasing urbanization theories, another explanation for the shift from goddess to male-centered religions in ancient times. We also see in Shlain a different version of the rise of abstract thinking; in Donald the primary cognitive shift or evolution was from mythic to theoretic thinking, supported by the development of written language, and it was first fully realized in the Greeks; in Shlain the shift was from imagery to abstract symbolism and it was first clearly expressed in Judaism. In general though, there is a consensus that the evolution of human thinking is intimately tied to the development of systems of representation.[153] (Recall the earlier discussions of cave art and language.) Though Bloom argues that humankind, in general, shows a history of ubiquitous violence and us versus them thinking and behavior, Shlain believes that it is male and left brain dominant cultures that show the greatest amount of intolerance and violence. Finally, whereas writers such as Watson see the key feature of the Axial Age as a "turning inward," away from sacrifice, external images, and ritual to find direction and meaning in life, Shlain highlights the shift from image to word as the key element.[154]

Although Shlain describes Judaism as predominately left brain and abstract, Judaism clearly includes other important features in its approach to life and mode of consciousness. The Judaic God may not have had a face, but He certainly had a personality. Further, He may have possessed a transcendent and abstract nature, but He was also repeatedly involved in the affairs of humans. *Yahweh*, in fact, was a combination of transcendent and personified qualities, and He was understood both through concrete narrative and metaphysical abstraction. On one hand, idolatry – the worship of images connected with polytheism – was condemned. *Yahweh* was beyond any concrete manifestation or image, and His presence was too terrifying and powerful to behold. Yet, *Yahweh* shows a variety of personified qualities, including jealousy, wrath, and anger, as well as compassion and a sense of justice. The Judaic *Bible* is filled with prophets who encounter God and hear His words and receive His messages and directions. There are numerous stories of these encounters. Although the true nature and form of God presumably can not be grasped or perceived by humans, the Judaic God is repeatedly interacting with His people, revealing features of his personality, and influencing human events throughout history. As with His Babylonian predecessors, He is the embodiment of order – in fact, in Judaism He becomes the abstract *Logos* of all existence - yet He is in a constant earthly fight in the world with the forces of chaos. The Judaic God is both transcendent and immanent - beyond human comprehension yet filled with human qualities and emotions -aloof and yet in the thick of things.[155] As Polak expressed it, the Jewish God is a "reconciliation of opposites," He is both loved and feared, strange and intimate, and mysterious yet revealing.[156]

Judaism is often described as creating a **"salvation history,"** a progressive view of history that eventually leads to salvation, but the road to salvation is a difficult and dramatic uphill climb. The story that emerges from the early history of the Jewish people, as contained in the *Bible*, is that the journey to the promised future is filled with struggle, repeated set-backs, human misery, and much violence. *Yahweh* on various occasions assists His people in their battles with their adversaries, but when His people do not maintain their loyalty to Him - when they break the covenant - they are the ones subjected to His wrath and are punished. (The story of Adam and Eve is the quintessential and archetypal example of disobedience and subsequent punishment.) *Yahweh* is a God of justice and compassion, and justice is repeatedly served in the saga of the Jewish people, but there is much sin, disobedience, and inhumanity along the way that needs to be rectified and overcome.

This image of the future as a difficult and often painful uphill climb has had a powerful impact on the history of Western thinking. Perhaps it is, in fact, a highly realistic and prophetic depiction of how the future of humanity has and will continue to unfold. This view of the future and time though clearly owes something to the ancient Mesopotamians and Babylonians with their emphasis on the struggle of order and chaos in the shaping of history.

Pivotal to the development of their salvation history were the writings of the great Jewish prophets ("One who speaks on God's behalf"), including Elijah, Ezekiel, Jeremiah, and Isaiah.[157] Presumably inspired by God, many of the Jewish prophets foretold of coming events in the future, in particular, pertaining to the future of the Jewish people. The Jews, of course, were not the only people in the ancient world who experienced revelations of the future; the Greeks, for example, often consulted "oracles" who presumably had divinely inspired visions of the future as well. But the Jewish tradition is especially associated with prophecy and revelation as an essential foundation to its beliefs about the future. A key feature of this mode of future consciousness is that the future is "revealed" or presented to the individual, whether in word or vision; the oracle or prophet does not actively reason or think out the future – they are more like a receptacle of knowledge of the future. According to Polak, the prophet is also important in Judaism in that he serves the role of a revolutionary, calling people to rise up, change their ways, and create a different world; the prophet challenges the status quo and makes everyone responsible for contributing to the creation of a better tomorrow.[158]

One of the most important prophets was Isaiah. Interestingly, it is a common view that there were actually two different writers who contributed to the book of Isaiah: A first Isaiah who began writing around 740 BCE and a second Isaiah who lived perhaps 200 years later. Watson states that the first Isaiah, in following the prophets before him, focused on inner spiritual and moral development and a turning away from the materialistic and sensual world. Further, Isaiah predicted an age of peace in the future if people followed the spiritual path of God, and even prophesized, according to some interpreters, the coming of a Messiah who would lead the Jewish people into the age of peace. As Watson notes, this description of the future gives history a linear and progressive quality.[159]

Armstrong contends that the Jewish belief in an absolute monotheism only appears in the writings of the second Isaiah. She states that the God of the second Isaiah had risen beyond whatever polytheistic elements remained in Judaic thinking and stood above the world and all creation. Isaiah saw the Judaic God as the creator of the world and he clearly expressed the view that this God is the only God – all other deities are false. It is this one God who conquered chaos in the past and will conquer chaos in the future. It is this one God who gives hope and purpose to the world.[160] As stated in the second Isaiah,

> "No god was formed before me,
> nor will be after me.
> I, I am Yahweh,
> there is no other savior but me."

A number of authors, including the prophets, contributed to the writing of the Judaic *Bible*. One set of writings within the *Bible*, attributed to the priestly tradition in Judaic thought, was compiled and finalized in the period 600 to 500 BCE, and has become known as the "P" collection or component of the *Bible*.[161] The famous opening chapter of *Genesis*, describing the creation, is generally believed to have been written by "P." In this description of God's creation of the world, God has become absolutely transcendent to the world and no image or concrete likeness of Him is possible. God wills or thinks the world into existence apparently out of nothing. The opening of *Genesis* is a creation of the Axial Age.

If the opening of *Genesis* solidifies Judaic monotheism, for it is a single God who has made everything, it also further reinforces the dualist dimension of Judaic metaphysics. God is clearly separate from His creation and the world of time. We have already seen that Plato, in describing a realm of eternal abstract forms that gives the world order, created a dualism of the eternal and the temporal. This same type of dualism emerges in Judaism, for God is eternal, non-physical, and the source of all order and creation. Further, the Judaic God is self-caused, whereas the world is dependent upon His existence. The dualist elements of Judaism and Platonism would come together and reinforce each other in Christianity. In understanding how this dualist metaphysics applies to the time and the unfolding of the future, the important point to see is that the flow of time was initiated and is being orchestrated from a separate and higher distinct realm of existence.

Still, the God of Judaism, though standing above creation, as noted, was routinely involved in the events of the world, and this involvement in the world clearly comes through in the idea of a Messiah that became increasingly more important in later Judaic thought. Throughout Jewish history, Yahweh repeatedly promised his people victory over their enemies and the establishment of a Jewish nation. The prophecy and promise of a Messiah, who would lead the Jewish people to final victory and salvation on the earth, was a personified expression of this general belief that with God's help and direction His people would triumph in the end. This prophecy of conflict and victory in the future can be compared with the Zoroastrian idea of a final battle between the followers of the good God *Ahura Mazda* and the followers of the evil spirit *Angra Mainyu*. And as Watson argues, the idea of a Messiah is probably Zoroastrian a well.

Aside from the growing importance of the Messiah, Watson and Polak describe other important changes that occurred in Judaic thinking during the 500 years preceding the birth of Christ. For one thing, the Messiah evolved from a human-like figure to a more supernatural and spiritual being, who not only promised victory on earth, but an eternal paradise as well. The future and the coming of a utopian paradise acquired a spiritual dimension – Polak refers to this change as the "eschatological shift." In general, new ideas on the future, of heaven and hell, of punishment and reward in an afterlife, and of Satan enter the picture after the Jewish people encountered Zoroastrianism during their exile in Babylon. Also, the idea of resurrection, another Zoroastrian concept, appears around 160 BCE. Further, Watson states that the Judaic *Bible* only acquired the status of divinely inspired text possessing absolute authority after 500 BCE. Although Shlain argues for an earlier date, Watson contends that the written word of the *Holy Scriptures* only become central to Judaic faith during the period 500 to 200 BCE - it was only by then that the pieces of the Judaic *Bible* were put together into an integrated and standardized whole.[162]

Christianity:
The Union of Opposites
and Augustine's Vision of Universal Progress

The story of the life of Jesus and his teachings is recounted in the first four books of the New Testament - the *Gospels* of Matthew, Mark, Luke, and John. All four of these disciples, who wrote their *Gospels* (ca. 60 to 110 AD) long after the life and crucifixion of Jesus, believed that Jesus was the promised Messiah sent by God to bring salvation, both spiritual and earthly, to humanity. Yet in spite of the rather rigid orthodoxy that would later emerge in Christian doctrine, the early history of Christianity was filled with numerous differences of opinion over the exact nature and identity of Jesus, what he meant by what he said, and his relationship with the one supreme God of Judaism. Christianity evolved over time.

The central doctrine of Christianity is that Jesus Christ was God incarnated. This idea that God could take on human form was a common belief in ancient religion and myth. God is humanized and made immanent.[163] The resurrection of Jesus - of God rising from the dead - which is one of the central "proofs" of the divinity of Jesus, was another common theme throughout ancient history as we have seen illustrated in the stories of *Osiris* and *Dionysius*. God has the power of life over death - God can transcend death. It is not that clear though whether Jesus ever explicitly claimed that he was God. He did reputedly say that he was "one with the Father," but he also refers to the Father as someone he serves and obeys. There was also great controversy over the resurrection in the time following the life of Jesus. Not all followers of his teachings believed that Jesus had risen from the dead. The debate among Christians over the divinity of Jesus and his resurrection continued for centuries after his death, and was not made part of official doctrine till the fourth century AD.

As its beliefs and practices coalesced and solidified in the following years, Christianity combined oppositional if not contradictory ideas and themes. First, consider the prophecy of the Messiah. Jesus of course was a Jew, and according to Matthew, a direct descendent of Abraham and Daniel. Although there is some historical dispute on this point, Jesus appears to have believed that he was the fulfillment of the prophecy of the Messiah - (as he states he was "sent by the Father") - but his vision and message, in important ways, differed considerably in spirit from the Judaic prophecy.[164] The Jews believed that the Messiah would lead them in an earthly battle against their oppressors and enemies and establish a permanent Jewish nation. Jesus, on the other hand, did not attempt to lead the Judaic people in a physical war against their Roman oppressors. Instead he preached love and forgiveness, even against one's enemies, and though he apparently believed in a coming earthly utopia, he emphasized a spiritual and "other worldly" salvation - a union with the "Father" in Heaven. Paul, in fact, came to especially highlight this spiritual meaning of salvation in his interpretation of the teachings of Jesus. Hence, the prophesized future of Judaism transformed from an earthly reward to a heavenly reward in Christianity and the road to salvation was through love, rather than war and violence.

Christianity added the earthly battle to its teachings in the final book of the *New Testament*, the *Revelation to John* (ca. 90-95 AD). As prophesized to John, Christ would return to earth in a "Second Coming" and lead believers in a great final conflict against non-believers and the forces of Satan. Hence, although Christianity begins with the idea of a spiritual salvation through love, it ends up combining this idea with the notion of a battle between good and evil that will result in both an earthly and spiritual victory. This final battle of Armageddon sounds very much like the prophecy of Zoroaster. In the "Final Judgment," those who believe in God are rewarded with both an earthly paradise followed by an eternal heavenly reward, whereas those evil souls and non-believers are damned to Hell and eternal punishment. Again, this sounds very much like the prophecy of Zoroaster.

These two visions of the future – of an earthly utopia versus a spiritual salvation – are according to Polak, two fundamental, yet disparate lines of thought that run through the history of Christianity, from its beginnings up through the Middle Ages. Polak sees the earthly utopian vision as a continuation of Judaic thinking, and contends that the Gospels of Mark, Matthew, and Luke, significantly reinforced by the book of *Revelation*, fall more in line with this version of Christianity. On the other hand, Polak sees the Gospel of John and the epistles of the Apostle Paul (3-67 AD) as emphasizing the spiritual vision of Christianity.[165] Watson agues that even Jesus, aside from his spiritual vision, anticipated the establishment of a "Kingdom of God" on the earth, and Jesus believed that he would rule in this new earthly kingdom.[166] As Polak states it, in the formative period of Christianity, the expression of "Kingdom of God" had both a spiritual and materialist meaning.

Early Christians believed that the coming earthly Kingdom of God was imminent. Jesus seems to have believed that it would occur either in his lifetime or soon thereafter; Jesus spoke with a sense of urgency regarding the future. With the death of Jesus and his reported resurrection, his early followers expected his return and the establishment of his earthly kingdom at any moment. According to Polak, Paul in his earlier writings seems to expect the return of Jesus very soon. Yet, as the years passed, Polak states that the tone of Paul changed, from urgency to patience. The values of faith and hope in the future – "of conviction in things not seen" – become paramount in Paul's writings.[167]

As Paul developed his views on the significance of Jesus, he also created a new vision of history and the future. Whereas in Judaism, the key event in the future was the anticipated coming of the Messiah, for Paul, the Messiah had come, marking a watershed point in human history. We were now entering a Post-Messiah period. A new covenant with Christ had been established and it was up to Christians to model the way of life that Jesus had exemplified. No longer emphasizing the imminent return of Christ and the creation of a Kingdom of

228

God on earth, Christians, in living the life of Christ, should look forward to a spiritual reward and eternal life in heaven. And further, as Watson argues, Paul universalized Jesus and the idea of the Messiah. No longer was the Messiah simply the savior of the Jewish people; Jesus the Messiah was now the savior for all humankind. Finally, for Paul, all of humankind was in need of salvation – we were all fallen from the grace of God – and it was only through Jesus – his life, death, and resurrection – that humankind had been saved. Humanity has only a hopeful future due to the intervention of God in the form of Jesus, the Messiah.[168]

The first combination of opposites described above, of an earthly utopia achieved through war versus a spiritual salvation realized through love and forgiveness (as well as faith in Jesus Christ), is connected to a second pair of contradictory themes – the masculine versus the feminine within Christianity. Although Judaic thinking attributed compassion and love to their God, there were equally strong elements of retributive punishment and outright violence connected with God. *Yahweh* evoked "fear and trembling" in both believers and non-believers. According to Shlain, Jesus preached a much more feminine set of values than the masculine values associated with *Yahweh* and Judaism. Jesus stressed non-violence, mercy, compassion, sacrifice, love, nurturance, kindness to the weak and sickly, and the equality of all humans - all feminine values. (Polak identifies love, forgiveness, justice, equality, mercy, justice, non-violence, and sharing as the "new" Christian values.[169]) For Shlain, Judaism, as expressed through such ideas as a judgmental God who stood on high, handed down abstract absolute laws, punished those who transgressed, and inflicted violence upon His enemies, was extremely masculine in its mindset and practices. If the worship of the goddess had steadily lost ground with the coming of male sky gods and male dominated social systems, Jesus represents a return of goddess values, albeit expressed through the voice of a male.[170]

In some respects the feminine values that Shlain identifies in the teachings of Jesus align with similar values in Buddhism.

As Watson notes, a common scholarly argument is that there is considerable overlap between the ideas of Buddha and Jesus. Both stressed an otherworldly attitude and an ethics of love, opposed violence, and renounced earthly satisfactions. Shlain, in fact, would agree that Buddha's teachings contained many feminine values.[171] The ideas of Buddha and Jesus emphasize a much less materialistic and much more peaceful and loving approach toward the creation of the future than the philosophy of power, greed, and conquest that has dominated much of human history.

Yet in other respects, Christianity, even in its earliest times, was not entirely feminine in tone. The crucifixion brought into the imagery of Christianity pain, suffering, violence, and death. The resurrection of Jesus, expressing the recurrent theme throughout Western mythology of life arising out of death, was connected with a male deity, the Father, who presumably raised Christ from the dead. The power of rebirth, such as in the story of *Isis* and *Osiris*, had been throughout the beginnings of ancient history generally associated with the female and the goddess. In the Christian story of the resurrection, that power has been usurped by the male sky god.

Even more so, after the life of Jesus, Christianity increasingly became male dominant in its thinking and practices. Within the emerging Christian world the male achieved and pretty much maintained a position of authority over the woman in sexual matters, and male controlled social hierarchies ruled the public and religious spheres of life.[172] According to Shlain, the feminine side of Christianity was progressively suppressed in the centuries following the death of Jesus. Shlain sees Paul, the primary architect of Christian religion, as greatly responsible for this shift in focus. Although Paul believed that the message of Christ and salvation was open to everyone and not just some chosen people, he established that the church hierarchy be run exclusively by men. Paul argued that the woman should be subservient to the man. Shlain states that although in his writings Paul elevated love as the greatest virtue above even hope and faith, Paul may not have practiced very well what he preached in his interactions with women. In describing the Holy Trinity, no room was made for the feminine side of God - the Father and the Son were clearly male and the Holy Ghost was identified with a gender neutral term. Mary, the mother of God, was relegated to a lower position in the Christian hierarchy.[173] So although on one hand, following the teachings of Jesus, Christianity professed a philosophy of love, forgiveness, and the equality of all human beings in the eyes of God, Christianity created a masculine deity, a male-dominant social and religious order, and a God who, in the Last Judgment after the final battle of good and evil, behaves as the stern unforgiving patriarch and condemns the souls of non-believers to Hell for all eternity.

Another combination of opposites that emerges in Christianity is between the intuitive and the rational. From a cognitive perspective, Shlain sees the values taught by Jesus as more right-brain than left-brain - more all embracing and holistic then divisive. Shlain connects left-brain thinking with analysis, literacy, abstraction, dualism, and social hierarchies whereas right-brain thinking he connects with intuition, concrete imagery, holism, and equality.[174] Christianity contains strong elements of both modes of consciousness.

First, let us consider the right-brain dimension of Christianity. The philosophy of Jesus offered an alternative to and escape from the left brain dominant rationality prevalent in Rome. According to Shlain, as Rome became more rational and literate, as its power grew, its people became more alienated, individualistic, and filled with angst. As Armstrong describes the Romans during the time of Jesus, they were conservative, pragmatic, action-oriented, and distrustful of change. They believed progress lay in a return to a Golden Age in the past and they were attracted to Greek rationalist philosophy as providing the answers to the fundamental questions of life. Because of their rationalist bias, the Romans initially saw Christianity as mad and irrational. Yet having been exposed to numerous cultures and different ideas, the Romans were also increasingly restless. Although they espoused practicality

231

and reason, many of them were drawn to the mystic rites of *Bacchus* (the Roman counterpart of *Dionysius*) and *Orpheus*. **Orphism**, a mystical cult going back to the Greeks, involved the practice of rituals that presumably would rid the self of evil. All told, feeling trapped in the confines of practicality, individualism, and reason, many Romans felt the need for redemption and salvation through the mystical. The right brain holistic philosophy of Jesus – of love, community, fellowship, and the heart – which shared some important features with Orphism, offered an alternative to Greco-Roman rationality, practicality, and extreme individualism.[175]

But again, after Jesus, Christianity integrates opposite elements into its philosophy, increasingly becoming more left-brain in its thinking and practices. Although the inspirational starting point of Christianity is the person of Jesus, who wrote nothing in his life and spoke in concrete metaphors and parables, the written doctrine of Christianity was created by Paul. Paul was a prolific and articulate writer and a grand theoretician – left brain qualities and strengths. In the battle in the fourth century AD between the Gnostic Christians, who were egalitarian, metaphorical, mystical, and intuitive, and the Orthodox Christians, who were dogmatic, rational, linear, literal, and guilt motivated, the Orthodox Christians won.[176]

As Christianity evolved in the centuries after Jesus, it incorporated various elements – often opposing elements of Greek philosophy – of both right and left brain thinking. Although Plato stood for the supremacy of reason, to recall, there was a mystical side to him as well. Early Christians, influenced by the ideas of Plato, attempted to combine both Platonic rationalism and mysticism with Christianity. They also incorporated Plato's dualism of matter and spirit into their religion. For example, Justin (100 – 165 AD) believed that Jesus was the incarnation of divine reason – of the Greek idea of the *Logos* of the world. (As stated in the opening lines of the *Gospel* of John, "In the beginning was the *Logos*, and the *Logos* was with God, and the *Logos* was God.") Clement (150 – 215 AD) was a strong advocate and follower of Plato, whom he believed was a prophet. Conceptualizing life in dualistic terms, Clement saw an ongoing conflict between the pull of passions and the discipline needed to contemplate and know God.[177] Clement also thought that Jesus was the *Logos*, and if one followed his practices and precepts, one would become God-like and in resonance with the divine *Logos*. Another early Christian writer, Origen (185 – 254 AD), thought that through contemplation the soul could advance in knowledge of God and transform into the divine. The Gnostic Christians, sounding very Platonic, believed that the physical world was an imperfect emanation of a perfect God and through intuitive (right brain) processes (as opposed to reason) could know God. In general, many early Christians thought that through the contemplation of God – His *Logos* incarnated and revealed within Jesus – one could liberate oneself from the body and connect with the absolute spiritual "One." This deprecation of the body and elevation of the spirit and mind was clearly Platonic.[178]

The rational versus the mystical is another opposition within Christianity. Orthodox Christianity identified with the rational and literate elements of Greek philosophy, whereas Gnostic Christianity identified with the mystical dimension of Greek philosophy. Although Gnostic Christianity was eventually defeated by Orthodox Christianity, in subsequent centuries, both the mystical and the rational aspects of Christianity would continue to flourish. Christian theologians, especially by the time of Scholasticism in the High Middle Ages, made great efforts to rationally prove the existence of God and articulate and defend Christian doctrine through reason and analysis. On the other hand, there was the contrary line of thinking in Christianity that God could not be captured through reason and could not be described in terms of earthbound human concepts. God must be "experienced" and this experience transcends normal human understanding. In particular, there are various mysteries, for example, the "Holy Trinity," that defy rational human understanding. This rejection of reason and rational categories of understanding as a way to know God has a long history, going back to early Judaism.[179]

The dualism of spirit and the physical body in Christianity points to another interesting combination of opposites in Christian thinking. The central doctrine of Christianity is that Jesus Christ is God, in some deep sense identical with the eternal transcendent God that created the world. Somehow God and man are united within the personhood of Jesus. This belief brings together the idea of immanence - that God is with us in the world - with the idea of transcendence - that God is beyond the world. In Christianity, God is both beyond the world and yet within the world. This doctrine of the divinity of Jesus Christ also connects man and God. As described earlier, the Hindu belief in the identification of *Brahman* with *Atman* is one that also unites individual souls with the universal soul. In Christianity this identification of the universal spirit with an individual human is limited to one person, Jesus Christ, but still, the dualism of God and humanity is overcome in the reality of Jesus Christ. Christ bridged the presumed gulf between God and humanity and the world.[180] All told, although on one hand Christianity emphasizes the dualism of the spiritual and the physical, it attempts to unite the heavenly and the earthly in the person of Jesus Christ.

Another combination of opposites within Christianity concerns its offer of spiritual salvation to all humanity, on one hand, and the tyrannical intolerance that emerged in its doctrine as the centuries went by. For Paul, the message of Jesus was for all humanity, and not just some chosen people. According to Paul, Jesus had come to save the world and not just the Jews. Yet, as Christianity transformed from a minority religious practice, persecuted by the Romans, into the official religion of the Roman Empire as established by Constantine around 330 AD, it aggressively attacked and pushed out all pagan practices and beliefs within Europe and the Mediterranean world. (This is ironical since many Christian beliefs, such as the virgin birth, are pagan in their origin.)[181] Christianity's professed love and openness to all people transformed into an increasingly aggressive effort to convert all people to its belief system. Its good news that God had sent His Son to save the world turned into an absolute "Truth" that negated all previous beliefs. The feminine and right brain qualities of love and inclusiveness became the masculine and left brain qualities of "us versus them" and the conquest of all non-believers. What was intended to unite actually generated much divisiveness. It is, in fact, a fascinating feature of monotheistic religions that although the idea of a single God is intended to envelop and unite, it invariably creates conflict and war. The "One" can not tolerate the "Other", and there always seems to be an "Other."

Of special significance to understanding the evolution of future consciousness within religion, Christianity attempts to fuse disparate concepts regarding the nature of time. For the evolutionary biologist Stephen Jay Gould, Christianity attempts to synthesize the idea that time is a sequence of unique events with the idea that time is lawful. The history of humanity as recounted in the *Bible* traces a story of distinctive and unique events, and in particular, the life, crucifixion, and resurrection of Christ as a singular and special event, never to be repeated again, that defines the direction of history. Yet, the *Bible* also expresses the cyclical and lawful theory of time as found, for example, in the *Book of Ecclesiastes*, where it is stated "that there is nothing new under the sun."[182]

According to Nisbet, Christianity combines the Greek idea of natural growth - that time involves the realization or actualization of what is potential - with the Judaic idea that history is guided and follows a necessary sequence. Both the Greeks and the Jews saw a teleological element to time, but to recall, Aristotle believed that the "telos" of change was inherent within nature. The Jews saw the "telos" of history as guided from God above. For the Christians, as expressed in the writings of Paul, the flow of events was both natural and necessary - intrinsic to the make-up of things yet determined by God. God had so designed the world that it would develop or unfold in a particular direction.[183]

The Christian theory of time connects with and attempts to synthesize both past and future and eternity and time. God's plan of salvation, *"oikonomia,"* through the death and resurrection of Jesus, though realized in time, was presumably pre-figured for all eternity in the mind of God. The crucifixion was foreseen by God in eternity. The conflict of good and evil and the triumph of God over evil is both an eternal pre-figuration and yet it is manifested and worked out, with great struggle, through time. (Recall Plato's comment that "time is the moving image of eternity.") Within this metaphysical scheme, past and future are connected as well. The meanings of past events become revealed through later events. Human sin and disobedience to God, such as in the story of the Garden of Eden, sets the stage for the eventual redemption of humankind through Christ. Judaic prophecies and the struggles of the Jewish people set the stage for the coming of the Messiah. Even the rebellion of Satan serves an eventual purpose, for without an evil one to tempt Eve leading to the "fall of man," there would be no need for the coming of Christ.[184]

A critical problem within early Christian thinking was how to reconcile the apparently contradictory beliefs that God is a one (monotheism) with the belief that Jesus Christ was God.[185] This contradiction was "solved" through the concept of the Holy Trinity. In 325 AD, in an attempt to reconcile various opposing camps of Christian thinking, Christian bishops at the Council of Nicaea established as official church doctrine the idea that Jesus Christ was God (as one of the three "persons" of the Holy Trinity), that God had created the world *"ex nihilo"* (out of nothing), and that because of inherent frailties and limitations, humanity and the world needed God and his eternal *Logos* to be saved.[186] The imperfect world required a perfect God. In a sense, the doctrine of the Trinity is an effort to synthesize the monotheistic and polytheistic in a mystical union.

Setting the direction for much of later Christian thinking, St. Augustine (354 – 430 AD) is the Christian theologian who is most well known for emphasizing the fundamental imperfections of humanity. Yet he also clearly articulated a linear and progressive view of human history and provided the theological foundation for Christian millennialism in centuries to follow. Augustine combines and synthesizes the antithetical themes of human sin and guilt with hope and the inspirational dream of eventual human perfection. Again we see in Christianity this effort to connect and unite opposites, and the writings of Augustine are a paradigm case. Augustine's impact on Christian philosophy has been immense; as Armstrong notes, next to Paul, Augustine was the most influential writer and thinker in Christian history.[187]

Augustine sets the stage for his theory of human history and the future in his doctrine of "**Original Sin.**" Since Adam and Eve disobeyed God's prohibition against eating the fruit of the Tree of Knowledge of Good and Evil, they fell from grace and innocence and were expelled from the Garden of Eden. This sin of Adam and Eve also irrevocably tarnished the souls of all their descendents - all humanity thereafter is born in a state of sin transmitted through generations from our ultimate parents Adam and Eve. Thus human history begins in a "**Great Fall**" and according to Augustine, because of this fall into sinfulness, humanity needs to be saved and redeemed by God. After the fall, humanity became a sick and suffering creature, a victim of its own freedom of choice, and helpless to do anything to change matters.[188] The guilt over human vanity, first articulated in the myth of Prometheus, is fully realized in Augustine. Humanity needs God to realize a better future - there is no other way.

Sex, the body, and women all acquire a bad name in the writings of Augustine and he uses his doctrine of Original Sin to support his negative views of physical sexuality and women. Augustine believed that Original Sin was passed on through the semen of the father. Sex therefore was the vehicle through which the sinfulness of humanity was transmitted. In his *Confessions*, Augustine describes how early in his life he was incessantly and powerfully tempted by his sexual urges. Sexual desire, an expression of the body, becomes the great adversary in Augustine's spiritual quest, and women were the source of this evil temptation. Of course, the ultimate temptress who first led humanity into sin was a woman – Eve. Influenced by the dualist ideas of Plato, Augustine sees the urges of the body, which he strongly associates with sexuality and women, as the lower and sinful reality of humankind, and the life of spirit, transcendent to the body and free of the carnal influences of women, as the ethically superior realm of existence. The feminine and the procreative power of the goddess are clearly denigrated and suppressed in Augustine.

Having established the fall of man, the resultant gulf between humanity's sinfulness and God's perfection, and the ethical dualism of the body and the spirit, Augustine develops his theory of the future of humankind as a rise from corruption and imperfection toward perfection and Godliness, all with the necessary involvement of God. Augustine clearly connects humanity's past with humanity's future. In his book *The City of God* he describes two alternative "cities" or ways of life for humankind. One is the "City of Man" ruled by physical desires, human choices, and self-love. It is the way of sin. The other city is the "City of God," a realm founded on the love of God where humankind follows the teachings of God. The "City of Man" leads to hell; the "City of God" leads to heaven. The unfolding of history is a struggle between the "City of Man" and the "City of God." The ultimate goal of humankind for Augustine is the abandonment and destruction of the "City of Man" and the complete realization on earth of the "City of God."[189]

Augustine strongly attacks the cyclical theory of time and instead argues that time is fundamentally progressive. He believes that God created time (the Platonic idea that eternity creates time) and set by design an objective linear direction to time. He believes that this developmental process within time is irreversible and controlled by God. Augustine expresses great confidence in a positive future for humankind, for he has faith in God's ultimate plan, believing that the historical development of humanity will culminate in a golden age of happiness on the earth. Synthesizing past and future, Augustine sees the developmental process of humankind as moving through a series of epochs or stages, advancing from the most primitive and infantile at the beginning of human history to the most elevated and mature at the end of time. Augustine aligns the six epochs in his history with different periods and significant events described in the *Bible* from Adam to Noah to Abraham and eventually to the appearance of Christ. Augustine believed he lived in the sixth epoch of human history. Though Augustine anchors his history to people and events in the Judeo-Christian world, he wants to include all of humanity in this developmental process – the "Unity of Mankind" doctrine in Augustine. All of humanity is moving forward. In the spirit of Christian openness, all humanity can be saved.

Although according to Augustine God sets the direction of time, time is not a simple and peaceful linear ascent. Augustine supports a conflict theory of progress and time. The developmental process of history involves a fundamental ongoing conflict between the two cities – of God and Man - and the forces of good and evil. (We have another paradox and combination of opposites here – God creates and controls time, yet time involves a conflict between opposing forces.) Augustine believed that the conflict of good and evil would continue through the sixth epoch, but would finally be resolved in the future in a seventh epoch or day. He predicts in the final resolution of the conflict the conversion of Jews to Christianity, the coming of the Anti-Christ, the culminating battle of Armageddon, the second coming of Christ, the destruction of evil, and the burning and renewal of the earth. He also seems to believe that the bodies of those humans who make it through the great conflagration will be transmuted and purified – "renewed in their flesh" – and that peace will be achieved on earth under the reign of a triumphant God. This utopian reality - the **"Millennium,"** or thousand year rule of Christ on earth - will precede the eventual ascension of souls into heaven and eternity on the eighth day and will be a time of total human equality, freedom, tranquility, security, and affluence.

On one hand Augustine describes the developmental history of humankind as an "ascension of mind over matter" - his dualist philosophy turned into a theory of progress. He states, "The education of the human race, represented by the people of God, has advanced, like that of an individual, through certain epochs or, as it were, ages, so that it might gradually rise from earthly to heavenly things, and from the visible to the invisible." Yet he also seems to believe that progress throughout history, and as most definitely realized on the seventh day, involves both material-earthly and spiritual advance. Human bodies are transformed and perfected and a society on earth is created that embodies all the ideal earthly social and political virtues. As Nisbet argues, Augustine's view of progress, as well as that of many Christian thinkers both before and after him, is natural and "worldly" as well as sacred and "other worldly." Again, quoting Augustine, "And by this universal conflagration, the qualities of the corruptible elements which suited our corruptible bodies shall utterly perish, and our substance shall receive such qualities as shall, by a wonderful transmutation, harmonize with our immortal bodies so that, as the world itself is renewed to some better thing, it is fitly accommodated to men, themselves renewed in their flesh to some better thing."[190] Hence, Augustine continues the dual themes of earthly utopia and spiritual salvation in Christianity.

Although Augustine created a grand history and future vision for all humanity, he also focused on the individual soul and individual salvation. In his general theory for all humankind, the future is set and there is a grand purpose to it all, but for each individual, the future will be a matter of choice. Augustine emphasizes the dimension of free will in determining one's future. Still he stacks the deck on this point, for the choice each individual has is between eternal damnation in hell and eternal happiness in heaven. Where is the choice in this? Polak contends that in the centuries ahead,

240

Augustine's emphasis on temptation, evil, eternal damnation, the inherent sinfulness of humanity, and free choice created an obsession with death and hell in the minds of medieval Christians.[191]

After Augustine, many Christians believed that there had been progress through human history and that a "golden age" lay ahead for humanity.[192] According to Nisbet, medieval Christians believed in both an earthly paradise and a heavenly paradise in the future. Polak though sees the spiritual and material visions as two contrasting and often competing lines of thought through the Middle Ages. For Polak, the spiritual vision tended to emphasize the idea of destiny, whereas the earthly utopian vision stressed humanity's ability to shape the future.[193]

Yet many early Christians, continuing through Augustine and into the Middle Ages, saw the Second Coming of Christ as imminent. The Apocalypse was around the corner and there was a sense of urgency connected with time.[194] Hence, although the Christians, especially as expressed in the writings of St. Augustine, believed in both spiritual and material progress and a more advanced and better world existing in the future, they did not have a sense of deep time in the future. The same was true about their view of the past. The world, according to the chronology of people and events described in the *Bible*, was not that old; according to most estimates it had existed a mere five thousand years. As Polak notes there were numerous and varied predictions throughout the Middle Ages regarding when "the world would end," but such predictions tended to be shortsighted. As one example, based on various references in the *Bible*, in the year 532 AD it was predicted that the world would end in another 271 years.[195] As far out as people could imagine into the future, there was the prophesized millennium of the City of God on earth, which would last but a thousand years after the Apocalypse and the battle of Armageddon.

Although there was considerable debate and theological controversy in Europe during the Middle Ages, in general, Christian ideas about reality and the future dominated

241

European thinking during this period.[196] The central concern about the future during this time was the coming "Kingdom of God," whether conceived in more earthly or spiritual – other worldly terms.

Interestingly, the expression "the Middle Ages" was only first used during the fifteenth century as a retrospective designation to cover the period between Roman times and what we would now call the Renaissance. In contemporary historical thinking, the Middle Ages is usually divided into the early Middle Ages (400 – 1000 AD), which encompasses the period in Europe referred to as the "Dark Ages," and the high Middle Ages (1000 – 1300 AD). In this chapter, specifically dealing with the growth of Christianity, I will end my discussion with the early Middle Ages and the "Dark Ages." In the next chapter, in describing the emergence of modern views of the future, I will begin with the high Middle Ages, which in fact, according to recent historical scholarship, is the actual starting point of Western modernism.[197]

In certain important respects, according to Watson, the "Dark Ages" in Europe were indeed dark. Compared to modern times, there was little sense of individuality; art, invention, and trade were significantly impaired, and the times were dangerous, unjust, and relatively unchanging. Illiteracy was high and the Christian church, in various ways, suppressed free and independent thinking and scientific and naturalistic inquiry. It was only through the word of God, as determined by Church officials, that the truth could be found. Because of the high rate of illiteracy and the scarcity of books, including the Christian *Bible*, not that many people actually read the *Bible*. In general, explanations of events in the world through natural causes were rejected in favor of explanations in terms of the will or purposes of God. In spite of some efforts to preserve them, many collections of books were burned and destroyed, and there was a general suspicion and antagonism toward the printed word.[198] To whatever degree people thought about the future, it was approached through theological dogma and blind faith. The emphasis on independent thinking, rationality, and naturalistic science inherited from the Greeks was repressed, if not lost in Europe, for countless centuries.

According to Shlain, after Augustine's moral attack on the woman, during the "Dark Ages" of Christian Europe, the feminine side of Christianity made a strong but temporary comeback. The worship and veneration of Mary increased. Cathedrals were dedicated to her. Numerous sightings of her were recorded. Still, Mary had been de-feminized, being robbed of her goddess power of sexual procreation. She was the "Virgin Mary." Also during the Dark Ages, the figure of Satan became more prominent. The image of a horned, serpent-headed and serpent-tailed red creature came into being - clearly an association between the animal nature within us and sin and evil. After the Dark Ages, when Europe entered a new age of learning, literacy, and "enlightenment," the feminine and the satanic were strongly tied together and the infamous and pervasive witch hunts spread across Europe and eventually to America.[199] Having begun with the teachings of Jesus, who preached love and the equality of all humans, Christianity had transformed into a worldview that deprecated women and saw sin, evil, and demons everywhere. This is part of the legacy of "the war of good and evil."

Islam:
Monotheism and Religious Conflict

Throughout history, monotheistic religions have had the tendency to preach and prophesize peace and togetherness on earth, yet to practice war and conquest. A case in point is the age old tension and conflict between Christianity and Islam - the two most influential and popular monotheistic religions in the world. Since the time of the Crusades, which lasted for roughly three centuries, when European Christians attempted to reclaim the Holy Land from the Islamic Empire, the Christian and Muslim worlds have been in a recurrent state of cultural conflict and antagonism and have on several occasions engaged in military confrontations with each other.

What is both fascinating and disconcerting about this religious and cultural opposition is that in many ways these two belief systems are very close in philosophy. In describing some of the main features of Islam below, I will frequently draw comparisons with Christianity to illustrate the connections and commonalities between the two religions.

Mohammed (ca. 570 – 632 AD), the founder of Islam, saw himself as continuing the prophecies and teachings of Judaism and Christianity – the third great prophet after Moses and Christ – and the culmination of their teachings. Mohammed lived in a time when the Arab world existed in a state of relative barbarism, where different competing tribes practiced violent retaliation against each other for perceived injustices, and the values of greed and egotism increasingly dominated human life. He believed that the Arab world needed to unify itself in terms of some central principle that transcended individual or local values and desires. This aspiration toward unification eventually led Mohammed to a monotheistic religious doctrine that not only provided a common ground for the different people of the Arab world, but also provided a theological system that explained and encompassed other religious systems, including Judaism and Christianity. For Mohammed, there was only one true God *"Al'lah"* (which literally means "the God") and all prophets and religious teachings point to this single God.[200]

Revelation played a critical role in the emergence of Islam, as it had in Judaism and Christianity. The traditional story is that Mohammed was awakened from sleep one night and felt enveloped by a divine presence, whereupon a voice commanded him to "Recite." Mohammed later identified this voice as coming from the angel "Jibril" or Gabriel, who was presumably speaking for God (or *Al'lah*). Literally overpowered by the divine personal presence, Mohammed began to recite the words Gabriel revealed to him. This recitation and Mohammed's subsequent efforts to record these divinely inspired words would continue, off and on, for the next 23 years and become the great religious text of Islam, the *"Qur'an."* For devout Muslims, the *"Qur'an"* is the revelation of God – of the nature of God – just as for Christians, Jesus is the revelation of God to humanity.[201]

The religious belief system expressed through the writings of the *"Qur'an"* is a thorough-going, extreme monotheism. There is only one true reality – *"Al'lah"* – and the entire world around us is but a manifestation and creation of *"Al'lah."* (In this respect, *Al'lah* is similar to *Brahman*.) *Al'lah* subsumes all and demands total obedience; submission and surrender to *Al'lah* are critical within Islamic religion. The term "Islam" means "to surrender" and a Muslim is "one who surrenders." Although originally Islam was a highly tolerant religion, since in Mohammed's mind, all religions are efforts to worship and understand *Al'lah*, Islam became increasingly intolerant of polytheistic beliefs and goddess deities. All other deities were pushed aside – there is only *Al'lah*. Mohammed rejected the Christian Trinity for God must be an absolute "One" – there is no plurality within God; hence, Mohammed sees Jesus as a prophet for it would make no sense to say that God had a Son. Watson argues that the Islamic God is closer to the Judaic God than to the Christian God, for *Al'lah* is more a god of might than a god of love; *Al'lah* is all-powerful.[202]

The principles of unity and tolerance were integral to early Islamic ethics. There was an emphasis on the oneness and the equality of all humanity. The brotherhood of man was an important guiding principle and initially women had equal rights to men. For Mohammed, one major goal of Islam was the development of a just and equitable society. It was part of the mission of Islam to eliminate oppression and injustice in human society.[203]

As Christianity emerged as the dominant religion in Europe, its religious leaders attempted to influence and direct the workings of human society and politics. Islam, beginning with Mohammed and continuing to the present day, perhaps even more strongly connected politics and religion. Islamic leaders believed that it was the will of *"Al'lah"* to enforce their religiously based ethical principles in their own lands and to spread such ethical principles in those lands that they conquered and assimilated.[204] In essence, the religion of Islam provided a basis for the ideal society, as well as an overall direction for the future of humanity.

Armstrong states that Islam did not begin as a militant religion; at first, there was no active effort to convert non-believers. As noted above, Mohammed believed in a philosophy of tolerance. But according to Armstrong, as the Islamic empire grew in the centuries after Mohammed, it became increasingly intolerant of non-believers, became more male dominant and repressive of women, and began to engage in aggressive conquest and conversion. Bloom, in fact, argues that contemporary fundamentalist Islam has turned aggression into a virtue. The Christian world has often thought and behaved in a similar manner; both religions have concepts of a just or holy war, where violence and killing is ethically and religiously justified.[205] For both religions, warriors are often promised heavenly rewards for killing the enemy. As a case in point, the Christian church, as a way to marshal support for the Crusades, offered "indulgences" (promissory notes to heaven) to knights who would participate in the retaking of the Holy Land back from the Moslems.[206]

In many other respects Christianity and Islam are similar as well. Both religions subscribe to the idea of one true, absolute, and transcendent God, though as mentioned above, Mohammed believed that he advocated a purer form of monotheism than the Christians and their doctrine of the Trinity. Still both Christianity and Islam see God as separate from and above the world. Hence, although monotheistic, Christianity and Islam are basically dualistic in their metaphysics and value systems, dividing reality into a lower material realm and a higher spiritual realm, and separating the moral realm into good and evil. Both religions see God as orchestrating the flow of events in the world. Further, both religions envision, as part of God's divine plan, a final battle of good versus evil and the reward of heavenly paradise to those individuals who believe in God and follow His will. (Both Christianity and Islam are Zoroastrian in this regard.) In fact, it is critical in both

Islam and Christianity that the will of God rather than human desires and aspirations determine a person's path in life. (Recall Augustine's allegory of the Cities of God and Man.) Extending back to their common heritage in Judaism, obedience to God is a central value in Islam and Christianity. Christians are supposed to follow the words of the *Bible*, and in particular, the teachings of Christ, who was the self-professed servant of the "Father." Both Christianity and Islam, though beginning as egalitarian religions, eventually created patriarchal systems. Both rejected imagery and idolatry, though perhaps Islam more strongly – for God was transcendent and beyond any concrete representation.[207]

As the futurist Wendell Bell has argued, it is frequently our similarities rather than our differences that produce conflict among us and the recurrent conflicts of Christianity and Islam are a good example.[208] Both Christianity and Islam polarize the world into good and evil, and both religions believe that they know the absolute truth and the absolute good. All other religions are false or mere approximations to the ultimate truth contained in their respective doctrines. Both religions are monotheistic and will not admit the existence of any other deities besides the "One" that their followers believe in. All other people, cultures, and belief systems embody elements of evil and need to be converted or assimilated. The intent of both religions is to convert the world to the one true faith. Since both religions, though professing peace as a central value, are associated with patriarchal and warlike societies, both Christian and Islamic civilizations have throughout their respective histories engaged in repeated war and conquest of other nations and people who did not subscribe to their beliefs. They have both pursued militant conversion in the name of religious enlightenment. If two groups of people are religiously intolerant, believe that they possess the only real truth, pursue global dominance, and are often militant in their interactions with other people, when these two groups meet, war is the natural outcome. As both religions evolved, they increasingly approached reality and other cultures in terms of an "us versus them" mentality.

In the centuries after Mohammed, as Islam spread across the Middle East and eastward through Northern Africa and into Spain, it encountered the teachings and ideas of many different cultures, both existing ones, as well as great past cultures. Through trade and conquest, Bloom's two primary integrative social forces, Islam created a thriving economic and intellectual network across the Middle East during the time when Christian Europe had sunk into the relatively illiterate, chaotic, and unproductive Dark Ages. The city of Baghdad became one of the great cities of the world and, at first, was very open and tolerant to the ideas and practices of non-Islamic people.[209] In the ninth century AD, one of the most significant intellectual events in the history of the world occurred – Islam discovered the ancient Greeks. Through the study of ancient Greek philosophy, science, medicine, and mathematics, coupled with their own genius and inventiveness, Islamic thinkers created one of the great intellectual cultures in human history. Armstrong describes this period as having features of both the European Renaissance and European Enlightenment. During this time, Islam produced a series of great scholars, philosophers, and scientists including Alkindi (813 - 880), Alhazen (965 – 1038), Avicenna (980 – 1037), Al-Ghazzali (1058 – 1111), and Averroës (1126 – 1198). Such philosophical thinkers and investigators of nature contributed significantly to the advancement of knowledge. While Medieval Europe had, to a great degree, closed its mind to pre-Christian traditions and culture, Islam opened itself to the multi-cultural heritage and intellectual wealth of the past and the present, and flourished. The lesson to be learned in this, building upon the model of Bloom, is that cultures that stay closed in both space and time stagnate, and those cultures that embrace and study both the past and other cultures move forward into the future. Also, during its apex, Islam was a culture of both faith and reason – of science and religion – attempting to synthesize these seemingly disparate elements in its philosophy and way of life. Yet it also should be noted that the conviction and enthusiasm of Islamic culture - that it possessed the all enveloping truth - motivated and supported its efforts to bring all human learning under the umbrella of its belief system. In the final analysis, faith in *Al'lah* and the teachings of Mohammed reigned supreme.[210]

Because of their importance and subsequent influence, several of these Islamic thinkers and the issues they discussed should be described in more detail. As Armstrong notes, after contact with ancient Greek ideas, an intellectual movement developed in Islam referred to as *"Falsafab,"* which was a synthesis of abstract science and philosophy and practical guidance in life; the intent was to live a philosophically enlightened life. In encountering the ideas of Plato and Aristotle, Islamic thinkers attempted to integrate their religion with the rationalist and naturalistic principles of the Greeks. Early on, Alkindi, though a student of the natural sciences, came to the conclusion that the revelations of the *Qur'an* took precedence over reason. Al-Ghazzali, whom Watson identifies as the second most important figure in Islamic history, also argued that the *Qur'an* and the model behavior of Mohammed were a sufficient basis for living the good life. But other Islamic writers during this time came to different conclusions. Alhazen, for example, embraced science and, incorporating ideas from the Greek atomists, created one of the great scientific works in the early history of the study of optics (which later would have a great impact on European science), presenting his ideas in a highly analytical and naturalistic format. Avicenna took the view that the universe was a rational and orderly system and although mystical revelation was important as a road to the truth, so was reason. In particular, Avicenna attempted to reconcile the ideas of Plato and Aristotle with Islamic religion and he developed rational proofs of the existence of God. Averroës, also attempting to synthesize Plato and Aristotle with the *Qur'an*, went even further, and articulated a highly rationalistic philosophy and theology, arguing that Greek reason and Islamic revelation were entirely compatible with each other.[211]

As a general point to make regarding these intellectual efforts, it is noteworthy that Islamic philosophers of this period struggled with bringing the rational-empirical and the mystical-revelatory approaches to life together into a synthetic whole. Armstrong describes it as an effort to merge monotheism with Greek philosophy and believes that the effort ultimately failed, allowing the mystical approach within Islam to life to maintain its dominance in the Moslem world. Further, after the great flowering of creativity, intellect, and imagination in the period being discussed, the Islamic world began to close itself off from outside influences, becoming more authoritarian and intolerant as a culture. But it is important to see that, for many of the great Islamic thinkers of the *Falsafab* movement, the scientific-rational and mystical-revelatory, which constitute two of the most influential approaches to the future and are usually seen as totally incompatible with each other, could be synthesized into a coherent whole.

The growth of Islamic culture became an increasing threat to the Christian world around the turn of the millennium. Not only was the Islamic world spreading across more geographical area, in effect, surrounding the Christian world on three sides from the west, south, and east as it spread up into Spain, but the ideas of Islamic writers were beginning to circulate through Europe. It was especially in the city of Toledo, Spain during the twelfth century that a large number of translations into Latin of works of Greek and Arabic philosophy and science were produced and distributed into other parts of Europe.[212] In particular, the writings of Avicenna and Averroës gained considerable attention and reintroduced Christian Europe to the pagan Greek philosophers and notably Aristotle. Christian theologians and thinkers increasingly felt the need to respond to the ideas coming out of Islam, but also indirectly to the ancient Greeks and the naturalistic and secular views of Aristotle. Although the Crusades, which began in 1095 and were spearheaded by Pope Urban II and his promise of heavenly indulgences, could be seen as opening the door to contact and exchange between Europe and the Islamic empire, according to Watson, very little new learning was transmitted as a consequence of this military confrontation. In fact, the Crusades intensified the bitterness and tension between Islam and Christianity.[213] Rather, it was through the spread of Arabic and Greek texts coming out of Spain that Europe rediscovered its heritage and was, as a consequence, shaken free of its culturally isolated dogmatic slumber. The rediscovery of the Greeks and the past through Islamic scholars and their efforts to reconcile reason and the *Qur'an* was one of the key events that triggered the rise of modernism in Europe. Christianity had been waiting for the Second Coming for a thousand years, looking toward a prophesized heavenly paradise in the hereafter, and the future meanwhile came knocking on the back door.

God, Religion, and the Future

Certain prevalent themes emerge in the historical review of myth, religion, and philosophy presented above. In following the logic of the Taoist *Yin-yang*, it seems that many major ideas can be described as reciprocal pairs – as complimentary and interactive throughout history. Though certain pairs of ideas, such as emotion and reason, the masculine and the feminine, good and evil, order and chaos, mysticism and rationalism, and monotheism and polytheism, can be seen as oppositional, following the philosophy of Taoism, what we take to be opposites actually are interdependent. Throughout the history of myth, religion, and philosophy such oppositional viewpoints and concepts are dynamically interwoven in theory, debate, and ways of life.

As a case in point, the idea of God is a weaving together of many reciprocities. God has been used as an explanation of reality. God is frequently the supreme archetype defining the cause, animation, purpose, and destination of humanity and the cosmos. In polytheism, gods and goddesses, forming an array of fundamental archetypes, provide the order, motive, force, character, and quality of the basic dimensions of life. Such conceptualized deities may create the universe from some primordial or chaotic base, or out of nothing through divine will. Reciprocally, just as the universe may begin in God, the universe and humanity may be heading toward a reunion with God at the end of time.

In the creation and animation of the cosmos, God may be seen as separate from the world, or God may be identified as the ultimate essence of reality, thus producing either dualist or monist views of reality. Gods and goddesses may be immanent and involved in the happenings of the world, or transcendent, standing outside of the world. The issue of transcendence versus immanence has been a point of significant and continued debate within the religious world - Judaism struggled with the question throughout its history, seeing value in both attitudes - and as in the case of Christianity, the attempt was made to synthesize the two views. In fact, as can be seen in this history of religion and myth, the themes of transcendence and immanence usually co-existed in various religious belief systems.

A recurrent issue throughout the history of religion is monotheism versus polytheism. A common belief, reinforced by monotheistic religions, is that polytheism is more primitive and as human history unfolded a shift occurred from polytheistic belief systems to monotheistic ones. Armstrong though notes that the opposite view has been defended – that monotheism came first and was later followed by polytheism.[214] At the very least, the earliest myths, according to Shlain and Bloom, involved both central male and female spirits corresponding to the Goddess Mother and the Male Hunter/Father, a fundamental reciprocity at the heart of the beginning of human myth and religion.

Even if monotheism embodied some kind of evolutionary advancement in human thinking, polytheism had repeatedly pulled the human mind in the opposite direction, even since the emergence of monotheism. Polytheism reasserted itself in ancient Egypt after the monotheistic reforms of Akhenaton. Armstrong describes in great detail the continual struggle within Judaism between polytheism and monotheism. Although

Hinduism identifies a single supreme deity - *Brahman* - it also embraces a vast assembly of lesser gods and goddesses, which include various different personae and manifestations of its central deity. Christianity has particularly wrestled with the issue, due to its belief in the divinity of Jesus, attempting to achieve reconciliation in the idea of the Holy Trinity which is both paradoxically a one and a many. (Hinduism has an analogous mystery in its Trinity of *Brahma, Vishnu,* and *Shiva.*) Further, Christianity throughout the ages has been populated with angels, demons, and the spirits of innumerable saints who have special powers concerning different aspects of life. The Holy Virgin Mary, in particular, though not officially part of the Godhead, has added a strong polytheistic element to Christianity. Monotheism and polytheism are often reconciled through spiritual hierarchies, with a central most powerful god at the top, such as *Zeus* or *Jupiter,* and many lesser spirits and deities with limited areas of power and influence below. Both Christianity and Hinduism have spiritual hierarchies.

The historical "tug of war" between monotheism and polytheism can be likened to the continued conflict within cultural and political human history of unification versus diversification, or as Empedocles would describe it, love and strife. Bloom describes the saga of history in terms of forces toward conformity and integration versus competing forces toward diversity and independence. As Armstrong notes, polytheism, which allows for the many, has been generally more tolerant and open, whereas, monotheism which elevates a singular one above everything, has been less tolerant and more militant. Speaking metaphorically, the "One" desires to assimilate the many, whereas the many wishes to break free of any constraints and express its diversity. Monotheism is motivated by the drive in the human mind toward unity and integration, whereas polytheism embraces the diversity of forces within the world. This interplay of unity and diversity - this *Yin-yang* - is a fundamental reciprocity within religious history.

Just as the Dionysian and the Apollonian, and the image and the word, are often contrasted and juxtaposed as distinct modes of consciousness, so are faith and reason. In his book The *Emergence of Everything*, Harold Morowitz highlights and distinguishes three forms of knowing God: through history, faith, and reason.[215]

Knowing God through history, Morowitz argues, clearly developed in ancient Judaism. The nature of God and His relationship with humankind was revealed through the stories of Adam and Eve, Cain and Abel, Moses, Noah, Abraham, and David. Yet much of human myth, even prior to Judaism, is a form of history, even if it is fanciful history. Knowledge of God through history is narrative and personified knowledge. This mode of understanding would correspond to Donald's notion of "mythic consciousness." And as it pertains to future consciousness, the future is understood as part of a great narrative that stretches back to the beginnings of time.

Next, Morowitz sees faith as a mode of knowing arising in Christianity, especially within the *Gospels* and the writings of Paul. (Faith though is also an integral part of Islam, in its acceptance of the divine origin and eternal truth of the *Qur'an*.) History presumably grounds human knowledge in fact, real or imagined; faith requires something more and something different. There are many different definitions of faith, but at its core, faith is belief and action without absolute proof. For Christians there is the expectation to trust in faith - it is a test of character and true belief. Yet, we have also seen that faith is an essential part of Judaism as well - the belief that God will honor His promises.

The significance of faith in understanding future consciousness is that the future is ultimately an uncertain reality. Although reason and science attempt to provide a sense of order and security regarding reality, as ancient myths had also tried to provide in earlier times, there is always an element of faith (belief without definitive proof) in all thinking about the future. Faith highlights commitment and determination in spite of the fact that human knowledge and action is contingent. Faith is an essential element in all human adventure.

Thirdly, according to Morowitz, there is reason as a mode of knowing God. Aristotle, for one, developed rational arguments for the existence of a "Prime Mover" – a first cause of the universe. The history of religion is filled with rational arguments for the existence of God and other metaphysical and moral principles. St. Thomas Aquinas believed that God could be known either through faith or reason, and believed that faith and reason supported each other. His *Summa Theologica* written in the thirteenth century is a paradigm example of rationalistic argument, and he presents a variety of logical arguments for the existence of God that in some cases are derived from Aristotle. As we have just seen, Islamic philosophers, who in fact inspired Aquinas, also attempted to understand God through reason, as well as revelation. Although Morowitz identifies science as the paradigm case of knowing through reason, the rational approach to reality goes back to ancient Greek philosophy and according to Shlain, to Moses and Judaism. Donald, to recall, sees the abstract and rational approach to life (clearly seen in the ancient Greeks) as the "theoretic" mode of consciousness and, in fact, the most evolved form of cognition so far realized in the human mind.

Another basic contrast that is frequently drawn in describing religious traditions is East versus West. In the East (for example, Hinduism and Buddhism) the future is a personal ascension to a higher realm of reality. In the West, deriving from the influence of Zoroastrianism, the future is generally seen as a cosmic and earthly conflict of good and evil forces, with good eventually triumphing. Individuals participate in this cosmic struggle of good and evil within their earthly lives. If the good within them triumphs – they ascend into a higher level of reality. The East does not have a vision of a great final conflict of good and evil in the future. The world may come to an end, as in the story of the Hindu god *Shiva* (the destroyer) bringing everything to an end in a great conflagration, but there is no sense of good or evil triumphing in a final universal destruction.

Shlain explains this fundamental difference of East and West in terms of the relative emphasis on alphabetic versus pictographic systems of symbolization. The West is generally more war-like than the East because the East retained more elements of pictographic representation in its written languages. (Consequently the Western vision of the ultimate future entails a great war.) Shlain also connects the extreme monotheism of the West with its heightened militancy and aggressiveness.

As a point of agreement, though, between the East and West, the future is seen as promising ascension to a higher level of reality. Yet even here there is a difference. In the West this vision of the future may be motivated by a fear of personal death. In the West, given our emphasis on individuality, we fear our personal death and hope for a continuation of our self (our soul) in Heaven. The promise in the East, for example within Buddhism, is an escape from the personal struggles of life by ascension to *Nirvana*, a breaking free of the "Wheel of Time" or the "Karmic Wheel," but there is no personal immortality – in fact, there is a transcendence of individuality and personhood. Fraser would argue, however, that in both East and West the motive behind their visions of the future is to somehow contend with and transcend the transience of time.

All told, we find the concrete and the abstract, the narrative and personified, the rational and the passionate, and prophecy and faith, all woven together in different combinations and versions throughout the development of mythology and religion. All of these modes of consciousness and ways of knowing not only influenced and structured beliefs about gods and goddesses and the origin of things, but also influenced ideas, secular and religious, about the future and the ultimate destiny of humankind.

What we find in examining ancient mythologies and early religions is that they unequivocally had many diverse and grand visions of the future – even the far distant future, as for example in Hinduism and the vast extended dreams of

Vishnu – and these visions, often embodied in stories with archetypal characters and themes, foretold of great coming battles, challenges, and eventual triumph, as well as a spiritual journey to some higher level of reality. These stories gave people hope for the future, as they continue to do today.

The futurist dimension to religious and mythological thinking is quite understandable. Myths and religions served the function of explaining reality and the scope of the cosmos. They provided visions, stories, and theories of the whole, including the nature of time, past, present, and future. Following Fraser, they provided a stable and meaningful structure for interpreting the passage and future direction of time.

References

[1] Bell, Wendell *Foundations of Future Studies: Human Science for a New Era*. Volumes I and II. New Brunswick: Transactions Publishers, 1997; Kurian, George Thomas, and Molitor, Graham T.T. (Ed.) *Encyclopedia of the Future*. New York: Simon and Schuster Macmillan, 1996; Loye, David *The Sphinx and the Rainbow: Brain, Mind, and Future Vision*. Bantam Books, 1983; Slaughter, Richard (Ed.) *The Knowledge Base of Future Studies*. Volume I. Hawthorn, Victoria, Australia: DDM Media Group, 1996. (a); Great Thinkers and Visionaries – Frontier Organizations - Alexander Chislenko Home Page : http://www.ethologic.com/sasha/thinkers.html.

[2] Fraser, J. T. *Time, the Familiar Stranger*. Redmond, Washington: Tempus, 1987, Page 95.

[3] Christian, David *Maps of Time: An Introduction to Big History*. Berkeley, CA: University of California Press, 2004, Chapter Seven.

[4] Donald, Merlin *Origins of the Modern Mind: Three Stages in the Evolution of Culture and Cognition*. Cambridge, Massachusetts: Harvard University Press, 1991, Pages 210 – 268; Watson, Peter *Ideas: A History of Thought and Invention from Fire to Freud*. New York: HarperCollins Publishers, 2005, Pages 49 – 50.

[5] Shlain, Leonard *The Alphabet Versus the Goddess: The Conflict Between Word and Image*. New York: Penguin Arkana, 1998, Chapter Four.

[6] Shlain, Leonard, 1998, Pages 47-50.

[7] Bloom, Howard *Global Brain: The Evolution of Mass Mind from the Big Bang to the 21st Century*. New York: John Wiley and Sons, Inc., 2000.

[8] Eisler, Riane *Sacred Pleasure: Sex, Myth, and the Politics of the Body*. San Francisco: HarperCollins, 1995, Pages 78 – 83.

[9] Eisler, Riane *The Chalice and the Blade: Our History, Our Future*. San Francisco: Harper and Row, 1987; Eisler, Riane, 1995.

[10] Watson, Peter, 2005, Pages 106 - 107.

[11] Watson, Peter, 2005, Page 107.

[12] Gell-Mann, Murray *The Quark and the Jaguar: Adventures in the Simple and the Complex*. New York: W.H. Freeman and Company, 1994, Pages 278-280.

[13] Hergenhahn, B.R. and Olson, Matthew *An Introduction to Theories of Personality*. 6th Edition. Upper Saddle River, NJ: Prentice Hall, 2003. See Chapter 16.

[14] Fraser, J. T., 1987.

[15] Hubbard, Barbara Marx, 1998; Nisbet, Robert, 1994: Watson, Peter *The Modern Mind: An Intellectual History of the 20th Century*. New York: HarperCollins Perennial, 2001, Pages 461-462, 771-772.

[16] Campbell, Joseph *The Power of Myth*. New York: Doubleday, 1988.

[17] Watson, Peter, 2001, Pages 461-462; Jung, Carl (Ed.) *Man and his Symbols*. Garden City, New York: Doubleday and Company, 1964; Campbell, Joseph, 1988.

[18] Jung, Carl, 1964; Hergenhahn, B.R. and Olson, Matthew, 2003. See Chapter 3.

[19] Brown, Donald *Human Universals*. New York: McGraw-Hill, 1991.

[20] Morowitz, Harold, 2002.

[21] Fraser, J. T., 1987, Pages 95-103.

[22] Franz, Marie-Louise von *Time: Rhythm and Repose*. New York: Thames and Hudson, 1978.

[23] Armstrong, Karen *A History of God: The Four Thousand Year Quest of Judaism, Christianity, and Islam*. New York: Alfred Knopf, 1994, Pages 14 – 17.

[24] Jaynes, Julian *The Origin of Consciousness in the Breakdown of the Bicameral Mind*. Boston: Houghton Mifflin, 1976.

[25] Armstrong, Karen, 1994, Page 98.

[26] Watson, Peter, 2005, Pages 99 – 101.

[27] Reading, Anthony *Hope and Despair: How Perceptions of the Future Shape Human Behavior*. Baltimore, Maryland: The John Hopkins University Press, 2004, Page 97.

[28] Bloom, Howard *The Lucifer Principle: A Scientific Expedition into the Forces of History*. New York: The Atlantic Monthly Press, 1995, Pages 19, 85, 183.

[29] Polak, Frederik *The Image of the Future*. Abridged Edition by Elise Boulding. Amsterdam: Elsevier Scientific Publishing Company, 1973, Page 10.

[30] Armstrong, Karen, 1994, Chapter Two.

[31] Campbell, Joseph, 1988.

[32] Armstrong, Karen, 1994, Introduction.

[33] Shermer, Michael *The Science of Good and Evil*. New York: Times Books, 2004, Pages 5-6, 31-40.

[34] Armstrong, Karen, 1994, Page xix.

[35] Armstrong, Karen, 1994, Introduction; Watson, Peter, 2005, Pages 513 – 514, 522 – 523.

[36] Watson, Peter, 2005, Pages 73 – 75.

[37] Polak, Frederik, 1973, Page 24; Watson, Peter, 2005, Page 90.

[38] Fraser, J. T., 1987, Pages 95-103.

[39] Sahtouris, Elisabet *EarthDance: Living Systems in Evolution*. Lincoln, Nebraska: IUniverse Press, 2000, Chapter Thirteen.

[40] Noss, David *A History of the World's Religions*. 10th Ed. Upper Saddle River, N.J.: Prentice Hall, 1999, Page 39.

[41] Armstrong, Karen, 1994, Pages 7-10.

[42] Noss, David, 1999, Pages 38, 41.

[43] Watson, Peter, 2005, Page 95; Noss, David, 1999, Pages 38 – 39.

[44] Noss, David, 1999, Pages 39 - 40.

[45] Franz, Marie-Louise von, 1978, Pages 46-47.

[46] Noss, David, 1999, Pages 41 – 42: Watson, Peter, Pages 88 – 89; Polak, Frederik, 1973, Page 24.

[47] Fleming, Fergus and Lothian, Alan *The Way to Eternity: Egyptian Myth*. London: Duncan Baird Publishers, 1997, Pages 22 – 45.

[48] Durant, Will *The Story of Civilization I: Our Oriental Heritage*. New York: Simon and Schuster, 1954, pp. 205-212.

[49] Fleming, Fergus and Lothian, Alan, Pages 24 – 25.

[50] Fleming, Fergus and Lothian, Alan, Pages 24 – 25.

[51] Fleming, Fergus and Lothian, Alan, Pages 76 – 81.

[52] Armstrong, Karen, 1994, Page 11.

[53] Shlain, Leonard, 1998, Chapter Seven.

[54] Watson, Peter, 2005, Page 103.

[55] Franz, Marie-Louise von, 1978, Pages 40 – 41; Fleming, Fergus and Lothian, Alan, Page 38.

[56] Durant, Will, 1954, Pages 205 – 212.

[57] Shlain, Leonard, 1998, Pages 59 – 61.

[58] Durant, Will, 1954, Pages 206 – 207.

[59] Watson, Peter, 2005, Page 115.

[60] Noss, David, 1999, Pages 43 – 44, 77 – 80.

[61] Watson, Peter, 2005, Page 115.

[62] Armstrong, Karen, 1994, Page 28.

[63] Solomon, Robert *The Big Questions: A Short Introduction to Philosophy*. 6th Ed. Orlando, Florida: Harcourt College Publishers, 2002, Page 328.

[64] Watson, Peter, 2005, Page 115.

[65] Watson, Peter, 2005, Page 116.

[66] Noss, David, 1999, Pages 89 – 90, 95 – 96; Armstrong, Karen, 1994, Page 29.

[67] Watson, Peter, 2005, Pages 116 – 117.

[68] Noss, David, 1999, Pages 92 – 93; Armstrong, Karen, 1994, Page 29.

[69] Hubbard, Barbara Marx *Conscious Evolution: Awakening the Power of Our Social Potential*. Novato, CA: New World Library, 1998, Page 30.

[70] Armstrong, Karen, 1994, Pages 31, 104.

[71] Noss, David, 1999, Page 125. It should be noted though that the Hindu trinity has been described in different ways. In another version, *Shiva* is the supreme deity and *Vishnu* and *Brahman* are faces or persona of *Shiva*. See Franz, Marie-Louise von, 1978, Page 61.

[72] Franz, Marie-Louise von, 1978, Pages 57, 70 – 71; Fraser, J. T., 1987, Pages 17 – 19.

[73] Watson, Peter, 2005, Pages 288 – 291.

[74] Fraser, J. T., 1987, Pages 17 – 19, 103.

[75] Shlain, Leonard, 1998, Pages 201 – 202; Watson, Peter, 2005, Pages 121 – 122.

[76] Noss, David, 1999, Chapters Six and Seven; Shlain, Leonard, 1998, Chapter Eighteen; Watson, Peter, 2005, Pages 114 – 115, 117 – 118.

[77] Shlain, Leonard, 1998, Chapter Eighteen; Armstrong, Karen, Pages 31 - 34.

[78] Armstrong, Karen, 1994, Page 33.

[79] Franz, Marie-Louise von, 1978, Pages 44, 64.

[80] Franz, Marie-Louise von, 1978, Pages 30 - 31.

[81] Noss, David, 1999, Page 180.

[82] Watson, Peter, 2005, Page 195.

[83] Noss, David, 1999; Smith, Huston *The World's Religions: Our Great Wisdom Traditions*. Harper San Francisco, 1991.

[84] Fraser, J. T., 1987, Page 98; Franz, Marie-Louise von, 1978, Page 37; Noss, David, 1999, Pages 253 - 254, 264; Universal Tao Center: http://www.universal-tao.com/tao/what_is_taoism.html; Chinese Cosmology and Metaphysics: http://www.kheper.net/topics/eastern/Chinese_cosmology.html.

[85] Noss, David, Page 253.

[86] Noss, David, Pages 253 - 254, 264 - 266; Franz, Marie-Louise von, 1978, Pages 32, 76.

[87] Noss, David, Pages 262 - 264.

[88] Noss, David, Pages 255 - 256.

[89] Shlain, Leonard, 1998, Page 179.

[90] Watson, Peter, 2005, Pages 120 - 121.

[91] Noss, David, 1999, Page 292.

[92] Watson, Peter, 2005, Page 192.

[93] Watson, Peter, 2005, Pages 311 - 312.

[94] Watson, Peter, 2005, Pages 112 - 114.

[95] Noss, David, 1999, Chapter Twelve.

[96] Polak, Frederik, 1973, Page 36.

[97] Noss, David, 1999, Pages 350-358.

[98] Donald, Merlin *Origins of the Modern Mind: Three Stages in the Evolution of Culture and Cognition*. Cambridge, Massachusetts: Harvard University Press, 1991, Pages 340 - 344.

[99] Shlain, Leonard, 1998, Pages 136 - 144.

[100] Simmons, Dan *Ilium*. New York: HarperCollins, 2003.

[101] Noss, David, 1999, Pages 43 - 53.

[102] Jaynes, Julian, 1976.

[103] Watson, Peter, 2005, Pages 125, 138 - 139; Polak, Frederik, 1973, Pages 27 - 30.

[104] Simmons, Dan *Olympos*. New York: HarperCollins, 2005.

[105] Noss, David, 1999, Pages 49 - 50; Fraser, J. T., 1987, Pages 100 - 101.

[106] Polak, Frederik, 1973, Page 28.

[107] Eisler, Riane, 1995, Chapters Four and Five.

[108] Franz, Marie-Louise von, 1978, Page 34.

[109] Sahtouris, Elisabet, Chapter Thirteen.

[110] Kirk, G.S. and Raven, J.E. *The Presocratic Philosophers*. Cambridge: Cambridge University Press, 1966, Chapters Six and Ten; Lombardo, Thomas *The Reciprocity of Perceiver and Environment: The Evolution*

of James J. Gibson's Ecological Psychology. Hillsdale, NJ: Lawrence Erlbaum Associates, 1987, Pages 19 - 21; Sherover, Charles *The Human Experience of Time: The Development of its Philosophic Meaning*. New York: New York University Press, 1975, Pages 11 - 13; Solomon, Robert, 2002, Pages 127 - 128.

[111] Edman, Irwin (Ed.) *The Works of Plato*. New York: The Modern Library, 1956; Tarnas, Richard *The Passion of the Western Mind: Understanding the Ideas that have Shaped Our World View*. New York: Ballantine, 1991, Pages 41- 47; Solomon, Robert, 2002, Pages 129 - 131.

[112] Armstrong, Karen, 1994, Pages 34 - 36.

[113] Armstrong, Karen, 1994, Pages 35 - 36.

[114] Tarnas, Richard, 1991.

[115] Armstrong, Karen, 1994.

[116] Watson, Peter, 2005, Pages 118 - 119.

[117] Tarnas, Richard, 1991, Pages 41 - 47.

[118] Shlain, Leonard, 1998, Page 202.

[119] Bloom, Howard, 2000, Pages 157 - 163.

[120] Bell, Wendell, Vol. II, 1997, Pages 14 - 16.

[121] Bloom, Howard, 2000, Chapters Fourteen and Fifteen and Pages 174 - 176; Jowett, B. *The Republic and Other Works by Plato*. Garden City, New York: Anchor Books, 1973.

[122] Polak, Frederik, 1973, Page 32.

[123] Watson, Peter, 2005, Pages 105, 118, 180; Polak, Frederik, 1973, Page 31.

[124] Kirk, G.S. and Raven, J.E., 1966, Chapters Two, Four, and Seventeen; Solomon, Robert, 2002, Pages 122 - 126.

[125] Watson, Peter, 2005, Page 128; Polak, Frederik, 1973, Page 32.

[126] Randall, John *Aristotle*. New York: Columbia University Press, 1960.

[127] Kosko, Bart *The Fuzzy Future: From Society and Science to Heaven in a Chip*. New York: Harmony Books, 1999, Introduction.

[128] See Bart Kosko (1999) for an extended explanation of how to formulate and apply a "fuzzy logic" that rejects the Law of the Excluded Middle. For Kosko an "either-or" logic does not apply to the real world because A passes into non-A and vice versa. In essence, this view goes in the opposite direction to that of Parmenides, accepting the reality of change and modifying one's logic to accommodate to ubiquity of change.

[129] Kirk, G.S. and Raven, J.E., 1966, Pages 187 - 196.

[130] Sabelli, Hector *Union of Opposites: A Comprehensive Theory of Natural and Human Processes*. Lawrenceville, Virginia: Brunswick, 1989, Page 3.

[131] Kirk, G.S. and Raven, J.E., 1966, Pages 324 - 330.

[132] Kirk, G.S. and Raven, J.E., 1966, Pages 332 - 336.

[133] Fraser, J. T., 1987.

[134] Franz, Marie-Louise von, 1978.

[135] Franz, Marie-Louise von, 1978.

[136] Nisbet, Robert *History of the Idea of Progress.* New Brunswick: Transaction Publishers, 1994, Pages 13 - 18.

[137] Nisbet, Robert, 1994, Pages 25 - 31.

[138] Watson, Peter, 2005, Page 141.

[139] Nisbet, Robert, 1994, Pages 37 - 43.

[140] Shlain, Leonard, 1998, Pages 132 - 135.

[141] Noss, David, 1999, Pages 412 - 413, 416 - 418; Armstrong, Karen, 1994, Pages 12 - 17, 62 - 65.

[142] Armstrong, Karen, 1994, Pages 17 - 18.

[143] Polak, Frederik, 1973, Page 38 - 43; Watson, Peter, 2005, Page 151.

[144] Armstrong, Karen, 1994, Pages 21 - 22.

[145] Polak, Frederik, 1973, Page 44.

[146] Watson, Peter, 2005, Page 162.

[147] Watson, Peter, 2005, Pages 108 - 109.

[148] Watson, Peter, 2005, Page 109.

[149] Armstrong, Karen, 1994, Pages 21 - 27.

[150] Shlain, Leonard, 1998, Pages 81- 85.

[151] Shlain, Leonard, 1998, Chapter Nine.

[152] Shlain, Leonard, 1998, Chapters One, Eight, Nine, and Ten.

[153] Donald, Merlin, 1991; Reading, Anthony, 2004.

[154] Watson, Peter, 2005, Pages 107 - 108.

[155] Armstrong, Karen, 1994, Pages 23 - 57.

[156] Polak, Frederik, 1973, Page 41.

[157] Armstrong, Karen, 1994, Page 405.

[158] Polak, Frederik, 1973, Page 43.

[159] Watson, Peter, 2005, Pages 111 - 112.

[160] Armstrong, Karen, 1994, Page 60.

[161] Noss, David, Pages 412 - 413; Armstrong, Karen, 1994, Pages 12, 63.

[162] Watson, Peter, 2005, Pages 158 - 162; Polak, Frederik, 1973, Pages 45 - 48.

[163] Armstrong, Karen, 1994, Page 83.

[164] Polak, Frederik, 1973, Pages 50 - 52; Watson, Peter, 2005, Pages 166 - 167.

[165] Polak, Frederik, 1973, Pages 50, 56 - 58, 62.

[166] Watson, Peter, 2005, Pages 169 - 170.

[167] Polak, Frederik, 1973, Pages 52 - 55.

[168] Watson, Peter, 2005, Pages 219 - 220.

[169] Polak, Frederik, 1973, Pages 52 - 53.

[170] Shlain, Leonard, 1998, Pages 213 - 222.

[171] Watson, Peter, 2005, Page 186; Shlain, Leonard, 1998, Pages 168 - 178.

[172] Eisler, Riane, 1995, Pages 29 - 32, 204 - 209.

[173] Shlain, Leonard, 1998, Pages 229 - 236.

[174] Shlain, Leonard, 1998, Chapters Three, Eight, and Nine, Pages 215 -223.

[175] Shlain, Leonard, 1998, Pages 225 – 229; Armstrong, Karen, 1994, Pages 91 – 92.

[176] Shlain, Leonard, 1998, Pages 237 – 251.

[177] Watson, Peter, 2005, Pages 182, 229.

[178] Armstrong, Karen, 1994, Pages 92 – 104; Shlain, Leonard, 1998, Pages 237 – 243.

[179] Armstrong, Karen, 1994, Page 114.

[180] Armstrong, Karen, 1994, Page 108.

[181] Watson, Peter, 2005, Pages 166 – 167, 225.

[182] Gould, Stephen Jay *Time's Arrow Time's Cycle: Myth and Metaphor in the Discovery of Geological Time*. Cambridge: Harvard University Press, 1987, Pages 10 - 11.

[183] Nisbet, Robert, 1994, Pages 48 – 49.

[184] Franz, Marie-Louise von, 1978, Page 82.

[185] Armstrong, Karen, 1994, Page 99.

[186] Armstrong, Karen, 1994, Page 107 – 119.

[187] Armstrong, Karen, 1994, Pages 119 – 124.

[188] Armstrong, Karen, 1994, Pages 123 – 124; Shlain, Leonard, 1998, Pages 248 – 250; Eisler, Riane, 1995, Page 23.

[189] Watson, Peter, 2005, Pages 231 – 232.

[190] Nisbet, Robert, 1994, Pages 54 – 76.

[191] Polak, Frederik, 1973, Pages 63 – 64; Watson, Peter, 2005, Page 231.

[192] Nisbet, Robert, 1994, Pages 77 – 86.

[193] Polak, Frederik, 1973, Page 62.

[194] Fraser, J. T., 1987, Page 104.

[195] Watson, Peter, 2005, Page 235.

[196] Polak, Frederik, 1973, Pages 61 - 78; Watson, Peter, 2005, Chapters 10, 11, and 16.

[197] Watson, Peter, 2005, Page 237.

[198] Watson, Peter, 2005, Pages 237, 242 – 252.

[199] Shlain, Leonard, 1998, Chapters Twenty-Six and Thirty-Two.

[200] Armstrong, Karen, 1994, Pages 132 – 135; Watson, Peter, 2005, Page 261.

[201] Armstrong, Karen, 1994, Pages 137 – 139; Watson, Peter, 2005, Pages 260, 266.

[202] Armstrong, Karen, 1994, Pages 149 – 152; Watson, Peter, 2005, Pages 262, 267.

[203] Armstrong, Karen, 1994, Page 135, 152 – 159.

[204] Armstrong, Karen, 1994, Pages 132, 156, 159.

[205] Armstrong, Karen, 1994, Pages 157 – 158; Bloom, Howard, 1995, Page 224.

[206] Watson, Peter, 2005, Pages 352 – 354.

[207] Shlain, Leonard, 1998, Chapter 27.

[208] Bell, Wendell "The Clash of Civilizations and Universal Human Values", *Journal of Futures Studies*, Vol. 6, No. 3, February, 2002.

[209] Watson, Peter, 2005, Page 276.
[210] Watson, Peter, 2005, Chapter 12; Armstrong, Karen, 1994, Pages 170 - 194.
[211] Lombardo, Thomas, 1987, Pages 41 - 46.
[212] Watson, Peter, 2005, Page 280.
[213] Watson, Peter, 2005, Pages 352 - 354.
[214] Armstrong, Karen, 1994, Page 3.
[215] Morowitz, Harold *The Emergence of Everything: How the World Became Complex*. Oxford: Oxford University Press, 2002, Pages 185 - 196.

CHAPTER FOUR

···===···

SCIENCE, ENLIGHTENMENT, PROGRESS, AND EVOLUTION

*"Give me a lever long enough and a place
to stand, and I will move the world."*

Archimedes

In this chapter I describe the emergence of secular and scientific theories of reality in the modern West and how these new theories were applied to thinking about time and the future. The central theme of the chapter is the emergence of a new progressive image of time, history, and the future of humanity. As I describe in the following pages, this progressive image is multifaceted, with many different interpretations and points of emphasis, and there are, as well, some noteworthy criticisms and counter-proposals that emerged in modern times. Overall though, the rise of modernity in the West was connected with a general shift to a new secular-scientific vision of progress. That is the focus of this chapter.

I begin with a history of the rise of modernism in the West, starting with the High Middle Ages and continuing through the Renaissance, the Age of Exploration, and the Reformation. I next examine one of the most significant defining events in the rise of modernity – the Scientific Revolution – and how it provided a new approach to the acquisition of human knowledge and predicting the future. Next, I describe the philosophy of the Enlightenment and, in particular, the idea of secular progress that crystallized and gained force within this philosophical movement. The idea of secular progress provided a general conceptual framework and set of ideals for understanding and directing the future. After discussing the Enlightenment, I review the ideas of Hegel and Marx, two of the most influential philosophical theorists of progress in the nineteenth century, though each put a particular unique spin on the idea of progress. Hegel and Marx are well known for developing the dialectical theory of change and applying the theory to both the history and future of humanity. Then I look at Romantic philosophy and its critique of the Enlightenment. Romanticism stands to Enlightenment philosophy as the Dionysian mindset stood to the Apollonian mindset in classical times. The Romanticists provided a different interpretation of the present, as well as a different vision of the future. In the final section of the chapter, I discuss one of the most important theories to emerge within modern science – Darwin's theory of evolution. The idea of evolution was seen by many philosophers and social theorists as providing a comprehensive scientific basis for understanding progress and change in both the natural and civilized world.

The Rise of Modernism

"In the past thousand years, and particularly in the past two or three hundred years, a transformation more rapid and more fundamental than any other in human history has taken place. A new threshold was crossed, leading to a fundamentally new type of society."

David Christian

During the seventeenth and eighteenth centuries, a popular wave of optimistic and progressive thinking spread across the European world culminating in the idea of **secular**

progress.[1] This new way of thinking, associated with the rise of **modernism**, derives from a series of historical developments beginning at least as far back as the High Middle Ages. Although this new way of thinking would challenge the dominance and validity of religious views in the West and provide a different approach to the future – in fact, the term "secular" means without association or connection to the religious or spiritual – certain key elements of its origins can be found in Western Christianity.

First let us get a quick overview of the social and philosophical transformation that took place during the rise of modernism. Modernism is both a philosophy and a way of life. It is a multi-faceted reality, connected with profound changes in technology, habitation, economy, politics, and culture, that has altered all aspects of human life especially over the last few hundred years. Although modernism began in Western Europe it has since spread across many areas of the world.

According to Steven Best and Douglas Kellner, the philosophy of modernism and secular progress arose as a consequence of a series of revolutions in thought and social organization, beginning with the Renaissance (ca. 1400 – 1500), and proceeding through the Age of Exploration and Colonization (ca. 1500 – 1800), the Scientific Revolution (ca. 1600 – 1700), the Age of Enlightenment (ca. 1700 – 1800), the emergence of capitalism and democratic states, and culminating in the Industrial Revolution (ca. 1750 – 1900).[2]

Walter Truett Anderson provides a succinct description of the change in mindset that emerged across this series of revolutions in thinking. The present increasingly was seen as the beginning of a new and different future rather than a repetition of the past and the decay and disintegration associated with it.[3] Modernism is hope for the future; modernism is forward thinking.

Shlain attributes the rise of modernism to the invention of the printing press in 1454 and the spread of literacy through Europe. Increased literacy stimulated great changes in science, philosophy, politics, and art, and in particular, instigated a reaction against superstitious, magical, and religious thinking in Europe. According to Shlain, five great abstractions of thought emerged in the literate west: imageless deities, abstract laws, speculative philosophy, mathematics, and theoretical science.[4] The last four abstractions were especially critical to the triumph of modernism in the West.

Along similar lines, Paul Ray and Sherry Anderson describe modernism as a cultural triumph over the authoritarian rule of medieval political and religious systems in Europe. They see the roots of modernism in European intellectualism, which grew as a consequence of the printing press and increasing literacy, burgeoning urban centers, the growing power of a merchant class, new economic and political systems, science, and the triumph of individualism. According to Ray and Anderson, Europe transformed from a God-centered world into a money and time-centered reality.[5]

Nisbet, on the other hand, traces secular modernism further back to the thirteenth century and the "Age of Inventiveness" (as he refers to it). During this period, humanity first strongly expressed the belief that nature could be mastered and controlled, a key theme within the theory of secular progress. Also during this period, we see the beginnings of the modern work ethnic and a significant rise in industrial and mechanical invention. Further, the philosophy of individualism became increasingly popular and there was a heightened interest in politics, economics, and society, all secular concerns as opposed to the otherworldly concerns of the Christian church.[6]

Finally, Watson, reviewing different theories of the rise of the West in modern times and pushing back the origins of modernism even further, argues that the key period that instigated the great transformation in thinking in Europe was 1050 – 1250, that the key event during this time was the rediscovery of Aristotle and his naturalistic and scientific philosophy, and that the central emerging theme was individualism.[7] Of special relevance to the history of future consciousness is that increasingly people came to believe in individual power and control over the creation of the future.

The historian David Christian, drawing on a vast and rich array of contemporary historical research, states that there is no single "consensual" explanation of the rise of modernism. Yet he does outline certain basic facts and conclusions that emerge from his review of a large range of books and articles.

First, contrary to the "Eurocentric" descriptions of the emergence of modernism, it was a global phenomenon, involving the contributions of nations and cultures around the world. Modernism first blossomed in Western Europe, but its causes were global and its consequent effects have been global. Still, according to Christian, the spark ignited in sixteenth century Europe during the beginnings of the Age of Exploration. Western Europe became the new hub of economic and informational exchange, connecting East with West, from the Americas to Asia, and benefited from the great flow of ideas and products that converged upon it from around the world. Western Europe took the lead in industrial production, technological development, and economic and military power and first "crossed over" into the modern way of life. Yet since the nineteenth century the philosophy, lifestyle, and technology of modernism has spread out across many other areas of the globe outside of Europe, in particular, North America, Australia, and much of Asia.

A second major point is that the central distinguishing feature of modernism is the accelerative growth of innovation in ideas, technologies, products, and social practices. World wide population growth accelerated, as did agricultural and industrial production, energy output and resource utilization, global commercial exchange, communication and information exchange, and the accumulation of knowledge. All these accelerative changes, which became especially pronounced in the eighteenth and nineteenth centuries, seem to have been caused by increasing innovation. Humanity became increasingly inventive during the rise of modernism.[8]

The increasing rate of innovation appears to be tied to two significant factors - exchange and competition. To recall, Bloom lists reciprocity as one of the two key forces that knit the modern world together, and conquest as the other major force. In the centuries preceding the rise of modernism, networks of commercial and information exchange evolved and spread across the Eastern Hemisphere, from China, to India, to the Islamic Empire, Africa, and Europe. The increasing rate of innovation, Christian argues, comes with increased sharing, cross fertilization of ideas, and in general, a building up of economic and informational reciprocities. Secondly, due to a widening sphere of potential markets and potential competitors, economic competition intensified. Competition stimulated innovation, which stimulated more competition and so forth.[9] The significance of competition as a driving force behind growth and change did not go unnoticed by writers and thinkers living during the rise of modernism. As I describe in this chapter, the idea that competition stimulates growth and evolution became a central theme within both economic and scientific theories of progress in the eighteenth and nineteenth centuries. The value of competition became a key theme in the Western modernist approach to the future.

Given the rapid economic, technological, and social growth occurring in Europe during the eighteenth and nineteenth centuries, it is not surprising that the philosophy that emerged in Europe at this time emphasized the theme of progress and advancement. A philosophy of growth emerged in a society that clearly was, in fact, growing by leaps and bounds. It is easy to be optimistic about the future when things are going well.

As can be seen from this brief introductory overview, there are many historical roots to modernism and, as Christian notes, varied interpretations of its causes. But it is clear that a significant and pervasive transformation took place beginning in Europe and then spreading around the world. The rise of modernism was, however, by no means a simple linear progression from one view of the world to another. There were surges forward, followed by roadblocks, counter-reactions, and temporary retreats. But there is a general pattern that emerges.

Let us begin the story of the rise of modernism in the High Middle Ages (ca. 1000 AD to 1300 AD). At the beginning of this period, the most important centers of new ideas and inventions were China, India, and Islam, rather than Europe.[10] But as Watson argues, it was during this period that a transformation in thinking took place in Europe, connected with the rise of individualism and the rediscovery of Aristotle, which would catapult Europe ahead of Asia in the centuries to come. Two key features of the modern West and its approach to the future are its emphasis on a secular as opposed to a religious vision of the future, and its emphasis on individual freedom and self-determination in the creating of the future, as opposed to the teleological and God-directed conception of the future contained in pre-modern Christianity. The emergence of these features of Western modernism can be traced back to the High Middle Ages.

Many factors contributed to the rise of individualism in the High Middle Ages. As Polak notes there was a general reactionary trend against theological dogmaticism and the power hierarchy of the church. Against the other-worldliness of Christianity, there was an increasing emphasis on life on earth; among the common people the themes of both social utopianism and utopian socialism became more attractive and powerful. Against the power structure of the church, there was a growing call for more intellectual freedom. As Watson recounts, in the Dark Ages there was less of a sense of individuality and of an individual inner life, but this changed in the High Middle Ages; more emphasis was placed on the self and emotional expression, there was a greater concern with privacy, and autobiographies became more popular as did literature and stories about love.[11]

One source of the growing individualism in Europe came from within Christendom itself. Papal and clerical authority was under attack and various "heresies" arose during this period challenging orthodox church doctrine. Of special relevance to the history of future consciousness, the famous mystic priest Joachim of Fiore (1132 – 1202) proposed a theory of history and the future that attacked the authority of the Papacy. According to Joachim, human history is divided into three periods: Early history as chronicled in the Old Testament – the age of Flesh and God the Father; recent history as described in the New Testament – the age of both Spirit and Flesh and God the Son; and finally, the future age of pure Spirit – dominated by God the Holy Spirit. Based on his reading of *The Revelation*, Joachim prophesized that the age of the Holy Spirit would begin in 1260 AD and predicted that monumental changes would occur as a result of passing into this new age. Human civilization would be transformed and, according to Joachim, the Christian Church would lose its power and disappear during the age of the Holy Spirit. (It would no longer be needed to control the spiritual lives of individuals.) Interestingly, Joachim and his followers came to believe that the Pope, in fact, was the prophesized Anti-Christ and the Vatican had become the modern Babylon (as described in *The Revelation*).[12]

But if heresies and criticisms of the Church expressed an independence of thought during this time, there was an equally strong counter-movement from within the Papacy to reassert its control and authority. Popes, such as Gregory VII (1020 – 1085), "the Julius Caesar of the papacy," Urban II (1042 – 1099), who initiated the First Crusades, and Innocent III (1161 – 1216), attempted to strengthen papal control and authority over upstart kings and secular rulers in Europe, as well as the general population. These highly authoritarian popes of the eleventh and twelfth centuries saw themselves as above anyone and everyone - (Innocent described himself as 'half-way between God and man") - and tried to enforce a centralized thought control over the people of Europe, using the threats of ex-communication and eternal damnation to keep both kings and common people in line. Out of these Papal efforts to control the minds and behavior of individuals arose a great inquisition in the following century. What is fascinating is that the strongest and most aggressive efforts of the Papacy to dominate European culture and the European mind occurred in conjunction with a growing dissatisfaction with and reaction against the authority of the church - absolutist centralized control and pluralistic individualism existed in a state of mutually escalating tension and conflict. Yet, as Watson notes, in the coming century the authority of the Papacy would plummet, never to return to the apex of power it had achieved in the eleventh and twelfth centuries.[13]

This conflict of authoritarian popes and the centralized church versus heretics, kings, and individual expression is highly representative of a general theme and trend that would run through the coming centuries. The growth of modernism, as manifested through the emergence of secular philosophy, mercantilism, and science repeatedly involved clashes between the Christian church and the newer ways of thinking and living. All along the way, Christianity, and Catholicism in particular, has repeatedly resisted modernism.

Another significant development, connected with the rise of individualism in the High Middle Ages, was the emergence of universities in cities and towns such as Paris, Bologna, Naples, and Oxford. In these early centers of learning, which over time became increasingly secular in their academic orientation, open debate, criticism and doubt, and the principles of logic and reasoning emerged as guiding principles of inquiry and study. The study of logic, in particular, was stimulated by the reintroduction of Aristotle and his works beginning around 1050 AD. In Paris especially, there was, over the years, significant growth in academic freedom of expression. Additionally, there was a burgeoning sense of optimism associated with these new centers of knowledge based on the belief that humans were capable of understanding the universe and mastering and controlling the world. (Such a belief system would be the cornerstone of the European Enlightenment six centuries later.) Overall, there was a shift from focusing on the past to forward looking and creative inquiry. And finally, it should be mentioned, reinforcing the point made in the previous paragraph, that once the ideas of Aristotle gained sufficient popularity and appeared to threaten the sovereignty and validity of Church doctrine, he came under attack from the church. Repeatedly, (for example, in 1231, 1263, and 1277) his books were banned and individuals who read or supported his non-Christian ideas were threatened with excommunication.[14] But in spite of such sanctions against Aristotle, his ideas would have a great impact on scholars in the universities and significantly contribute to the undoing of the authority of the Christian church.

In this regard, it was at the University of Paris in the thirteenth century that a momentous meeting of minds took place that would have a great effect on the further development of European thinking, and it involved Aristotle. The famous German theologian Albertus Magnus (1193 – 1280) brought his young new Italian student St. Thomas Aquinas (1225 – 1274) to Paris to teach him Christian theology but also Aristotelian philosophy. Albertus firmly believed that Christian thinking could be integrated with Aristotle. He argued that there were, in fact, multiple paths to the truth – scripture, logic, and empirical observation. For Albertus, Aristotle provided a way to approach the truth through logic and naturalistic observation.[15] Aquinas, inspired by his mentor and teacher, would continue and expand greatly on this line of thinking.

Christian Scholasticism in thirteenth-century Europe reached its intellectual apex in the writings of St. Thomas Aquinas. In his masterwork, the *Summa Theologica*, Aquinas created a unified and comprehensive philosophy of God, reality, and humanity. In the centuries that followed, Aquinas became recognized as the greatest and most influential thinker in the history of Christianity. Although his Christian philosophy was founded on faith in the revelations and stories of the *Bible*, he also acknowledged and defended the central importance of reason and natural observation as well. One of Aquinas's main goals in writing the *Summa Theologica* was to reconcile the rationalism of Aristotle with the faith and revelation-based authority of the Christian church. In achieving this philosophical synthesis and reconciliation, he officially approved and opened the door within Christendom to reason and naturalist observation as legitimate paths to the truth.

In many respects, Aquinas embraced Aristotle. He supported many of Aristotle's main theories and invoked and used many of Aristotle's central concepts and arguments. Contrary to previous Christian thinking which had downplayed the importance of the natural world, Aquinas saw great value in understanding and appreciating nature – clearly an Aristotelian sentiment. Aquinas supported the exercise and development of the intellect – the human capacity of thinking should be reinforced rather than repressed. In opposition to St. Augustine, Aquinas saw our present existence on earth as important and valuable, rather than evil and inferior. In essence, Aquinas assimilated the rational and naturalistic optimism of Aristotle into Christianity, or perhaps, as some would say, he assimilated Christianity into Aristotelianism.[16]

There were important things happening at Oxford as well. According to Watson, Oxford was particularly strong in mathematics and the natural sciences. Science, of course, in the modern sense had yet to emerge as a significant social and intellectual movement, but two teachers at Oxford, Robert Grosseteste (1168 – 1253) and Roger Bacon (1214 – 1294) anticipated and would later influence the development of science and technology. Both Grosseteste and Bacon were well acquainted with Aristotle. Grosseteste, based on his extensive study of Aristotle, appears to have understood how induction and deduction, as complementary forms of logical reasoning, apply to the study of nature. In addition, Grosseteste outlined the significance of analysis in scientific observation and even provided a clear description of hypothesis testing and science experimentation. In essence, Grosseteste anticipated the fundamentals of modern scientific methodology and epistemology – the set of procedures and activities that would provide an alternative pathway to knowledge than that offered by tradition and authority or revelation. If Grosseteste described some of the main principles of the scientific method, Bacon followed through and attempted to practice these principles. Though Bacon is also known for being a mystic who was interested in the occult, alchemy, and astrology, he was foremost an advocate of science as the pathway to the truth. Bacon was very familiar with the great accomplishments of Arab science and attempted to practice the experimental method in his multifarious investigations of nature. He strongly attacked the blind acceptance of authority and believed that science would provide the way not only to gain mastery over nature, but to transform the world in the future. He was, in fact, the great futurist visionary of the High Middle Ages. Bacon predicted such modern technological developments as the microscope, telescope, eyeglasses, automobile, submarine, flying machine, and steam ship.[17]

Due to the opening of the European mind to scientific and secular approaches to knowledge and truth, a philosophical dualism (a "double truth universe") emerged in the West that to the present day still exists.[18] Aquinas and Albertus, among others, were willing to accept the rational and naturalistic ideas of Aristotle and the ancient Greeks alongside the mystical, mythological, and faith based ideas of Christianity. Aquinas thought that the core ideas and beliefs of these two approaches were compatible or consistent with each other, but the methods are clearly different. Aquinas hoped for a peaceful co-existence. In the centuries after Aquinas, however, the tension between the religious and the secular-scientific worldviews would grow as new ideas emerged in the latter approach that clearly seemed to challenge Western religious doctrine. Not only did a new vision of reality emerge in science, but a new vision of the future developed, and ultimately the value and validity of faith, revelation, and myth came under attack. But the tension, as noted, still remains, for religious thinking about reality and the future has not gone away; it co-exists along side the secular and the scientific. Contrary to the dream of Aquinas, the modern West lives in a dualistic world, a house divided against itself – with two contrary views of reality and the road to the future.

Even if the High Middle Ages achieved a high level of intellectual and theological order in the writings of Aquinas and other scholastic thinkers, Europe, in the time of Aquinas,

278

lagged far behind both the Islamic Empire to the southeast and China and India to the Far East economically, technologically, and socially. A vast network of trade and exchange from the Middle East to China had emerged by 1200 AD, and both China and Islam were centers of commerce, industry, science, technological innovation, and intellectual activity. As noted in the previous chapter, the power and presence of Islam, in particular, was clearly felt in Europe, and the re-introduction of Greek philosophy and science to European thinking was due to Islam. While Islam and China had formed vast political and social empires, Europe, though dominated by Christianity, was politically fragmented throughout much of the Middle Ages.

Yet, in one significant respect, Europe and Asia were similar. Though at that time the biggest cities in the world were in Islam, China, and India, the overwhelming bulk of the world's population was still rural and pastoral rather than urban. One of the central trends that emerged with the rise of modernism was accelerative urbanization (a trend that is still continuing today); one could say that urbanization is both a cause and result of modernization – urban centers intensify innovation and provide for all the perceived benefits and opportunities of modern life. In the Middle Ages though, pastoral groups regularly unsettled agricultural and urban communities across the globe; (the spreading reign of terror and conquest by Genghis Kahn and the pastoral Mongols is the most dramatic example from this period). The balance of power was clearly not with the cities, but this would change in the centuries ahead.[19]

If thirteenth-century Christian Scholasticism was in many ways a high expression of an Apollonian mindset and a tradition-bound social order, the "calamitous" fourteenth century in Europe was filled with "sound and fury," Chivalric romanticism, and a heightened Dionysian quality. It was an era of passion, of love and war, of the Great Plague or "Black Death" and forebodings of catastrophe, and in particular, an intensifying reaction against the perceived repression and corruption of the

Christian church.[20] Shlain describes fourteenth- century Europe as dominated by a "hunter-killer" mentality.[21]

At a more global level, after the sustained period of technological and economic development that had occurred in Islam and the Far East for several centuries, progress significantly slowed down around the world. Population growth and production took a down turn. China, in particular, became less open to the outside world. This change would be significant. Though China could have led the way into modernism, perhaps much earlier than in Europe, historical events took a different course. In the next two centuries, while China stayed closed off, Europe opened its doors further and would "set sail" in the Age of Exploration, exposing itself to new ideas, cultures, products, and resources beyond its borders. This was critical to the rise of modernism.

One factor that significantly contributed to the decline of both China and Islam as economic powers was the Great Plague. Although the plague spread throughout Europe as well as Asia and the Middle East, the great trade routes of Islam and China were significantly and disproportionately affected by the plague. European trade routes were not as greatly disrupted. Hence, moving into the fifteenth century Europe was able to capture more of world trade and commerce, which would be a key catalyst behind the flowering of the Renaissance in Italy.[22] And more generally, as Bloom and Christian would argue, the degree of openness and exchange within a culture is a critical variable in determining the rate of growth and development for that culture. It was in the fourteenth century, in particular, that Europe started to become the new global hub of interaction and trade.

Contact with new ideas and cultures would have a significant impact on Europe. It instigated increasing openness to alternative modes of thinking and raised skepticism among Europeans regarding their Christian-dominated social order. (This, of course, had begun with the introduction of the ancient Greeks and Islamic science and philosophy in the High Middle Ages.) Opposition against the authority of the Christian church continued to increase in the coming centuries, but the church stayed entrenched in its ways. According to Shlain, the Papal-dominated church of the fifteenth century was totally corrupt, authoritarian, filled with "sins and vices," and would not reform.[23] First with the Renaissance, and then with the Reformation in the following century, the domination of the church, as well as the medieval way of life, was challenged and ultimately unsettled and transformed.

One could say that the promise of Christian religion, at least as it was espoused and practiced by the medieval church, was left unfulfilled. The "Second Coming," though repeatedly anticipated and predicted throughout the Middle Ages, had not arrived as expected; the Black Death seemed to many people to be the end of the world, but clearly not as foretold in the *Scriptures*; and the church, which was the supposed moral leader for humanity, was both internally degenerate and externally embroiled in war and perpetual violence. It was time for a change.

Another significant factor in the shifting of power in Europe was the steady rise of the commercial and merchant class. This change is particularly important in understanding the Italian Renaissance (1400 to 1500 AD). Powerful city states, such as Florence and Venice that were controlled by a wealthy merchant class had developed in Europe in the fifteenth century. Whereas royalty and the church had ruled in earlier centuries, now there was a shift of power from the "other-worldly" to the worldly. Thus, the philosophy of secular progress that later emerged in eighteenth-century Europe was rooted in a world where the social power structure had already been changing from the spiritual to the secular and economic in the preceding centuries. One of the major seeds in the growth of power in the economic sector was the Renaissance city states of Italy. This growth of commercialization was a significant factor in the increase in innovation and the rise of modernism – economic competition stimulated creativity. As Europe transformed in the fifteenth and sixteenth centuries into a society dominated by commercial power structures and commercial values, it laid the foundation for the modern revolution.

Although the artistic creativity of the Renaissance is usually highlighted as the distinctive feature of the era, Watson argues, based upon recent scholarship and study, that the revolutionary nature of the Renaissance was founded upon an economic and commercial transformation. International trade accelerated and blossomed in the Italian city states. The philosophy of capitalism gained increasing power and influence – the free market became a central principle of this new philosophy. There was a revolution in banking and finance. A wealthy class arose which valued the accumulation of material possessions and money. Contrary to the spiritual and other-worldly values of Christianity, this new wealthy class saw meaning and purpose in life through economic growth and material accumulations. But the wealthy class also valued education and literacy and became great patrons of the arts, supporting the elevation of the artist as a new cultural icon. The Renaissance was foremost a great capitalist transformation.[24]

Technological advance also significantly contributed to the Renaissance. Two of the central values of the Renaissance were invention and imagination, and the period showed great technological innovation. In the previous century, the precise quantification and measurement of space and time significantly advanced. The compass, mechanical clock, gunpowder, and the printing press all emerged as significant technological developments in European life. The creation (or discovery) of the principles of perspective (geometrical projection) was another noteworthy event of the period. The first printed book in Europe, that can be dated, was produced in the year 1457, and printing caught on very quickly thereafter. The number of secular and popular books increased dramatically, challenging the dominance and monopoly of religious texts of previous centuries. Interestingly, almost immediately, and as a reaction to the distribution of objectionable works, the Church began the public censorship of various books.[25]

Another significant feature of the Renaissance was the rise of humanism. Humanism provided an alternative philosophy to the spiritualism and other-worldly attitude of the Christian church. Humanism emphasized the importance of individuality and human dignity, earthly existence and values, and the arts, and found its inspiration in classical philosophy and literature, such as in the Greeks. The Italian scholar Francesco Petrarch (1304 - 1374) is frequently identified as the founder of humanism, and along with Dante, one of the fathers of the Renaissance. A poet, who extensively wrote and romanticized on love, as well as the inner life of the human mind, Petrarch, though a devout Christian, found great philosophical inspiration in the ideas of ancient Greece. Perhaps the first person to recognize and label the "dark ages" as a period of intellectual decline, he foresaw a great turning point in the future prospects of humanity.[26] The most well-known of the humanists was Desiderius Erasmus (1466 - 1536) who, although ordained as a priest, was highly critical of the traditions, superstitions, and intolerance associated with the Christian church. Erasmus, like others before him, attempted to reconcile the ideas and writings of classical literature and philosophy with the Christian doctrine, but ended up being labeled a heretic. He emphasized individual judgment and conscience over the abstract formalism of Christian doctrine and scholarship and because of his individualistic and critical philosophy was identified as the person who "laid the egg that Luther hatched."[27]

If the rediscovery of Aristotle played an essential role in the intellectual achievements of the High Middle Ages, the rediscovery of Plato was equally important to the Renaissance. The ideas of Plato did much to support and reinforce the visionary and aesthetic philosophy of the Renaissance. Although Plato emphasized the eternal and mental realm over the physical and temporal realm, in the Neo-Platonism of the Renaissance, spirit, beauty, and order were seen in everything, and earthly life was consequently viewed as having great value, as well as being intelligible to the human mind - the world was beautiful and it made sense. Further, and quite significantly, as

Plato had identified beauty with the good, Renaissance artists and thinkers saw beauty as morally valuable. Art both informs and pleases the human mind, and wisdom was described as the synthesis of beauty and enlightenment. Overall, the Renaissance expressed an aesthetic and moral vision of the future.[28]

According to Watson, the rise of humanism and the flowering of the Renaissance reinforced and further amplified the growing individualism in the European mindset. The growth of capitalism stimulated individual competition; the emergence of humanistic philosophy intensified self-consciousness and individual conscience; and the fascination, if not worship, of the artistic genius contributed to the increasing interest in the uniqueness of different people. Fame and individual glory became important values. Individual achievement became an important goal in the life of many people. In fact, virtue was re-conceptualized as an individual achievement built on reasoning and planning. The image of the "Renaissance Man" arose – a person of education and varied interests – who possessed a "self-conscious optimism" that life possessed an intelligible order that could be grasped and understood and directed toward positive ends.[29] All in all, the Renaissance greatly contributed to the modern view that the future can be self-determined through individual effort and talent.

Perhaps the most famous example of the "Renaissance Man" was Leonardo da Vinci (1452 – 1519). Considered one of the greatest artists and creative geniuses of all time, Leonardo combined interests in painting, human anatomy, architecture, optics and perspective, engineering, and the study of nature. He synthesized in his work the scientific, technological, and naturalistic with the artistic and humanistic, as, for example, in his detailed studies and drawings of the human body and his work in vision, optics, and the principles of perspective. Leonardo is also a supreme expression of the Renaissance value of human inventiveness. Of special relevance to the evolution of future consciousness, in his "Notebooks" can be found innumerable designs and drawings of mechanical and engineering devices which anticipate technological inventions in later centuries. The first documented design of a mechanical robot can be found in his "Notebooks". There are also drawings for the construction of a helicopter, tank, submarine, mechanical calculator, car, and solar energy device. In many respects, Leonardo captures the essence and values of the modern mind; his interest and creative inventions in technology, his forward looking attitude, his individualistic lifestyle, his fascination with the wonders of nature, and his great and varied accomplishments all point toward modern values, and away from the past.[30]

The Renaissance rose up as a dissident counter-culture, an attack on tradition and authority, and a rejection of the philosophy and way of life of the Middle Ages. The writers and artists of the Renaissance called for a "return to the Golden Age" of ancient Greece, a culture filled with secular and non-religious ideas and ideals. With the Renaissance came a renewed belief in the individual, in human freedom and human ability, in reaction to the perceived repressiveness of the Middle Ages. If the Scholastic era was Apollonian, then the Renaissance was Dionysian.

The Renaissance though was a prelude to modernity rather than its beginning. In several important respects it was more regressive than progressive. For one thing, the period of the Renaissance was a time of extreme subjectivism (a necessary accompaniment of its heightened individualism), in some ways much more passionate and Dionysian than rational and Apollonian, and filled with magical belief systems and occult practices. When modernity does arrive, the magical, the occult, the religious, and the passionate would be replaced with the rational and the secular. Moreover, because the Renaissance rejected the mindset of the Middle Ages, the idea of progress as defended in the writings of St. Augustine and other Christian writers was abandoned. Instead, the Renaissance accepted a cyclical theory of time – the goal was to return to a higher civilization represented in the Greeks rather than building upon the accomplishments of the Middle Ages. They were not moving forward; rather they were returning to something better that had existed long ago.[31] In spite of these pre-modern qualities to the Renaissance, what the Renaissance did accomplish was to oppose and help to further unsettle the authority of the Church, through both humanistic and economic individualism and the elevation of secular values, and the Renaissance was just the beginning of this revolt.

Two important historical events occurred in the sixteenth century that would significantly impact the growth of modernism in Europe. The countries of Western Europe began to aggressively and competitively explore and conquer the Americas. The Age of Exploration began. Second, the authority of the medieval Church was overturned in the Protestant Reformation. Let us first look at the Reformation.

With literacy spreading throughout Europe after the introduction of the printing press, the Protestant Reformation, led by such figures as Martin Luther (1483 - 1546) and John Calvin (1509 - 1564), openly and decisively challenged the authority and sovereignty of the Roman church. Luther's rejection of the church as an intermediary between individuals and God further reinforced the growing individualism of the West. Instead of having to follow the dictates and directions of the church, which presumably spoke for God, Luther emphasized the importance of inner faith and piety and argued that individuals should read the *Bible* for themselves. Although it might seem paradoxical, given the emphasis that the Christian church placed on Biblical doctrine, the church hierarchy did not want the common people to read the *Bible*; the church wished to maintain control over how to read and interpret their holiest book. When Luther translated the *Bible* into German so that the general population in his homeland could read the *Bible* for themselves, he struck a great blow for individual religious freedom.[32]

The Reformation has been described as a "great renewal of religious commitment" and a "moral cleansing" of the vices of the church. According to Nisbet, there was a renewal in the belief in progress, which began in the writings of Jean Bodin (1530 - 1596). Bodin, in fact, attempted to combine the cyclical and linear theories of time, arguing that humanity did not begin in some Golden Age, but rather in a state of primitiveness, and through a process of growth, decay, and renewed growth, had progressed to its present state of development.[33]

Yet, the Reformation, like the Renaissance, was in important ways not a progressive movement forward. Luther was decidedly anti-intellectual and pro-faith, in spite of his professed desire that individuals read and study the *Bible*. He believed, as did Calvin, in pre-destination, and both of them were highly superstitious, believing in witches, demons, and evil spirits. Calvin, in fact, was decidedly deterministic about the future, believing that whether one would go to heaven or hell was pre-determined from birth (due to God's omniscience) for everyone. Both Luther and Calvin repressed the rights of women and Calvin, in particular, sought to establish a male dominant authoritarian hierarchy of rule within Christianity. In general, Calvin argued for the strict enforcement of moral rules and supported the centralized control of his new church over the behavior of individuals. Further, the Reformation and Papal Counter-Reformation (which included the Inquisition) led to new religious wars and excessive violence and murder. (The infamous witch-hunts were a product of the Reformation and Inquisition.) Specifically regarding the Counter-Reformation, the Catholic Church reacted to Luther and the Reformation in a manner similar to the way it had previously reacted to new ideas and threats to its authority – it became more entrenched, rejected Luther and the rise of Protestantism as heretical, and once again banned and censored whatever books it saw as objectionable. In general, although as Watson states, the Reformation was associated with increasing tolerance, there were also many counter-efforts at the same time to re-assert authority, including those of Calvin. Although humanists such as Erasmus opposed the anti-intellectualism of Luther, according to Shlain, it was the authoritarian power of Calvin and his rejection of worldly concerns that stalled the continued rise of secular philosophy and ended the free flight of the Renaissance. One should add that the widespread murder and torture connected with the Inquisition and the witch hunts, instilling fear in so many people across Europe, did not help matters either. [34]

Still, Europe, coincident with the Reformation, was embarking upon a great geographical and commercial adventure in the sixteenth century. According to Christian, after the economic and social regression of the fourteenth century, a new wave of economic development emerged across Europe and the most significant factor in this upturn was the bridging of the Atlantic in the sixteenth century. The first truly global commercial network of exchange came into existence linking the Americas to Europe and the rest of the world. Products from the Americas, including silver, tobacco, and a variety of new foods found their way across through Europe and across the Eastern Hemisphere and, in turn various products, animals (the horse for example), and quite destructively, for the indigenous people, new germs found their way to the Americas. The hub of economic exchange shifted from the Middle East to Western Europe. Again, following Christian's logic, exposure and increasing exchange of information and commercial products were critical stimulating factors behind a new wave of innovation in the sixteenth and seventeenth centuries. As a prelude to the rise of modernism, Western Europe became the center of a rich global "open system" of trading and interaction that for the first time encircled the world.

Watson also concurs that the discovery and exploration of the Americas was of immense significance in the economic, social, and intellectual rise of Europe. It was the final key factor (in the period of 1050 to 1500 AD) that moved Europe, and in particular, Western Europe, ahead of the rest of the world. Not only did the Americas provide tremendous new resources and products, as well as a growing market for European goods, its discovery had a significant intellectual impact on the Western mind. The discovery of these new lands and new people had been totally unexpected and it altered the European mind's conception of the earth, of space, and of culture and history.

How did the American Indians fit into the grand scheme of things? As Watson notes, encountering these new people and their cultures stimulated the beginnings of an evolutionary theory of culture.[35]

It should also be noted that during the same period that Western Europe was exploring and settling the Americas, it was also pushing outward in other directions, exploring Africa, reaching India, and culminating with the historic voyage of Magellan, circling the globe. Western Europe not only settled the Americas, but conquered and colonized many other parts of the world, bringing diverse other peoples and cultures into its economic and political web. Modernity begins with Europe's economic and political globalization of the world.

With the Age of Exploration, European nations entered into a period of fierce competition as first the Americas and then other parts of the world were explored and often conquered in the process. Competition among Spain, England, France, and Portugal fueled innovation in navigational and military technology. And following Bloom's logic, just as the creation of economic reciprocities helped to knit the world together, the wave of conquest and colonization pulled the resources of many parts of the world together under the control of Western Europe. Europe became the center of military and political power as it became the center of trade and exchange.

Christian argues that England was especially primed to enter into this new era. The rural population of England had been over the previous centuries increasingly pushed into employment in urban areas to survive. They were being pulled into the web of commercial exchange centered in the cities. "Proto-industrialization" was developing across England during the seventeenth century as a prelude to the Industrial Revolution a century later. Competitive capitalism had become the dominant economic system throughout England.

England would become the first nation to clearly move into the modern era. Technological innovation accelerated in England during the seventeenth century. Cities became the unequivocal centers of power. Capitalist commercialism reinforced and solidified secular and competitive values. Wealth and production escalated as England utilized the resources and populations of the many diverse lands that were swallowed into its growing global empire.[36] England, in particular, was primed for the final significant event in the rise of modernism that "lit the fuse."

The Scientific Revolution

"Copernicus, Kepler, and Galileo put in place
the dynamite that would blow up the theology
and metaphysics of the medieval world.
Newton lit the fuse.

Neil Postman

"This most beautiful system of the Sun, planets, and comets, could only proceed from the counsel and dominion of an intelligent and powerful Being."

Isaac Newton

While the Reformation and Counter-Reformation were spreading across Europe, something less violent, less boisterous, but ultimately more earth shattering was emerging in the minds of men. Nicolaus Copernicus (1473 – 1543) published his famous theory that the earth was not the center of the universe but actually revolved around the sun. Galileo Galilei (1564 – 1642), following the lead of Copernicus, discovered the moons of Jupiter and the rings of Saturn, formulated the beginnings of modern physical science in opposition to the authority of Aristotle, and openly challenged the authority of *Genesis* as a correct account of the origin and nature of the universe. And Johannes Kepler (1571 – 1630), co-inventor of the telescope, discovered the laws of planetary motion, further reinforcing Copernicus' theory, as well as setting the stage for Isaac Newton's grand scientific synthesis later in the century. Though the historical development of science can be traced through the ideas of the ancient Greeks, the investigations and studies of Islamic scholars, and the writings of Grosseteste and Roger Bacon, among others, modern science was born in the sixteenth and seventeenth centuries. The Scientific Revolution found rational and mathematical order within the changing world of nature and time, and ultimately, without having to resort to the hypothesis of God.[37]

In the centuries ahead modern science would increasingly challenge not only the authority of the Christian church, but almost all religious beliefs and practices worldwide. Science began to formulate a new explanation and understanding of nature and reality. Eventually it overturned both mythological and religious stories of the origins of the universe, as well as traditional histories of nature and humankind. Further, it provided a new way of thinking and a new method for investigating nature – the scientific experiment. Finally, it laid the theoretical seeds for a new view of the future – the theory of secular progress. Although there were some notable connections between Western religion and science, as well as ancient Greek philosophy and science, a distinctively new belief system and approach to reality emerged in the Scientific Revolution during the sixteenth and seventeenth centuries.

Watson asks why the Scientific Revolution occurred in Europe as opposed to Asia or the Middle East. According to Watson, the rise of individualism (encouraging freedom of thought and inquiry), the increasing emphasis on quantification and precision, and the materialistic and competitive nature of Western capitalism all contributed to creating a favorable climate in Europe for the beginnings of science. Although there were as many scholars in Islam or China as in Europe during the period of the High Middle Ages, the former two societies had centrally controlled intellectual cultures, whereas European intellectual culture was more open, individualistic, and critical. All in all, it seems clear that the philosophy and practice of freedom of inquiry and individualistic competition were critical factors in stimulating the growth of science in Europe.[38]

The spirit of openness and freedom of inquiry associated with the rise of modern science would eventually lead to tension and conflict between science and the Christian church. When Copernicus first published his theory of the heavens, the church did not react critically to it. In fact, church leaders responded favorably to Copernicus's theory, treating it more as a valuable and interesting system for making astronomical calculations and predictions than as a theory that made claims about the nature of reality. But as time went by, the implications of Copernicus's theory became clearer and more unsettling. According to Copernicus's heliocentric theory, humanity was no longer at the center of the universe, a theory that seemed to contradict certain passages in the *Bible* about the nature and creation of the universe. Eventually, the Christian church found the ideas of Copernicus objectionable and when, in the seventeenth century, Galileo aggressively defended Copernicus, arguing that the *Bible* contained errors concerning the cosmology and the heavens, the church branded Galileo a heretic, forced him into recanting his views, and imprisoned him.[39] As had frequently occurred in the past, the Christian church aggressively attempted to suppress ideas that challenged its sovereignty and hold on the official truth.

One of the most important features of the theory of Copernicus was that it exposed a deep egocentricity and narrowness of point of view in humanity's conception of the universe. Because we observe the sun, the stars, and the planets relative to our position on the earth, it "appears" as if the sun circles around the earth and the stars and planets rotate in the sky. It appears that we live at a stationary center point in the cosmos. Yet, this appearance is due to a limited and local perspective on the nature of things. As science advanced in the centuries ahead, humanity would discover still other ways in which our view of nature and the universe was limited and egocentric. To recall from earlier chapters, a critical feature in the evolution of consciousness has been the movement from the egocentric to ever expanding vistas in space and time. The theory of Copernicus was a highly significant step in this ongoing evolutionary process.

As modern science progressed it would challenge and overturn many traditional and common beliefs about reality. But what is important to see is that these new ideas derived from a new emerging methodology for studying and understanding nature. The origins of the new philosophy of scientific method, as well as the modern secular notion of progress, can be found in the great early promoters of science, Sir Francis Bacon (1561 – 1626) and Rene Descartes (1596 – 1650). Bacon first articulated the empiricist philosophy of science and Descartes became the great advocate of the rationalist theory of science. Both openly questioned all the beliefs passed on to them from tradition and the past, and both supported a progressive view of history and the future. In particular, they both saw science as the way to continued progress in the future.[40]

The contemporary scientist, E. O. Wilson believes that Francis Bacon was the critical figure in the rise of modernism and modernism's supreme philosophical expression in the Age of Enlightenment.[41] Bacon proposed that all past beliefs which were ungrounded in fact or reason should be rejected. He referred to these unsubstantiated beliefs as "idols of knowledge." He proposed a new method that would liberate humanity from these idols – a method that was rational and based upon fact. Specifically, he formulated the principle of **induction**, where knowledge should be based upon generalizations of observed facts. Instead of consulting religious authority or basing one's beliefs on faith, one should directly investigate the natural world as the method for gaining knowledge.[42] Bacon called for a "Great Instauration" – "a total reconstruction of the sciences, arts, and all human knowledge, raised upon the proper foundations".[43]

Bacon though did not see science and scientific knowledge as an end in itself. Rather, science should serve as a foundation for the improvement of humanity.[44] Through understanding nature, one could control nature, and through the control of nature one could improve human life. Bacon was a pragmatist – for him "knowledge is power." To quote, "Knowledge and power come to the same thing for nature cannot be conquered except by obeying her."[45] This pivotal idea that humanity could improve natural reality, as well as human society, through the application of scientific knowledge was at the core of the secular approach to the future and the concept of secular progress.[46] Through human reason and scientific principles, the future was something that could be positively directed. Progress could be achieved through scientific methods. This new philosophy was an incredibly powerful idea that eventually transformed the Western world.

Instead of focusing on spiritual improvement and spiritual ends through spiritual means, Bacon was proposing the use of secular or scientific means to achieve natural or secular ends. In fact, Bacon envisioned an ideal world ruled and controlled by science and scientists in his utopian book, *The New Atlantis*. This vision was not specifically set in the future, but it was clearly not an otherworldly or spiritual reality.[47]

The great scientist, mathematician, and philosopher Rene Descartes proposed a second method for achieving scientific knowledge. Where Bacon focused on the generalizations of observations and facts, Descartes focused on rationality or reason. Descartes, like Bacon, wished to free himself of the false and ungrounded beliefs of the past, and decided to begin by doubting everything he believed, including the existence of an external physical world. In his famous insight "I think therefore I am" he found something he could be certain of – his own existence as a thinking being - and proceeded from this starting point to deduce a variety of other conclusions. Descartes argued that he would only believe what he could form "clear and distinct ideas" about and what could be rationally deduced through reason. (Note that "I think therefore I am" is a rational deduction – the conclusion logically follows from the premise.) Descartes believed that scientific knowledge should be built upon rational deductions.[48]

Though Descartes wished to identify a sound methodology for the acquisition of knowledge, it is important to see that his starting point is doubt. Bacon also begins from a critical attitude regarding traditional beliefs. As part of the growing freedom and openness of inquiry in sixteenth century Europe, philosophical skepticism was becoming an increasingly influential idea. Descartes' contemporary, the French essayist Michel de Montaigne (1533 - 1592), was especially noted for his skeptical attitude regarding all human beliefs and customs. In the spirit of scientific empiricism and naturalism, Montaigne was especially critical of supernatural and other-worldly ideas, and in resonance with Bacon's view of the function of human knowledge, Montaigne argued that knowledge, instead of being used as "the preparation of man for a safe death," should be used to improve our earthly life and existence.[49]

Returning to Descartes, he also helped to clarify the concept of natural or scientific laws, proposing that the universe could be completely described in terms of mathematical laws. (Galileo and Kepler had formulated their physical and astronomical laws in terms of mathematical equations.) Wilson sees this insight as supporting the idea of the Enlightenment that all human knowledge can be unified - in this case through mathematics. Further, anticipating Newton, Descartes argued that the universe was fundamentally a machine; hence Descartes is the beginning of the modern **"mechanistic"** model of nature. The mechanistic model would provide a key idea in theories of secular progress. Just like a man-made machine can be constructed to serve human goals, both nature and human society could be viewed as machines and manipulated to serve human ends.[50]

Morris identifies Descartes as the beginning of the idea of secular progress. Not only did Descartes clarify many important concepts of early science and attempt to distinguish scientific knowledge and scientific ends from superstitious and unscientific ideas of the past, he also entertained the idea of natural evolution and progress. He suggested that the universe had evolved, due to the laws of nature, from a primordial chaos but he ultimately rejected this hypothesis because it seemed to him to contradict *Genesis*. He did believe, as did Bacon, in future human progress through the advancement of science.[51]

As noted earlier, Western Europe in the sixteenth and seventeenth centuries had become the hub of economic and informational exchange for the world. Western Europe also explored, and in many cases conquered, many different regions of the world, and thus was exposed to numerous and diverse cultures. This access to different ideas and products loosened the monolithic hold of the Christian Church on people's minds. Bacon, Descartes, and Montaigne are illustrative of the skepticism that entered into the minds of Western Europeans regarding their own cultural traditions and beliefs. But further, not only did the European mind become more open toward different views of the world, it also became more competitive in assessing and judging the validity of these different views. Christian argues that the Western European mindset had moved increasingly toward a competitive approach to life and away from a tributary and obedience-driven way of life. Openness to different ideas lead to competition among ideas for there was no longer just one view dominating the scene. On what basis was one to evaluate and decide among different points of view? Bacon and Descartes provided methods for evaluating and comparing different knowledge claims – through direct observation and reason. Both Bacon and Descartes were highly critical of many common beliefs of popular and traditional culture. Ideas could no longer simply be upheld and supported because of religious, cultural, or royal authority. Ideas had to be subjected to the methods and scrutiny of science. According to Christian, beliefs were now tested and debated in the "market of ideas." Science was born in a world of openness, exploration, and competition.[52]

Aside from induction and deduction and the use of mathematics, one other key feature of scientific methodology needs to be identified and described. According to Watson, one of the most influential ideas in the history of humanity is the "scientific experiment." Scientific ideas (hypotheses or theories) can be empirically tested through experimentation. In essence, if a scientist has a question concerning nature, a scientist can ask nature for an answer. Through the direct manipulation of nature, in the form of an experiment, a scientist can observe what effects appear to follow and thus gain an understanding of the workings of nature. Experimentation would become a central and defining activity in modern science, giving science, according to Watson, a democratic quality. If scientists disagree on some issue, the question can be put to the test through experimentation. Whereas previously, differences of opinion were settled through consulting and interpreting authoritative texts, through logical argumentation, and frequently through intimidation and suppression, disagreements could now be (hopefully) resolved empirically in a fair and objective fashion by manipulating and observing nature.[53] Attempting to understand nature, as well as control it through the use of experimentation represents a significant advance in human thinking and behavior and without question had a great impact on how humanity approached the future. At least within the scientific community, as we move forward into contemporary times, predictions about the future and decision making concerning the future increasingly has become grounded in experimental data and results.

Interestingly, in the midst of increased openness and competition among different points of view, one theory emerged that came to dominate early thinking in science. The Scientific Revolution culminated in Sir Isaac Newton's (1642 - 1727) theory of mechanics, motion, and gravitation which seemed to provide a comprehensive scientific explanation of the physical universe. It appeared that nature could be completely understood through reason, mathematics, and generalizations of observable facts - the dream of Bacon and Descartes. Newtonian science provided the theoretical foundation for the Industrial Age and inspired the philosophy that human society could also be modeled on unifying and comprehensive scientific principles and controlled through the application of these principles.

Newton described the physical universe as discrete, solid objects of matter moving through empty space. Material objects influenced each other through material forces. (In many ways

Newton's physics is similar to the theory of the Greek atomist Democritus.) The motions of objects and the effects of physical forces were governed by stable laws of nature. The universe, as a whole, behaved deterministically and the motions of all physical objects, earthly and heavenly, could in principle be calculated out indefinitely into the future.

Further, in his physics Newton transformed the concept of time. To recall, the ancients usually associated time with deities, archetypes, and concrete dimensions of reality (such as the rising and setting of the sun or the changes in seasons). Newton proposed an abstract and absolute concept of time - totally disconnected from any concrete manifestations. Absolute time flows throughout the universe independent of specific events in nature.[54] Newton's concept of time is one example of a general trend in science to describe the world in the most abstract terms, devoid of personification, cultural or personal bias, or concrete metaphors or associations.

Also, in one important respect, Newtonian physics connects the heavens and the earth. Kepler, in his discovery that the planets moved around the sun in elliptical orbits, had demonstrated that the heavens were not "perfect," in contrast to the accepted belief that heavenly bodies presumably all moved in circular orbits - the circle being the "perfect" geometrical form. The dualism of the Middle Ages had separated, in a Platonic fashion, the imperfect and corruptible earth from the perfect and eternal heavens above.[55] Building on Kepler's insight and discoveries, Newton demonstrated that the same physics that applies to the earth also applies to the heavenly bodies above. His laws of mechanics united and comprehensively covered all of observable nature. Newton "demystified the heavens."[56]

Yet Newton also maintained a strong Platonic element in his thinking. Newton believed that the laws of nature presumably existed since the beginning of time. The laws of nature are permanent and stable. Early scientists, such as Newton, believed that the laws of the universe had been created by God and imposed upon nature from "above." Order was stamped upon the natural world of flux. The contemporary scientist and cosmologist Lee Smolin sees Newton as following Plato, in this regard, postulating eternal laws created and dictated by an eternal creator. What is eternal gives order to time.[57]

The central metaphor of the Newtonian view of the universe was the clock. Both Descartes and Newton replaced the earlier idea that nature was like a living organism – filled with spirits - with the idea that nature was a machine.[58] Nature was de-personalized and objectified. The mechanical universe had been set and ordered at the beginning of time by God and, like a clock or perfect machine ticking away at a regular and predictable rhythm, moved in a totally lawful way. Increasingly, during the Industrial Era, human society was modeled on the metaphors of the clock and, more generally, the smoothly running activities of a deterministic machine.[59]

Hence, although science brought with it a spirit of open inquiry and a rejection of authority, science also created a new system of belief that emphasized order, lawfulness, and integration in nature, as well as in human society. We will see that this same philosophical dualism of freedom and openness versus order and unity would permeate through other aspects of secular modernism as well.

The principles of science were not only applied to physics and astronomy, but all other dimensions of nature as well. Andreas Vesalius (1514 – 1564) and William Harvey (1578 – 1657) made significant scientific advances in the study of human anatomy and physiology, including Harvey's epochal discovery and explanation of the circulation of blood and the pumping action of the heart. The father of microbiology, Anton van Leeuwenhoek (1632 – 1723) with the newly developed microscope discovered a whole new realm of the very small – protozoa, bacteria, and many other microscopic forms of life - again demonstrating that our everyday view of reality was very limited.[60]

Although most early scientists believed in the Christian God, there was a growing sense that with the emergence of science a revolution in thinking was taking place. When the Royal Society of London was founded (1660 or 1662), its purpose was to defend and support the "new experimental philosophy." Newton would become President of the Society in 1703. According to Watson, by the time of Newton, a great shift in intellectual standards had occurred. Theology had been pushed out of its central position in academia and instead of providing the standard against which beliefs were judged, science had now become the new standard against which theology was judged.[61]

At this point, let us look more closely at the concept of scientific or natural laws, a pivotal idea in the new way of thinking and theory of order that emerged in science. The scientific concept of natural laws has great relevance to the view of time and the future that emerged in modern science.

The modern scientific concept of a law of nature derives from the ancient Greeks. A law of nature is a general pattern of change. To recall, for Heraclitus, change was pervasive through nature. Plato, following Heraclitus, believed that within the natural world all was in flux. For Plato, stability and permanency was only to be found in the eternal forms. To whatever degree there was order in the physical world of time, it derived from the order given to it from the eternal realm of forms. We have already seen how Newton applied this Platonic idea to his concept of laws of nature.

However, Heraclitus suggested that although everything in nature changed, it changed in accordance with a certain pattern, law, or rhythm - the "*Logos*," the logic of change. Aristotle in his concept of "formal cause" followed Heraclitus. For Aristotle every object had a "form" to how it changed. Also, according to Aristotle, there were general or abstract patterns of change in nature. If not interfered with, all acorns grow (change) into oak trees, human embryos develop into adult humans, and water runs downhill. The forms of change for Aristotle were teleological involving the actualization of potentials within things toward natural ends ("final causes"), but he did reject Plato's idea that order and form existed in a realm separate from physical nature. Further, Aristotle

300

rejected the separation of order and change. For Plato, order applied to what was eternal, static, and unchanging. Aristotle believed that there could be orderly change and this "form" of things resided in nature. Modern science would take up this idea of laws of change and make it central to the scientific description and explanation of nature.[62]

Within science, a law of nature is conceptualized as a regularity of change in nature. Although things constantly change and move about, things change in a predictable way. Modern science begins with the discovery of a variety of physical laws. For Galileo, when objects fall to earth they accelerate in velocity according to a general and universal formula. For Kepler, the planets moved in their orbits around the sun in accordance with three basic laws of planetary motion. According to Newton, the acceleration of an object upon impact is proportional to the ratio of force applied and the mass to be moved. For every action, there is an equal and opposite reaction. The Scientific Revolution found order within the changing world of nature and time.

What Newton kept from Plato was that this order in nature was imposed by an eternal reality, yet in the coming centuries science would move closer to the Aristotelian mindset, abandoning the notion of a separate realm of eternal order. The order in nature somehow directly derived from nature itself. This insight was critical in Darwin's formulation of his theory of evolution.

Because science adopted the concept of lawful change, all change in nature was presumably determined and predictable from natural laws. This is the concept of **lawful determinism**. The specific flow of events is determined by general natural laws. Connected with the idea of lawful determinism was the concept of **mechanistic causation**. Each individual event in nature is an effect, totally determined by specific antecedent causes. Combining the two ideas, it is the laws of nature that determine what particular effect will follow from what particular cause. Given a particular cause, you can predict the effect. As it is frequently stated, cause-effect relationships in nature are lawful.

The mechanistic notion of causality adopted within science is often contrasted with the teleological view of change. The theory of mechanistic causation implies that the past determines the present; the teleological view of change implies that some future event or intended future purpose determines the flow of events in the present. As we saw in the previous chapter, religious and mythic views of the future were frequently teleological – the flow of events into the future was guided or controlled by the purposes or intentions of deities. Science challenged the teleological view of change, and consequently the teleological view of the future. There is no "intended" future that sets the course for present events.

Since from the perspective of lawful determinism all change in nature is determined by natural laws, there is no chance, free will, or unpredictability. We may not know all of the laws, but, if we did, we could predict, in principle, everything that would happen to the end of time.[63] The following famous quote from Pierre Simon de Laplace (1749 – 1827) sums up this view:

"We may regard the present state of the universe as the effect of its past and the cause of its future. An intellect which at any given moment knew all the forces that animate matter and the mutual positions of the beings that compose it, if this intellect were vast enough to submit that data to analysis, could condense into a single formula the movement of the greatest bodies of the universe and that of the lightest atom; for such an intellect nothing would be uncertain; and the future just like the past would be present before its eyes."

This absolute determinism for all of nature espoused within early science conflicted with Christianity's belief that the individual souls had free will. Early scientists, such as Descartes, attempted to combine the determinism of science with the idea that humans possessed individual freedom. For Descartes, physical matter obeyed the lawful determinism revealed through natural science. On the other hand, Descartes believed that the human mind was non-physical and consequently free. Yet, for both many scientists and philosophers, Descartes's dualistic solution to the problem of free will in a deterministic world of nature had problems, for there is no clear way to understand how the mind – possessing free will and an immaterial existence – could influence a deterministic physical reality – the body.

In spite of the various important differences between early science and Western Christianity, described above, there were also notable connections and similarities. The belief that nature was lawful was based on the Christian belief that God had created a lawful and rational universe that obeyed a *Logos* determined by God.[64] Early scientists believed that they were discovering the laws set down by God. Kepler believed that he saw a connection between the form and dynamics of the solar system and the Holy Trinity.[65] As indicated by the quote at the beginning of this section, Newton believed that the beauty and orderliness of nature must be due to a supreme being. The idea of progress, as it evolved in the Age of Enlightenment, had its beginnings and its inspiration in Western Christian thought, in particular in St. Augustine's vision of the universal linear progress of humankind.

Enlightenment and the Theory of Secular Progress

"Thus we surpass all the times that have been before us; and it is highly probable that those that will succeed, will far surpass us."

John Edwards

"The negation of nature is the road to happiness."

John Locke

In this section I describe the growth of the theory of secular progress during the Age of Enlightenment. In particular I highlight the development of new social, political, and economic ideas during this period and connect these ideas to the emerging theory of secular progress. I also examine, continuing the discussion of the previous section, how the growing power of science increasingly challenged religious doctrine and authority and how this conflict played into the evolution of the theory of secular progress.

The historian Robert Nisbet asserts that progress is the most important idea ever developed in Western Civilization.[66] He identifies five basic premises behind the modern secular theory of progress: The value of the past, the superiority of the West, the worth of economic and technological growth, faith in reason and science, and the importance of life on earth. The philosophy of secular progress as it developed in the modern West assumed that progress is cumulative, building upon the accomplishments of the past; that the West should take a leadership position as the most modernized culture and society in the world; that economic and technological developments facilitate advances in all spheres of human reality, including morals, psychology, and social-political organization; that reason and science, over faith, revelation, and religious doctrine, are the preferred modes of inquiry and understanding to advance human society; and that worldly concerns are at least as important as other-worldly values.[67]

Nisbet though does not believe that the idea of progress begins with the modern era. Nor does he believe that the idea of progress has a secular origin. Rather, as we saw in the previous chapter, the idea of progress arises in a religious and specifically Christian mindset, notably within the writings of St. Augustine. (There were also clear indications of the idea in Greco-Roman thinking, such as in Lucretius, where the idea has a secular and naturalistic quality.) To recall, Christianity adopted a linear view of time, which it inherited from Judaism and ultimately Zoroastrianism. Further, early Christianity viewed linear time as progressive, leading to the Second Coming and the ascension of all deserving souls into heaven. The focus of this progressive view was more spiritual and otherworldly than secular, but even this is only a half-truth, for many Christians, including Augustine, believed that "Providence" was guiding humanity toward a better world in this natural reality as a prelude to a greater reward in the next.

In Nisbet's most general definition of progress, he presents a broad and abstract formulation that captures both the secular and the spiritual elements of the idea. He states that progress throughout Western history has meant a movement from the inferior to the superior. He notes that this belief in a progressive direction to history meant two things: first that human knowledge grows or advances across time, and second, that humans are moving forward along various dimensions of improvement, including moral and spiritual development and overall happiness. The general belief in progress implied a movement toward human perfection. Earlier religious writers on the idea of progress highlighted spiritual criteria of advancement, whereas later modern writers highlighted more secular criteria of improvement, but the general idea of improvement from what was inferior to what is better captures what is basic to all the different versions of the idea across Western history. As Nisbet sums it up, "...the idea of progress holds that mankind has advanced in the past – from some aboriginal condition of primitiveness, barbarism, or even nullity – is now advancing, and will continue to advance through the foreseeable future."

What happened in the period of 1600 to 1800 is that the Augustinian notion of progress became increasingly secularized and connected with science, rather than with *Scriptures* and Providence. Rather than the *Bible*, science became the means to progress. But this did not happen suddenly, and many writers combined religious ideas with scientific ideas in formulating their view of progress. Even early scientists such as Descartes, Kepler, and Newton did not immediately abandon religion in favor of science – they attempted to synthesize the two perspectives.

A notable example of early efforts to combine science and religion within a theory of progress is contained in the writings of Gottfried Wilhelm Leibniz (1646 - 1716). Leibniz was a philosophical and mathematical genius, unequaled in sheer intelligence and breadth of interests, co-inventor of calculus along with Isaac Newton, who influenced diverse areas of science and philosophy and anticipated the modern relativistic conception of space and time.[68] Leibniz proposed a theory of **universal progress**. He argued that following from the **"principle of plenitude"** (that everything that can be will be) the whole universe should show infinite progress as it moves into the future. In essence, following an Aristotelian line of thinking, progress was the actualization of the infinite potential of the universe. From within a Christian framework, Leibniz argued that the infinite progress of the universe is the realization of the perfection and beauty of God - the Creator of the universe.[69] For example, he states, "...To realize in its completeness the universal beauty and perfection of the works of God, we must recognize a certain perpetual and very free progress of the whole universe, such that it is always going forward to greater perfection."

Although the theory of progress developed by Leibniz invokes a supreme God and reflects his Christian philosophy, he does break free of the narrow vision of reality bequeathed from the Middle Ages. Leibniz was very aware of the advancing discoveries of science - of the potential vastness and intricacy of the universe beyond what was visible to the naked eye (he was particularly fascinated by the discoveries of Leeuwenhoek - and his expansive vision of the entire cosmos in a state of progressive development anticipates twentieth century evolutionary cosmology.

A notable example of an idea that combines science and religion in a theory of progress is **Puritan Millenarianism**. The Puritan Revolution of the seventeenth century greatly influenced many early scientists, including Isaac Newton, and the Puritans had a strong progressive philosophy of history. They believed that a divinely created universal law (Providence) was in operation in the history of humankind that eventually would lead to a Golden Age or "millennium." They also thought that the pursuit of human knowledge, especially scientific knowledge, would accelerate the arrival of the millennium. They saw a strong connection between scientific progress and spiritual progress.[70]

Yet the free spirit of critical inquiry that began in the seventeenth century would create increasing tension and difficulties between science and religion as we move into the eighteenth century. Descartes in fact had set the stage by doubting everything that could not be proven, and Bacon, in arguing that scientific truth depended on observation of facts, would likewise undermine the claims of religion. Both rationalism and empiricism as philosophies of knowledge evolved throughout the eighteenth century and as the implications of these two approaches to knowledge became clearer, secular philosophy increasingly became more at odds with religious belief.

The secular approach to the future not only derived its inspiration from science, but it also drew upon the whole history of rational and empirical philosophy stretching back to the ancient Greeks. As science took hold in modern Europe, secular and critical philosophy clearly emerged again as a pursuit separate from religion and theology. In Medieval Europe, Christian thinking dominated philosophy, for example within Christian Scholasticism, but the expression "rational enlightenment," often used to describe the philosophy of the Enlightenment, refers to the emancipation of both science and philosophy from religion and theology. Although secular philosophy did not directly involve the scientific method of experimentation, academic philosophers of the Enlightenment would reinforce the belief that the future and, more generally, all reality could be understood and predicted through rational and empirical methods, rather than through metaphysics, prophecy, and mystical revelation.

The impact of secular philosophy on the modern concept of progress is especially significant in the emergence of social and political philosophy. Niccolò Machiavelli (1469 – 1527) was one of the most famous early modern writers to apply secular thinking and philosophy to issues of politics and social order. In his well known book *The Prince* (1513) Machiavelli outlines rules and strategies for the maintenance of the political state which, in historical retrospect, have frequently been seen as ruthless, manipulative, and unethical in nature; Machiavelli is remembered for his expression "The ends justify the means." But Machiavelli saw himself as a realist who was simply attempting to describe how to most effectively govern and run a political state. For Machiavelli, politics should be based on a realistic understanding of the nature of humans and their behavior. Further, politics should not be subordinated to religion but stand as an independent discipline with its own principles and laws. Machiavelli is often seen as the starting point of modern political philosophy.[71]

A second major philosopher who did much to determine the future course of political philosophy was Thomas Hobbes (1588 - 1679) who is best known for his major philosophical work *Leviathan* (1651). Hobbes not only saw political philosophy and politics as independent of religion, he, in fact, believed that a strong secular government and political order was needed to counter-act the negative effects of religion. Civil law took precedence over religious doctrine. Hobbes was highly critical of the religious conflicts, atrocities, and fanaticism of his day, and he believed that a strong central government was needed to bring order to his turbulent world. Although Hobbes was critical of the destructive aspects of religion, his call for a strong central government to maintain order in the world was ultimately based on a rather pessimistic view of basic human nature. According to Hobbes, humans are inherently selfish and hedonistic and prone to war as a means to secure what they want at the expense of others. War is natural to humans, and ethics is reduced to human desire: What is good is simply what we desire or want, and what is evil is what we hate or avoid. Hence without a strong government to control the selfishness and violence in humans, all would be conflict and chaos, and in the final analysis many or most of us would not get what we want. A strong authoritarian central control protects us against each other.[72]

What is especially noteworthy in both Machiavelli and Hobbes, aside from their philosophical emancipation from religion, is their argument for strong centralized government and control. Both are often seen as having rather negative views of human nature, and their political philosophy, to a great degree, follows from their ideas on human psychology. In the evolution of social and political thinking on the nature of secular progress, two different central themes emerged: Progress was associated with both increasing social order and increasing freedom and individuality. These ideals (compared to Bloom's conformity versus diversity) not only seem contradictory, but would recurrently lead to war and conflict among nations in the centuries ahead.

On the other side of the philosophical continuum, the great British philosopher John Locke (1632 - 1704), one of the great inspirational starting points for the Enlightenment, emphasized individual human rights, freedom, and self-governance, and the right of the general population to determine the legitimacy of those who rule them. A key theme that emerged during the Enlightenment was a questioning of all kinds of authority, religious or secular. According to Locke, there are no "divine rights" that only the privileged few possess; rather, all men are equal and no one is above the law. People have the right to challenge the authority of their government when its leaders fail, through their actions, to serve the public good. Locke also strongly argued for toleration among diverse peoples and defended freedom of religion. He is well known for his defense of the natural rights for life, liberty, and property. Contrary to Hobbes, Locke believed more in the inherent goodness and rationality of people; rather than war, he argued that humans, by nature, use reason to solve or resolve problems and challenges. Because of this more optimistic view of humans, Locke argued for limited power in government claiming that people have the natural capacity and moral character to determine their own lives. Hence, whereas the Christian church of the Middle

Ages, following Augustine's view of the inherent evil nature of humans, emphasized strong central control, Locke is the first of the great liberal thinkers who took the opposite view that individuals have the right to control their own destiny.[73]

As Watson argues, the writers of the Enlightenment had a strong interest in understanding human nature and their views on progress and how to improve human society were based upon their ideas about human psychology and the human mind. Locke was highly influential in his development of a comprehensive theory and description of the human mind in his book *An Essay Concerning Human Understanding* (1689), considered the modern starting point for empiricist psychology and philosophy. It is noteworthy that Locke's *Essay* and many subsequent works delving into psychology, the mind, and human nature attempted to provide a scientifically informed picture of humanity independent of religion. Secular theories of political and social order were intimately connected with secular and scientific theories of human nature. According to Watson, pessimistic psychologies, such as in Hobbes and Machiavelli, provided the seeds of later conservative and authoritarian political philosophies, whereas more positive psychologies, such as in Locke, provided the foundation for the development of political liberalism.[74]

A second great philosopher and contemporary of Locke, and one who championed the importance of freedom in his political writings, was Baruch Spinoza (1632 - 1677). Along with Locke, he is also considered one of the major inspirational sources of the Enlightenment. Whereas Locke was an empiricist, Spinoza was a rationalist; in fact, in ways he was the supreme rationalist of modern times, believing that all of existence, including human nature and God, could be deduced and comprehended through reason. In totally rejecting the authority of traditional religion (including both Judaism and Christianity), Spinoza eschewed miracles, the supernatural, and the afterlife and argued in his great philosophical work, *The Ethics* (1677), that God and the universe were ultimately the same thing; Spinoza was a monist and a pantheist. Spinoza embraced the ideals of science and believed that all of nature could be understood; he modeled his *Ethics* on the principles of mathematical proof and deduction, and within *The Ethics* created both a comprehensive psychology and moral theory based on reason and a naturalistic perspective. Whereas philosophers and scientists alike, up to Spinoza's time, may have criticized certain features of traditional religious doctrine or religious practices, Spinoza attempts to completely break free of religious authority, substituting reason and science as the final arbiters of truth. For Spinoza, the search for knowledge must be a totally democratic process with no special interest group determining what is deemed acceptable.

According to Watson, Spinoza created the modern world. He integrated theology (a rationalist version), science, psychology, ethics, and politics into a coherent whole. In his *Tractatus Theologico-Politicus* (1670) he outlined his political views. According to Spinoza, and contrary to Hobbes, not only do humans possess a fundamental need to help each other but it is the function of government to help people to realize their potentials, in particular, their capacity for reason. Governments can not control people through fear. As with his views on religion and the quest for knowledge, Spinoza attacked tyranny and repression. For Spinoza, as was the case with Locke, freedom is a critical value in human society.[75]

Spinoza has been identified as the founder of modern Biblical criticism.[76] As we move into the eighteenth century, we find increasing skepticism towards the validity and morality of Christian doctrine and practices. As we have seen, secular writers in the sixteenth and seventeenth centuries attempted to emancipate both political philosophy and ethical theory from religion, increasingly turning toward science as a foundation for creating normative ideals and direction for humanity.

The philosopher and essayist who brought this secular re-orientation to its apex and culmination, and who is often identified as the father of the Enlightenment, was Voltaire (1694 - 1778). A great admirer of Newton, Voltaire argued that human society needs to be reconstructed based on science, reason, and observation. He attacked all forms of absolute authority and dogma, religious and secular, and defended various civil liberties including freedom of religion. He did not believe that God determined human destiny and came to totally reject religion as a structure that could provide beneficial guidance in life. A cynic, skeptic, and satirist who critiqued Leibniz's optimistic vision of the world and universal progress in his well known satire *Candide* (1759), Voltaire penned the famous line "If God did not exist, it would be necessary to invent him."[77]

Central to science and the new secular philosophy of the Enlightenment was a new theory of the nature and acquisition of knowledge. As we have seen, Descartes and Bacon, early on, articulated the principles of rationalism and empiricism as fundamental to the new theory of knowledge. Two philosophers of the Enlightenment are especially important in understanding the new epistemology, or theory of knowledge, as it further evolved in the eighteenth century. They are the Scottish philosopher David Hume (1711 - 1776) and the German philosopher Immanuel Kant (1724 - 1804). Hume carried the philosophy of empiricism to its logical and most dramatic conclusions, whereas Kant attempted to synthesize rationalism and empiricism into a consistent and coherent philosophical framework. Hume's two most noteworthy philosophical works are *A Treatise of Human Nature* (1739) and *An Inquiry Concerning Human Understanding* (1748). Kant's most famous philosophical work is the *Critique of Pure Reason* (1781). Hume is generally considered the greatest empiricist philosopher – in fact, even "the greatest philosopher to write in English," whereas Kant is regarded as the greatest German philosopher of all time. According to Kant, it was Hume who "woke him from his dogmatic slumber" and stimulated him to write the *Critique of Pure Reason* in response to the highly skeptical ideas and conclusions espoused by Hume. [78]

Let us begin with Hume. His basic philosophical starting point was that knowledge either derived from perceptual sense impressions or reason. If a statement or belief could not be supported through either sensory observation or reason, then the belief, according to Hume, did not constitute real knowledge. From this starting point, Hume demonstrated that beliefs such as the existence of an external world, cause - effect relationships, the existence of a self (Descartes' presumed indubitable starting point), and the existence of God or anything metaphysical, could not be definitively supported or proven either through reason or sense experience. These beliefs were simply habitual beliefs or thoughts and did not constitute real knowledge. It is clear that Hume's conclusions undercut any rational or empirical attempts to prove the existence of God, but they also undercut the idea that even science could prove anything. The postulation of "scientific laws" based on generalizations of facts cannot be proven for we can never be sure that the "law" will hold through subsequent observations. Although Hume's reasoning provided a basis for rejecting religious beliefs on the grounds that they could not be proven either rationally or empirically - which philosophers of the Enlightenment would embrace - Hume's ideas also revealed a real philosophical weakness in the presumed certainties of science. For Hume, beliefs about laws of nature are contingent. As we will see, Hume's skeptical critique of science would eventually have an impact on the evolution of the philosophy of the Enlightenment. Yet, in the enthusiasm of the first century of the Enlightenment, Hume's skepticism regarding science did not significantly undermine the secular agenda of the time.

Kant, on the other hand, aware of the skeptical conclusions of Hume, attempted to demonstrate that science did have a solid epistemological basis. Kant's response to Hume was to argue that there was a set of necessary "categories of human understanding" that science assumes in its investigation and conceptualization of nature (what Kant called "synthetic a priori" knowledge). These categories of understanding cannot be questioned since all human thinking, as well as all human experience, assumes these categories as its starting point. The categories of human understanding only apply to the world of sense experience, and if we attempt to apply them to what exists beyond sense experience (for example God), they generate antimonies or contradictions. Hence, there is no rational or empirical way to demonstrate the existence of God, because God lies beyond the realm of meaningful and intelligible human experience. Instead, for Kant, God belongs to the realm of faith. By setting boundaries to the limits of science, Kant made room for the importance of faith.

Kant's conclusion that science contains certain empirical knowledge because it assumes unquestionable categories or concepts in its formulation has not stood the test of time. The fundamental concepts of science have changed since the time of Kant. It does not appear that the human mind is somehow welded to a set of unchangeable categories of thought. Yet Kant's conclusion that science or any form of human understanding presupposes some set of concepts in making sense of the world has become a highly influential idea in modern intellectual history. The problem is that this conclusion opens the door to **subjectivism** - all humans understand reality through conceptual categories that filter and organize experience, hence objective knowledge is impossible. All human knowledge is from a conceptual point of view.

The skeptical conclusions and implications connected with the philosophies of Hume and Kant had their most immediate impact on the legitimacy of religious, mythological, and metaphysical belief systems. Although many illustrious philosophers, theologians, and religious figures throughout history had attempted to prove the existence of God and embraced all kinds of metaphysical beliefs, such as the existence of angels, demons, heaven, hell, and higher spiritual realms, Hume and Kant philosophically demolished the rational credibility of these beliefs and arguments. Consequently, the way was opened for a total rejection of religious and metaphysical ideas in formulating a vision of reality and the future.

Science and secular philosophy broke free of religion and religious notions of progress in the eighteenth century. The great French economist Jacques Turgot (1727 - 1781) provided the first clear expression of a purely secular concept of progress. Turgot identified the ultimate objectives or ideals of progress as knowledge, freedom, and economic growth.[79] Emphasizing these three central goals of progress is noteworthy because they are all secular in nature. Turgot does not include anything spiritual on his list of the fundamental objectives of progress.

We have already seen that knowledge - in particular scientific knowledge - was strongly connected with the idea of progress in Bacon and Descartes, among others. Even the Puritans connected the advancement of knowledge and science with progress. Yet early science exclusively dealt with the physical world rather than the spiritual realm, hence the advancement of scientific knowledge could be described in entirely secular and non-spiritual terms.

The importance of freedom as a second central ideal reflects the spirit of individualism in modern Europe in the eighteenth century and is indicative of the growing opposition to authority, political or religious. We have already noted how freedom was a key political and social ideal in the philosophies of Locke, Spinoza, and Voltaire and how the rise of individualism in the West actually goes back to the High Middle Ages and the Renaissance. By the end of the eighteenth century freedom and liberty would become the battle cries of the American and French Revolutions, and individualism, freedom, and democracy would go on to become central ideals or values of modernism. But individualism and freedom are fundamentally secular ideals, and in fact, as we have seen, have often been at odds with the authority of organized religion.

The third ideal - economic growth - is decidedly materialistic. To suggest material advance as a fundamental criterion of progress is clearly aligned with the materialist mindset of science rather than with the spiritual mindset of religion - it is secular rather than otherworldly. Just as freedom was a central theme in many theories of secular progress, economic and materialist advance was another key goal identified in secular theories of progress.

Turgot would have a significant influence on the father of modern economic theory, Adam Smith. Smith drew a strong connection between freedom and economic growth in his theory of progress. Smith identified freedom as the means to economic advancement, just as Bacon and others had identified scientific knowledge as a means to the material improvement of the human condition.

It is noteworthy that Turgot also formulated a general historical theory of progressive change that was fundamentally secular in nature, and connected his vision of the future with his theory of the past. He argued that historical progress was cyclical, involving alternating periods of barbarism (chaos) and rationality (order) - progress was not a steady linear flow upward. Each cycle of chaos and order brought humanity further along to a higher stage of social and economic development. In addition, each cycle moved humanity more toward individual freedom and away from centralized, authoritarian control. In general, for Turgot, history advanced in stages (an idea we saw in Augustine), and he combines in his theory of change, both progressive and cyclical views of time.

Hence, although reason and science, as well as cumulative growth, were emphasized in many early visions of secular progress, theories of historical change during this period did not always see progress as smooth, steady, or even peaceful. Turgot presents a view of oscillating order and chaos, reminiscent of Babylonian mythology and the philosophy of Empedocles. His view is also suggestive of the Zoroastrian - Christian idea of the war of good and evil, except now good and evil are interpreted in secular terms. In the previous century, the great Italian philosopher and historian Giambattista Vico (1668 - 1744) formulated a grand cyclic theory of progress that described alternating periods of growth and decay, of order and chaos in human history. For Vico, the ongoing conflict was between the primitive impulses of individuals and the progressive realization of a harmonious and civilized social order among humans. Although Vico mixes Biblical and secular-naturalistic ideas in his history, he does see an overall progressive evolution of humanity, civilization, and even religious doctrine and practices.[80] Another example of cyclical order and chaos was presented by The French bishop and scholar Jacques-Bénigne Boussuet (1627 - 1704) who, in his *Universal History* (1681), described history in terms of the rise and fall of empires. As Nisbet points out, the ideas of cyclic change, stages of development, violence and conflict, and both cumulative and revolutionary change were often important and central concepts in theories of progress during the seventeenth and eighteenth centuries.[81] As we will see, all these ideas and themes would continue to influence the evolution of the philosophy of secular progress in coming centuries.

Another point to highlight regarding these early theories of progress during the Enlightenment is that coincident with their development, historiography was also taking on a decidedly secular flavor as well. Generally, the study of history had been dominated by Christian visions of the past, but in the eighteenth century, as philosophy and theories of progress became more secular, scientific, and naturalistic, history also broke free of religious influence. This is important because secular theories of progress and the future, such as in Turgot and Voltaire, grounded their ideals and visions in theories and interpretations of the saga of the past. Secular theories of progress required secular theories of the past. Not only did secular views of history break free of religion, but in the case of Edward Gibbon's (1737 - 1794) highly influential *The History of the Decline and Fall of the Roman Empire* (1776, 1788), religion, and in particular Christianity, is seen as interfering with human progress.[82]

If Turgot and others separated progress from the spiritual and religious, Marques de Condorcet (1743 - 1794), in his classic philosophical statement of the Enlightenment *Sketch for a Historical Picture of the Progress of the Human Mind* (1795), clearly set science and progress in opposition to religion.[83] Wendell Bell refers to Condorcet as 'the first futurist." Condorcet was a highly influential figure in the evolution and articulation of the idea of secular progress. Echoing the individualist theme we have been following throughout European history since the High Middle Ages, Condorcet declared himself to be the adversary of all forms of tyranny, which in his mind included royalty, nobility, political monarchies, and the priesthood. Religion, he believed, was based on superstition and an ignorance of nature and thus was the enemy of progress which, for Condorcet, meant increasing freedom. Similar to Bacon, he envisioned the ideal society of the future as one ruled by science and reason[84] and it was only through science and reason that humanity could be liberated from the closed-minded tyranny of religion. Progress having first emerged as an influential idea within Western Christianity, had now not only emancipated itself from religion, but literally turned against religion in the writings of Condorcet.

Condorcet saw no limit to the perfectibility of humanity. Developing a secular theory of history to support his secular theory of the future, Condorcet argued that there had been ten stages thus far in human history, with the French Revolution ushering in the beginning of the newest and potentially most "glorious" period in human advancement.[85] Condorcet's concept of progress, though highlighting the importance of science, reason, and liberty, was broad in scope. He hoped and expected that there would be improvements in the future in the arts, morality, human intelligence, physical health and abilities, and of course science. It is noteworthy that he includes morality in this list for morality in the past had been strongly associated with religion. Yet, given the continued criticisms that had been raised against the church regarding its own moral behavior, it is understandable that Condorcet would see religion not as the foundation of morality but perhaps the reverse. Religion had led to immorality, including war, persecution, corruption, greed, and the suppression of the rights of human beings. Hence, it is important to see that religion and myth, though once having provided a basis and justification for morals as well as for social justice, were now rejected as legitimate and valid foundations for morality. The secular approach and, in particular, the use of reason, according to Condorcet, would provide a new and better basis for providing moral direction in the future.

If what was good, that is Christian religion, had now become bad, the reverse that what was bad had now become good, perhaps had also occurred. This is the argument of Dinesh D'Souza in his book *The Virtue of Prosperity*.[86] According to D'Souza, a new set of fundamental human values emerged during the Enlightenment. He notes that during the period of the fourteenth through the seventeenth centuries, Europe was ravaged by war and chaos in the name of religion and that the people of Europe suffered from both a scarcity of food and resources, as well as disease. In the minds of the architects of the Enlightenment, D'Souza states, the idea began to emerge that there must be a better way to live than under the dominion and influence of religion. The quest for virtue and the perpetual wars of good versus evil had produced violence and immense human suffering. Hence, the thinkers of the Enlightenment, believing it was time for a change, substituted material and commercial gain and self-interest for religious virtue as the central goals of society. (We should note that this change was beginning as early as the Italian Renaissance.) This was a significant shift, both ideologically and socially. Not only during the Middle Ages, but as far back as the ancient Greeks, economic trade and technological development had been seen as inferior to the pursuit of virtue. But we have seen that throughout the Middle Ages, and accelerating with the rise of modernism, commerce and technological innovation became increasingly powerful and central as guiding forces in human society. D'Souza states that as early as the sixteenth century, Machiavelli had abandoned the goal of virtue in political and social affairs, replacing it with self-interest and personal power. The classic case, which we will come to momentarily, is Adam Smith, who founded his whole economic theory on the free pursuit of self-interest. Following D'Souza's interpretation, the founders of the United States took the ideas of Smith, as well as Bacon and Locke, and created the first true secular society based on **"enlightened self-interest,"** establishing a clear separation of the power of religion from the operations of the state. Therefore, what had once been considered vices from a Christian and spiritual standpoint - namely self-interest and material wealth - had been turned into the central values of a modern society.

The themes of freedom, self-interest and self-determination, economic and material advance, and power over nature through the application of science, expressed in the writings of the philosophers of the Enlightenment, are intimately connected with social, political, technological, and commercial changes that were occurring in Western Europe during this period. Western Europe had become the new commercial hub of global exchange. Industrial production in Europe was steadily rising and would eventually surpass China and Asia in the early nineteenth century. With the development of the modern steam engine by James Watt in the 1760's, industrial energy production skyrocketed. Agricultural production also dramatically improved in the eighteenth century. Factories, with highly organized and efficient systems of manufacturing, sprouted and grew throughout Western Europe.[87] It appeared to many observers of the time that humanity was gaining control over nature through science, technology, and industry. The competitive and individualist philosophy and practice of capitalism increasingly drove economic and commercial development. Mass education and literacy increased which undermined the authority of political tyrannies and religion. Educated people in Western Europe learned about science and its principles of empirical and rational inquiry. There were political revolutions and changes as well, in the name of liberty, human equality, and democracy. Overall, not only were there significant changes in beliefs and philosophy, there were significant and resonant changes in ways of life as well.[88]

In his writings, Adam Smith (1723 - 1790) captured and crystallized many of these social-economic trends. He articulated the central economic theory that would explain and justify the growth of the modern secular society, and he tied this economic theory to the concept of progress. In his highly influential book *The Wealth of Nations* (1776) Smith argued that if individuals were allowed to freely pursue their own self-interest through the creation of products and services that they could sell for profit and monetary gain, the overall effect would benefit the public good. In his mind, competition among producers for the sale of their products to the public would cause steady improvement in the products, as well as control the prices of such products. This epoch-making formulation of the philosophy of **capitalism** was founded upon a clear connection in Smith's mind between freedom, material gain, and the idea of progress.

Smith believed that there was a natural progressive movement, revealed within history, toward the advancement of society and the growth of wealth. This "law of progress" was not interpreted within a religious framework, but rather within a secular and naturalistic framework. He described this general process as the "natural progress of opulence" which leads to increasing happiness for everyone. Smith believed that if individuals were given the "natural liberty," which for him meant "economic freedom", to pursue their own goals and self-interests they would benefit, - through the "invisible hand" of free competition, - the overall public good of society. For Smith, increasing human freedom facilitated the natural social and economic progress within history.[89]

For Smith, as well as many other advocates of secular progress, economic, industrial, and technological development would produce social, political, and moral advance. Hence, both individual and social virtue would be served. Secular progress would cure all social ills, such as crime, disease, poverty, and mental disorders. Progress was the royal road to human happiness.

Not only did the philosophy of capitalism provide an economic and materialist justification for secular modernism, it also provided a clear alternative to the idea of divine Providence as the cause of progress in human history. From Smith's perspective, the operation of capitalism, which would involve competition among producers, leads to progress without some divine hand guiding the process. As the idea of secular progress evolved in the eighteenth and nineteenth centuries, various philosophers, social theorists, and scientists, began to formulate naturalistic explanations for the mechanism of progress. Instead of turning to the teleological explanations of progress provided by such religious thinkers as St. Augustine, they developed scientific explanations and used these secular ideas to support their social and political philosophies of how to direct and guide the future. Progress not only had secular goals, but secular causes.

Although Smith is known for emphasizing economic self-interest and economic competition in his theory of progress, he saw his philosophy as having a moral focus. In place of the tyranny of business monopolies controlling the economy, he argued that the general population, through consumer judgment and demand, should control economic development – economic growth should be founded on a democratic process. Further, Smith's ideas reflected the emerging popular view that the modern commercial society was a new stage in human progress – a positive advance over previous human societies. Wealth was not inherently evil, but was built on productivity and exchange, that is, hard work and labor, and mutually beneficial reciprocities. Finally, Smith believed that social justice was critical to modern society, and supported some degree of government intervention to insure that the benefits of economic growth were distributed throughout society. As both Watson and Nisbet argue, Smith never separated economics from social and ethical concerns and values.[90]

As D'Souza argues, the ideas of Smith, as well as those of Locke and Bacon, were critical in the creation of the United States of America. D'Souza contends that the United States was the first true secular society. Watson, in a similar vein, sees the "invention of America" as the concrete realization of the principles of the Enlightenment. Although colonial America had a much more pragmatic bent than Europe, the leaders of the American Revolution supported and adopted many of the main philosophical principles of the Enlightenment.

Thomas Paine (1737 - 1809), in his famous book *Common Sense* (1776), which did much to ignite the American Revolution, strongly argued for the right to rebellion in order to realize freedom and independence from English domination. Paine believed in the ideals of progress and human improvement, strongly defended human rights (he opposed human slavery), argued for such contemporary ideas as free public education and minimum wages, and is generally considered one of the modern founders of political liberalism. He attacked all forms of organized religion and rejected monarchial government. A Deist, he wrote that "My own mind is my own church." Benjamin Franklin (1706 - 1790) and Thomas Jefferson (1743 - 1826), two of the central architects of the American Constitution and Declaration of Independence, worked into these pivotal documents as central values the principles of freedom, equality, and prosperity. Both Franklin and Jefferson were very interested and active in science and critical of traditional religion (though Franklin did have some mixed thoughts on this point). Jefferson is particularly noted for his strong support of separation of church and state, though he believed, as had Locke before him, that fundamental and unalienable human rights were derived from God. As Watson recounts, Jefferson was a great defender of the United States and was very optimistic about his country's future. According to Watson, America surpassed Europe in political development for, in spite of the European Enlightenment proclaiming the values of human freedom and individualism, eighteenth-century Europe was controlled by authoritarian monarchies. In America, there were no established dominance hierarchies to overcome, and thus democracy grew and flourished with greater ease. Unprecedented new freedoms emerged in the United States, especially as declared in the Bill of Rights, which would be a major inspirational source for the French in the creation of their Declaration of the Rights of Man. All in all, the American people, united in a spirit of common destiny, had a powerful sense of a promising and better future. The United

States was the great experiment of the Enlightenment and the philosophy of secular progress.[91]

In coming to the end of my review of the eighteenth century, it seems clear that the philosophy of secular progress, at least in the minds of many of its advocates, had severed any remaining ties with religious thinking. Although there were still some nineteenth century philosophers and social theorists such as Hegel, Herder, and Lessing in Germany, who showed a strong Christian influence in their visions of progress, a relatively autonomous and comprehensive secular philosophy of human life, progress, and the future had emerged in human thinking in modern Western Europe by the year 1800.

In summary, in numerous ways secular modernism challenged the authority of the religious view of life. One main difference between the secular view of the future and earlier religious views was that secular modernism saw the future as something that could be understood and controlled through reason, science, modern economic practices, and industry. As noted earlier, religious and mythological views often saw the future as something revealed and often under the control of supernatural or spiritual powers.[92] Secular modernism empowered humanity, conveying the message that, rather than follow the dictates of authority or tradition, individuals should pursue freedom; at last people were the architects of their own destiny.

Concerning the issue of truth, science and the philosophy of the Enlightenment challenged the validity of religious, mystical, mythological, and magical approaches and beliefs. The foregone conclusions and certainties of religious revelations and prophecies regarding the future were questioned, rejected and replaced by scientific laws and principles and rational and empirical methods.

The biologist Kenneth Miller states that the conflict between science and religion as it emerged in modern times was framed in extremist and absolutist terms. According to Miller, science presented itself as offering a complete explanation of reality – in materialist terms – thus excluding any need for divine or spiritual forces. The question of whether science can provide a complete explanation of the universe is still being debated today, but as Miller quite readily admits, every time some critic of science states that science will never be able to explain some feature or dimension of reality, history invariably proves the critic wrong. The growth of science over the last few centuries has been a steady and tenacious drive to turn the inexplicable into the explicable.[93] Further, as Galileo first realized and suffered for in the end, scientific ideas frequently contradict the views espoused in religious explanations of reality. Whether science, in the final analysis, will be able to explain everything is a question yet to be answered, but clearly religious explanations have suffered repeated defeats and contradictions at the hands of science. This historical pattern of contradiction and retreat began with the Scientific Revolution and continued into the Enlightenment.

Another major point where secular modernism challenged religion was on the issue of values. Values identify the ideal or preferred direction for the future. Secular and scientific thinking brought with it the view that values and ideals could be arrived at through reason and human dialogue, rather than through divine authority and revelation. If we trace the history of ideal visions of the future in the modern West, there is a definite shift in focus from religious justifications of ethics and morality to rational, materialist, and democratic justifications.[94] Humanity no longer followed a script or set of values created by the gods; rather humanity through reason, science, debate, and discussion became the creators of the script and the ideals for the future. This is the view espoused by Voltaire and Condorcet, among others. We have also seen that in the minds of economic theorists such as Adam Smith, free enterprise and economic development would create a happier, more ethically advanced world than would a social system of religious authoritarian control.

The concept of secular progress entailed defining growth or improvement in the human condition in terms of values derived from empirical or rational criteria, rather than religious or spiritual sources. Secular ideals of progress included material wealth and improved living conditions, the control of nature, the advancement of scientific knowledge, the evolution of technology, economic freedom, democratic rights and participation in government, and greater opportunities for education. Many of these values became central ideals in the creation and development of American politics and social philosophy and the American way of life (though it should be noted that at least for some of the founding fathers some of these values were justified in terms of beliefs about God).

As one final important shift, science and secular modernism provided a new story for humanity, overturning that found in the *Bible*. Stories by definition have temporal extent; they relate a series of events that occur over time. Stories will also usually have a direction in the form of a plot and a climatic resolution. Western religion provided both an explanation and narrative of the origin and development of humanity and the world and a set of visions and predictions of the future grounded in its historical narrative – it connected past and future usually in the form of mythic drama. Science, beginning in the eighteenth century, began to piece together a new story of our creation and the evolution of the world. (More will be said on this in subsequent sections.) Secular histories began to appear. This new set of stories had a different plot and identified natural forces, rather than supernatural or spiritual ones, in explaining human history. Enlightenment philosophy, inspired by the promises of science and technology, presented a new secular vision of the possibilities of tomorrow built upon these secular histories. Further, Enlightenment philosophy identified a set of secular values that gave history a progressive direction. Aside from Turgot and Condorcet, Bernard de Fontenelle (1657 – 1757), in his *A Digression on the Ancients and Moderns* (1688), identified the cumulative growth of knowledge as a fundamental trend within human history; Voltaire wrote a general history, *Essay on Customs* (1756), which highlighted the improvement and "enlightenment" of the human mind; and William Godwin (1756 – 1836), husband of Mary Wollstonecraft and father of

Mary Shelley, in his *Enquiry Concerning Political Justice and its Influence on Morals and Happiness* (1793) strongly argued that individuality and the growth of freedom was the key dimension of historical progress and predicted its inevitable further advance into the future.[95]

Central to the new story of humanity was the idea of natural progress. As we have seen Smith believed that progress in human history occurred through natural forces without the need for divine guidance or intervention. The idea that there was a **natural law of progress** was pursued and investigated into the nineteenth century, notably in the writings of social thinkers like Auguste Comte (1798 – 1857) and Herbert Spencer (1820 – 1903). Both believed that there was an inherent tendency in nature towards progress.[96] For both Spencer and Comte the law of progress was inseparable from the linear flow of time, yet whereas Spencer saw natural progress moving toward increasing freedom and individuation, Comte saw something very different.

Herbert Spencer argued for a philosophy of extreme liberal individualism. Instead of any type of authoritarian or centralized control on the behavior of individuals, Spencer believed that social order should arise through voluntary cooperation rather than government coercion. Spencer was the supreme advocate of the Enlightenment philosophy of freedom and individualism. Spencer connected his social philosophy to a general cosmological principle, which he referred to as the universal "developmental hypothesis." In some important ways anticipating Darwin and contemporary evolutionary theory, (Spencer coined the expression "survival of the fittest"), Spencer argued that the universe as a whole moves from the homogeneous to the heterogeneous – from lack of form to increasing differentiation. At the human level, he saw a general trend from the static, authoritarian, and monolithic to the diverse, pluralistic, and individualistic. In the spirit of Heraclitus, everything for Spencer was process and motion. Unequivocally supporting the idea of progress, Spencer argued that both natural and social change were directional

327

and developmental. There is a "beneficent necessity" that inexorably moves the cosmos, humankind included, toward greater individuality, freedom, and diversity. If chaos is identified with lack of structure and form, then for Spencer the "developmental hypothesis" implies that the general direction of time is increasing order arising out of chaos - but order in the sense of structure and differentiation rather than conformity and uniformity. Progress is also moral, for according to Spencer, evil is due to some type of deficiency in humans or nature, and as the universe evolves, all evil will disappear - what Spencer refers to as the "evanescence of evil."

Auguste Comte, though emphasizing the growth of order in his theory of progress, has a diametrically opposed interpretation of progress and the nature of order. Comte is remembered for creating the discipline of "sociology", a term that he coined. As Newton had developed a scientific explanation of the physical world that empowered humanity, through technology and industry, to manipulate and control physical matter, it should be possible, according to Comte, to develop a **"social physics"** that would describe the laws of human society and empower humanity to shape and direct the social world. In essence, Comte was applying the logic of science to human society and following through on the argument of the Enlightenment that science could be used to improve the human condition. As Watson notes, Comte, among other pioneering social scientists in the nineteenth century, wanted to explain the growth of modern society (a scientific explanation of social history) and apply this knowledge to politics; it was not enough to describe and explain progress - this knowledge must be used to create a better world.[97] Interestingly, Comte believed that nineteenth century Europe was in a state of "spiritual anarchy" with the church having lost its control and authority - an apparent confirmation of the victory of secularism over theism but with negative consequences. Further, Comte believed that individualism had become the "disease" of the West - another confirmation of the success of the Enlightenment, but again interpreted as a

328

negative result. Whereas the architects of the Enlightenment, such as Condorcet and Smith, had strongly connected progress with freedom and individualism, Comte took the opposite stance and connected progress with increasing order – an order of regularity, connection, and organization. (For Newton, order in nature meant regularity and uniformity.) In Comte's mind what Europe needed was more stability and order, not more freedom and individuality. In Comte's mind the natural law of progress produced increasing organization in the world rather than anarchistic individualism.

The contemporary futurist Virginia Postrel defines a "technocracy" as a rationally controlled and managed society, based on the idea that the behavior of humans and social organizations can be scientifically predicted, and hence directed toward some focused set of goals in the future.[98] This clearly appears to be Comte's ideal and vision of progress. According to Nisbet, there were numerous other nineteenth century social thinkers, such as Rousseau and Saint-Simon, who believed that under the banner and justification of progress, human society should be controlled, directed, and organized.[99] Claude Saint-Simon (1760 – 1825) argued in a manner resonant with the philosophy of Francis Bacon that human society should be organized along scientific principles. A great believer in future progress, Saint-Simon contended that "The golden age is not behind us, but in front of us. It is the perfection of social order."[100] For all these writers, progress was connected with increasing social order, in the sense of organized coordination and uniformity, and this progressive order needed to be imposed on human society - a seemingly opposite message to the original ideal of freedom in the Enlightenment.

As I introduced earlier, the philosophy of the Enlightenment actually bequeathed to posterity two apparently contradictory ideas regarding the nature of secular progress. On one hand, freedom and individualism were central values of the Enlightenment – a consequence of the centuries old battle against the perceived repressive authority of both the church and royalty. Yet the Enlightenment also embraced science as the road to truth and early science described nature as a deterministic and orderly reality, subject to laws that governed its behavior. (Note the parallel between Western religion and Western science – in the former case God ruled the heavens and the earth, in the latter case, natural laws ruled the universe.) If science is applied to the world of humanity, the implication is that there are discoverable laws that describe human behavior and that an understanding of these laws would empower humans to control human affairs just as humans had learned to control the processes of nature. Although it might seem paradoxical, Condorcet, the great defender of individualism and freedom, believed that the future of human society was scientifically predictable.

Further, the Enlightenment embraced reason as the appropriate method for discovering the truth, but rationality, as for example practiced in mathematics and logic, yields singular truths rather than many different truths or perspectives. The points of views of different individuals ultimately are insignificant – what matters are the singular and unequivocal truths revealed through reason and science. Hence, in this sense, reason and science are tyrants – there is only one correct view of reality. As Christian points out, one of the central goals of science has been to discover abstract truths that are universal and independent of cultural or individual bias and point of view.[101]

This unity of opposites – of universal order and individual freedom – clearly shows itself in the growth of modern nations in the eighteenth and nineteenth centuries. Although the tyranny of despotic royalty and religious authority was presumably challenged and overthrown in the democratic and scientific revolutions connected with the Enlightenment, the long term result has been that modernized governments regulate and control individual human behavior much more so than in the past. Citizens of modern society are monitored, policed, subjected to legal rules and regulations, and obligated to participate in many government controlled practices such as mandatory taxes and education. Many of these new forms of surveillance and control have been implemented, presumably to protect the rights of citizens and ensure for the public welfare, but the overall effect has been heightened national power and regulation over citizens. People of the Middle Ages were neither watched nor controlled any where near as much as in modern times. Yet, modern nations also support, to various degrees, a host of individual human rights, individual participation and democratic input in government, freedom of religion, and freedom to pursue personal and economic goals.[102] Although there are clearly cases in modern times where excessive government control has negated human freedom, such as in Nazi Germany and the Soviet Union, the overall trajectory of modern history seems to have been a co-evolution of both government regulation and individual freedom and rights.

Immanuel Kant, who believed in historical and moral progress, did not see a paradox in the evolution of human freedom and the growth of government and social order. Kant viewed individual human beings as autonomous, rational, and free, and he saw progress as the advancement of freedom and reason. He did see, however, a fundamental clash between individualism and selfishness and the need for social community and human communion. Yet according to Kant, humans, who possess an "unsocial sociability", can find ways to advance the cause of freedom within the bounds of social order. Humans can be motivated to enter into social cooperation and collaboration if their individual needs are satisfied in the process. For Kant, the overall purpose of human social advancement is to create laws and institutions which will maximize individual power and freedom. So instead of seeing government and social organization as suppressing human freedom, as for example in authoritarian and centralized regimes of control, Kant saw the ideal social organization as serving the needs of both the individual and the community.[103]

The ongoing *Yin* and *Yang* of social order and individual freedom, of unity and diversity, of the whole and the parts, has been a central issue in the development of human society.

Bloom highlights this conflict in his theory of historical change. For writers like Kant, the belief was that these two forces could be synthesized and made mutually compatible. Other theorists and philosophers emphasized one factor over the other, seeing an inherent contradiction between these two dimensions of human life. This difference of opinion is itself a *Yin-Yang*; the idea that social order and individuality are incompatible versus the idea that social order and individuality are reciprocal realities. We can see in the philosophy of the Enlightenment these different points of view, and consequently a variety of interpretations of what constitutes progress in human history, as well as what direction to take in the future.

The issue of social order versus individual freedom serves as a good starting point for considering another idea that became very influential during the eighteenth and nineteenth centuries. This idea was utopianism. The term "utopia" literally means "no place" or "not a place." Sir Thomas More (1478 - 1535), a humanist thinker of the Renaissance, first used the term in his fictional book *Utopia* (1516), which described an imaginary society in which everything was morally perfect and harmonious; "utopia" was a well ordered society. More's intent was not to imagine some perfect society of the future, but rather to satirically critique the customs and practices of his own time.[104] Yet More's utopia, although peaceful and cooperative, was also static and boring. On the other hand, toward the end of the sixteenth century Francis Bacon wrote *The New Atlantis* in which he imagined an ideal society built on the principles of science.[105] In Bacon's ideal society, as well as in Saint-Simon's utopian vision two centuries later, scientists rule society, producing both social order and continued scientific advance.

Saint-Simon was only one among many modern thinkers who created utopian visions. With the coming of modernism and the Age of Enlightenment, many writers began to envision ideal societies that presumably could be realized in the future through the application of the principles of science, reason, and secular values. The optimism of the Age of Enlightenment led many people to believe that humankind could create ideal or, at least, much better societies in the future. These imaginary ideal or perfect societies of the future were usually referred to as "utopian." Utopias were projections and predictions of ideal societies as imagined through someone's eyes, but they were also proposals and calls to action – they were intended as seeds of revolution and reform.

According to some writers, utopian visions went through a significant evolutionary development from their earliest expressions, such as in More's *Utopia,* to later formulations such as in Condorcet's *Sketch for a Historical Picture of the Progress of the Human Mind.* Initially these ideal human societies were simply imagined as hypothetically existing in some other place than the society and world in which the writer lived. What the Scientific Revolution and Age of Enlightenment brought into the picture was the view that these ideal societies could be seen as existing potentially in the future. The shift in focus was from "another place" to "another time."[106] Throughout human history there have been stories and fables of ideal societies or worlds that existed in the past (the myth of the Golden Age), but with the coming of modern times the ideal societies were now imagined in the future.

The futurist Warren Wagar offers a somewhat different but compatible assessment, arguing that pre-modern utopias were static and a-historical, whereas modern utopian theories were dynamic and historical, describing how humanity would progress in time to achieve a more ideal society. Modern utopian visions provided future histories.[107]

Though there is some element of truth in these generalizations, the first view ignores the historical fact that pre-modern religious thinking did contain stories and predictions of more ideal societies in the future. St. Augustine clearly believed in the future advancement of humanity on the earth, and throughout the Middle Ages there were many advocates and followers of the vision of millennialism – that an ideal human reality, lasting a thousand years, would be achieved on earth with the second coming of Christ. Wagar's generalization is also limited because Augustinian and millennial thinking, in fact, did describe a process of moving from present times to envisioned ideal states. While these historical or dynamical processes did involve supernatural and spiritual forces, still the ideal worlds envisioned were described as a result of a developmental process.

Wendell Bell in his *Foundations of Future Studies Vol. II* provides a historical review of the evolution of utopian thought.[108] Bell, like Wagar, sees a real value in examining utopian images of ideal societies. As Wagar states it, the study and consideration of utopian thought is "normative future studies". Utopian thought assumes some set of prescriptive values that the utopian writer thinks should be followed and realized in the future. Utopias are normative or prescriptive visions. As Bell sees it, in examining different utopian theories, we are able to see how different value systems, which by definition are normative and prescriptive, could hypothetically be realized in human society. Utopias are thought experiments of the ideal.

Other writers see utopianism as counter-productive, archaic, and dangerous. Since More seemed to imply by the use of the word that such a society did not and perhaps could not exist, the ideal of social and human perfection is perhaps unrealistic. Leszek Kolakowski has stated that "Utopia is a disparate desire to attain absolute perfection; this desire is a degraded remnant of the religious legacy in nonreligious minds."[109] If utopias aspire to perfection, such ideal states are impossible, for human reality is fluid rather than static. Augustine could imagine an ideal perfect world because he saw the temporal world eventually coming to an end in the ultimate fulfillment of God's plan. He believed that there was a perfect moral order determined by God. But can humans ever achieve perfection? Can perfection, without recourse to some absolute authority such as God, even be defined?

Since a utopian vision is an ideal, as well as a call for action, a central question throughout history has been how to realize the prescribed ideals of utopian visions. Writers of the Enlightenment called for a change in thinking, from being superstitious and irrational to becoming more scientific and rational. They also called for equality and human freedom. Yet how does one realize the ideal of freedom? Given the perceived authoritarian and repressive rule of royalty and the church at the beginning of the modern era, the only way to achieve this goal was through open rebellion and revolution in order to overturn authoritarian regimes, which is what occurred repeatedly in Europe and America beginning in the late eighteenth century.

Nisbet argues that there are two central themes within modern theories of progress (what I have noted as the two apparent contradictory messages of the Enlightenment): increasing freedom on one hand and increasing power and order on the other.[110] Further, Nisbet notes that if social perfection, however defined, was the stated goal of a progressive or utopian image, then within human history perfection has often been sought through violent and revolutionary means. People have fought to free themselves from oppression, but people have also fought to bring order and control to a "chaotic" situation. Again, progress both in theory and practice, has not turned out to be a peaceful and steady advance. It has been "punctuated" by revolution, upheaval, and violent transformation.

In the writings of Comte and Spencer we saw that there was a growing belief in the nineteenth century that progress was a fundamental law of nature; consequently writers and revolutionaries would often justify whatever means were necessary to achieve perfection through the presumed "law of progress." Nisbet cites Karl Marx as another utopian and progressive thinker who attempted to justify the call for rebellion and revolution through the law of progress.[111] If progress is the way of the world, then we should go after it, whatever the means. As we move through the nineteenth and twentieth centuries, there have been great wars and excessive human violence committed in the name of progress – whether it has been to achieve greater freedom or greater law and order. Aside from wars among themselves, European nations often conquered and subjugated more "primitive" cultures in the name of progress and the advance of civilization. The modern story of secular progress, in fact, starts to sound somewhat similar to the story of the religious wars of the fifteenth and sixteenth centuries, which were presumably waged in the name of God and virtue. E. O. Wilson has asked whether the dream of perfection and order through science and reason was the fatal flaw of the Enlightenment,[112] but the dreams of perfection and order go back much further in human history. In fact, the secular ideal of progress owes a great deal to the earlier religious view of progress. And throughout history, human perfection, however defined, has often been sought through violent and disruptive means.

Whatever may be the flaws or contradictions inherent in the theory of secular progress and the philosophy of the Enlightenment (and I will discuss this point in more depth later in this chapter), secular modernism became the dominant belief system and way of life in the West in the nineteenth and twentieth centuries. In fact, modernization has steadily spread across the globe in the last two centuries. To various degrees, many areas of Eastern and Southern Asia have adopted capitalist economies, cultivated and developed high tech industries, embraced science, assimilated Western popular culture, commercialism, and consumerism, and worked toward democratic systems of government. Further, as modernism has spread outward from Western Europe it has conquered and destroyed the indigenous cultures and economies of many non-industrialized nations.[113] Modernism has been progressively conquering the world.

Modernism has come to stand for many different things. Best and Kellner list mechanical metaphors, deterministic logic, critical reason, individualism, the search for universal truths and values, political and social justice, human emancipation, unifying schemes of knowledge, and an optimistic belief in human progress as the key themes of modernism.[114] According to Ray and Anderson, modernism has created the present world - a world of equality, freedom, justice, human rights, democracy, industrialization, urbanization, commercialization, analysis, control, science, efficiency, and the compartmentalization of life – a world in which time is equal to money.[115] As can be seen, there are both positive and negative dimensions to the rise of modernism, and elements of both increasing order and increasing freedom embodied within it.

Perhaps more than anything else, modernism is associated with the triumph of science. The modern world, in many respects, is a creation of science. The central conviction of the Enlightenment was that reason and reason alone should guide humanity into the future. God was dethroned and replaced by science as the "locus of knowledge and value." "Scientism" became the new God.[116] Secular modernism was the story of Prometheus retold – knowledge of fire was stolen from God thus making humanity equal or perhaps even superior to God.

Hence, with modernism comes a renewed and evolved human hubris. A new story and new vision emerged that valued competition and self-interest, as well as the domination and control of nature. Modernists embraced Bacon's idea that knowledge is power. Capitalists, industrialists, and technologists alike all valued the practical and profitable applications of knowledge. Modern industry and technology, modern agriculture, and our modern economy all rely upon the systematic use of scientific knowledge to enhance productivity, efficiency, profit, and control.

Control over nature, and in particular, the capacity to influence and direct the future, requires the ability to predict the consequences of our actions within the natural world. What science contributed to this goal of modernism was an unequivocal demonstration that the future to some degree could be predicted. The scientific theories of Galileo, Kepler, and Newton, among many others, provided a basis for making exact predictions about the behavior of many natural phenomena. Although the theories of science have evolved and been modified along the way, there has been a steady increase in the exactitude and range of predictions that are repeatedly confirmed through observation and experimentation. Within

337

science our predictive power has expanded into the areas of physics, chemistry, biology, geology, astronomy, and even to some degree the human and social sciences, economics, psychology, and sociology. The philosophy of determinism does seem to apply to a great deal of nature. There are ongoing debates in contemporary science as to the limits of determinism and predictability in nature, but there is a vast arena of complex and intricate phenomena that can be predicted based upon deterministic laws and principles.[117] Although, many futurists want to emphasize the element of possibility regarding the future, there is no question that to a significant degree we can predict many things about tomorrow. Science has demonstrated this general feature of nature, and our modern industry and technology functions because of nature's predictability. Without some level of deterministic order and predictability in nature, our efforts to influence the future would be pointless.

A good way to conclude this section is through looking at E.O. Wilson's analysis and defense of the philosophy of the Enlightenment. In his book *Consilience: The Unity of Knowledge*, Wilson presents an overview of the goals and strengths of the Enlightenment vision of the future, arguing that fundamentally they "got it mostly right."[118]

Wilson believes that the basic tenets of the Enlightenment were that the universe was lawful and could be understood through science; that all human knowledge could be united through a set of fundamental scientific laws - laws that gave order to nature; and that through understanding and applying these laws of nature the potential for infinite progress in humanity could be realized. For Wilson, science is religion liberated from the constraints of dogma. Enlightenment philosophers and scientists had a passion to demystify the world, a thrill of discovery, a central belief in the power of reason, and a strong commitment to education. Wilson thinks that the great goal of the Age of Enlightenment and the West's greatest contribution to the world was the idea that secular knowledge (science and rational philosophy) could facilitate and drive the evolution of human rights, ethical and moral advancement, social development, and human progress. Wilson, along with the futurist Wendell Bell, sees Condorcet as one Enlightenment philosopher who clearly articulated and supported this secular view of the future. Wilson notes that Condorcet, among others, saw human progress as an inevitable expression of the laws of nature. (Recall the idea of the "Law of Progress" in the writings of Smith, Spencer, and Comte.) Thus the lawful process of nature is the engine of growth and change and the doorway into tomorrow. By understanding and controlling this process we will create a better future for humanity. The presumed gods and supernatural forces of existence are no longer seen in control. Through science and reason, humanity has become empowered.

Hegel, Marx, and the Dialectic

"The goal, which is absolute Knowledge or Spirit knowing itself as Spirit, finds its pathway in the recollection of spiritual forms as they are in themselves and as they accomplish the organization of their spiritual kingdom."
"The more conventional opinion gets fixated on the antithesis of truth and falsity... [yet] each is as necessary as the other; and this mutual necessity alone constitutes the life of the whole."

Georg Wilhelm Friedrich Hegel

"The mode of production of material life conditions the social, political, and intellectual life process in general. It is not consciousness of men that determines their social being, but, on the contrary, their social being that determines their consciousness."

Karl Marx

The German philosopher Georg Wilhelm Friedrich Hegel (1770 - 1831) created one of the most comprehensive and grandiose theories of reality, time, and the cosmos ever produced in Western history. Hegel attempted to synthesize in one philosophical system the diverse wisdom and teachings of all past traditions and articulate a scheme of thought that would describe in main outline the past, present, and future of all humankind and the universe. He was also clearly a philosopher of progress, believing in the inevitable advancement of humanity and reality as a whole, but he emphasized spirit and consciousness over the secular and material in his general theory of reality and progressive change.

Hegel's impact on Western thought has been significant though his influence has waxed and waned over the last two centuries.[119] He inspired a whole generation of German thinkers and philosophers, including Karl Marx, as well as many British and American philosophers in the late nineteenth century, but he has also been severely criticized as obscure, obtuse, illogical, and politically authoritarian by many noteworthy philosophers in both the nineteenth and twentieth centuries, such as Arthur Schopenhauer, Søren Kierkegaard, and Bertrand Russell. To say the least, Hegel's ideas have been highly controversial.

The first thing to understand about Hegel's philosophy is that everything in the cosmos is in motion – there is no stasis. All is flow. Hegel is Heraclitian. Second, everything is in a state of becoming. Nothing is complete unto itself. Everything is moving toward fulfillment and realization. There is no "being" – there is only "becoming." Third, there is a direction to the universal process of becoming. This direction, which defines the nature of progress, is toward the realization of the Universal Spirit or God. God is the ultimate goal of the universal process of becoming. The universe is the becoming of God.

Hegel explained the process of becoming and the nature of progress through the concept of the **dialectic**. The dialectic is the logic of change – the *Logos*. According to Hegel the cause of progress is the dialectic – it is the engine or motive force of change. Further, according to Hegel, the pattern of progress is dialectical. Historical change moves dialectically. For Hegel, the dialectic is how things change and why they change. In particular, Hegel invokes the dialectic to explain why history moves in a progressive direction toward the realization of God.

In the concept of the dialectic, Hegel synthesized the circular and linear theories of time. He proposed that time has an oscillatory form of growth. The idea of the dialectic implies that change moves from an initial thesis to its antithesis (its opposite), and then to a synthesis of the two polarities. Each new synthesis in turn becomes a new thesis which will produce its opposite and a new cycle of growth will begin. History therefore moves forward by encompassing more and more reality, progressively circling outward to form greater and greater wholes. History both spirals and advances. Hegel believed he observed the dialectical process throughout all of human history, where trends and ideas swing toward one direction, then in the opposite direction with elements of conflict, and eventually to a progressive synthesis.[120]

In the dialectic, Hegel rejects both the Law of Identity and the Law of the Excluded Middle. First, he believes that that everything is born with its own inner contradiction - that is every thesis contains its antithesis. Everything contains its opposite; hence, "A" is equal to both "A" and "Not A." This belief contradicts the Law of Identity. We have already encountered this kind of logic in the Taoist *Yin-Yang*: *Yin* contains *Yang* and *Yang* contains *Yin* - everything contains its opposite. Hence, for Hegel, any emergent reality instigates or produces its opposite as a natural consequence of its own existence - it creates a mirror image of itself. All realities are born with implicit divisions. In the second phase of the dialectic, Hegel rejects the Law of the Excluded Middle, that is, the logic of "either-or." Once opposites are generated, these opposites seek synthesis and unity. A synthesis combines realities that seem mutually exclusive. Again, using the *Yin-yang* to illustrate this point, although *Yin* and *Yang* are "opposites," they are united in the *Tao*. Similarly, for Hegel, reality is "both/and" rather than "either/or."[121]

Hegel's theory of the dialectic involves two complementary forces. First, the dialectic implies that growth involves conflict,

341

a view of history and time we have seen previously expressed in the writings of numerous philosophers and religious thinkers. Heraclitus, for one, presumably said that "the father of all things is war," Zoroastrianism and Christianity saw history as fueled by the conflict of good and evil. Hegel, in this tradition, in fact, sees war as a necessary element in progress.[122] Opposition is a necessary component of change. Second, Hegel also argues that opposites seek unity and synthesis. In complementation to division and pulling apart, there is a force toward coming together which fuels the second phase of the dialectic.

Recall that the ancient Greek philosopher, Empedocles, had proposed that "Love and Strife" equally direct change. Hegel's dialectic is in a sense a more modern version of Empedocles; there is a force toward unity (love) and a second force that produces difference (strife, conflict, and opposition). To draw a parallel with contemporary physics, the modern cosmologists Fred Adams and Greg Laughlin propose that the complementary attractive and repulsive forces in nature generate the evolutionary pattern of change in the universe, and in modern social theory, Robert Wright has hypothesized that the complementary processes of cooperation and competition produce social change.[123] And Bloom has argued for the dual processes of reciprocity and conquest and conformity and diversity. In all these cases there are dual forces of togetherness/ unity and opposition/division that create change.

The dialectic is an oscillation of synthesis/thesis (love) and antithesis (strife), generating a spiral of expansive growth into the future. For Hegel this oscillation of synthesizing love and opposing strife produces a progressive motion in time. Does love make the world go round? For Hegel, it is both love and hate. But the world does not simply go round, it moves outward and upward through this oscillatory process.

As with his logic and theory of reality, Hegel's explanation of change also bears certain similarities with Taoist philosophy. Within Taoism, the eternal continuance and rhythm of time (the *Tao*) is maintained by a cycling of *Yin* and *Yang*. For Hegel, there is also an oscillation within time between thesis and antithesis. Using Taoist terminology, Hegel's dialectic is a moving back and forth between unity (as expressed by the *Tao*) and plurality (as expressed by *Yin* and *Yang*). Because of the oscillatory and oppositional nature of change, progress for Hegel is not a smooth, direct line forward. Growth or progress is "give and take," back and forth, up and down, unity and disruption.

We have already encountered similar views of change in previous Western writers, such as Vico, Boussuet, and Turgot – progress is oscillatory, with alternating periods of integration and disintegration. Again, what Hegel articulates in his dialectical theory of change is a synthesis of the linear and circular models of time, a new *Yin-Yang* of sorts that leads to progress. Does time move in a circle or a linear direction? For Hegel the answer is a *Yin-yang* – it moves both ways.

Hegel believed that progress is inevitable. In his theory of the dialectic, he stands with other philosophers, such as Comte and Spencer, who thought that there was a natural law of progress. Built into the very fabric of reality is the force of progress, which for Hegel is the dialectic. Yet Hegel goes beyond a theory of naturalistic necessity and views progress as having a purpose or *telos*. The *telos* of history is the realization of the universal spirit – God – through the dialectic. Time or becoming is a process of the purposeful self-discovery and self-creation of God.

According to Hegel, there is a purposeful "impulse to perfectibility" within history. The general direction in history toward the realization of God produces increasing ethical and logical perfection and absolute truth and absolute freedom. Hegel believed in the possibilities of perfection and absolute truth and viewed God, as had many religious thinkers of the past, as the realization of these ideals. In general, God is the "Absolute' relative to which everything is measured and relative to which everything is moving.

Yet within Hegelian philosophy, God is not separate from nature and the world, but rather the evolution of the cosmos is the means by which God becomes realized and self-conscious.[124]

Hegel, like Spinoza, is a pantheist - the universe is God. Each new synthesis in history, according to Hegel, is an incomplete perspective on the whole, and in instigating its own opposite, the original perspective is balanced by its mirror image. The new synthesis that arises combines both earlier perspectives - broadening the view of the whole. This process of an ever expanding perspective on the whole continues toward God and the absolute truth - the conscious synthesis and realization of all perspectives and the whole.

Since Hegel believed that all perspectives on the whole are limited (except for God's), he viewed all human belief systems as culturally and historically constrained and relative. What we see and what we understand is always from a particular point of view in space and time and is naturally colored and influenced by our localized perspective.[125] Just as Kant had seen all human understanding structured in terms of categories that filtered and organized consciousness, Hegel sees the human mind structured and filtered by history and culture.

What might seem paradoxical in Hegel's thinking is that he saw his own philosophy as somehow providing a universal and all-encompassing vision of reality, in spite of the fact that according to his very philosophy all human belief systems are historically and culturally bounded. Hegel presents a relativist theory of human knowledge but excludes himself from it. Hegel desired to capture the whole.

In fact, Hegel's overall philosophy is exceedingly holistic. It is the whole which is ultimately most real and true. Every finite thing finds its reality within the whole and all finite things are steps toward the realization of the whole. God or the ultimate whole is also that which is ethically and logically perfect. God encompasses and resolves all contradictions and provides an absolute benchmark for all ideals. God is the Absolute Truth and the Absolute Good. Reality, truth, logic, and the good are all anchored and defined relative to this ultimate whole.

Progress for Hegel is holistic. God provides the unifying direction toward which everything is moving. Since God or Universal spirit is the absolute whole, encompassing everything, progress is movement toward the whole. Although Hegel describes progress as moving toward greater freedom, an idea we have seen expressed in many philosophers of the Enlightenment, Hegel ultimately sees progress as an absolute integration of everything into God. Thus Hegel is usually seen as siding with those theorists who believed that progress was toward increasing order and integration.

Hegel is often referred to as a German idealist, since he believed that mind creates reality. Ultimate reality is therefore mind – in particular the conscious mind of God - and the end result of all progress is the realization of this absolute mental reality. God is pure thought thinking about itself. God is also pure self-consciousness, since God is everything and consequently there is nothing beyond God for God to think about. In essence, the nature of God is God contemplating God. God is God's Idea of God. Hence, the entire motion of progress is toward a purely mental and spiritual reality – a theory we have seen expressed throughout many earlier religious and philosophical traditions.

Hegel is also seen as both a rationalist and a romanticist – a synthesis of opposites - since on one hand, he views history as a rational process, which is the logic of the dialectic working its way out through time, and yet, on the other hand, he views history as filled with conflict and passion.[126] To recall, Hegel sees war as a necessary component of progress. Since Hegel saw his philosophy as encompassing and synthesizing all past philosophies and points of view, through the logic of the dialectic, it is not surprising that there are both Apollonian and Dionysian dimensions in his theory.

Nisbet argues that Hegel follows the general line of thinking in nineteenth-century Germany in viewing progress primarily in holistic or group terms. It is the whole that evolves. In particular, according to Nisbet, German thinkers emphasized the importance of the national state. Hegel, for one, believed that all conflicts inherent in the dialectic – between the secular and the spiritual and the individual and the state, for example - would be fused or resolved within the perfect realization of the state, which he saw occurring in his native Germany.[127] Hegel stated, "The German spirit is the spirit of the new world. Its aim is the realization of absolute Truth as the unlimited self-determination of freedom ..."[128]

From this quote we can see that Hegel connected the evolution of his native Germany, a nation state, with the full realization of human freedom. For Hegel, a nation is a spiritual entity and the embodiment of ethical ideals. According to Hegel, "The state is the Divine Idea as it exists on earth".[129] It is the national state that supports the full expression of human identity and individuality. Hegel sees history as leading to increasing individual freedom, but freedom as realized in the context of community and the whole - that is, for Hegel, in the context of the national state.[130] Without the state there is no freedom. And pushing this holistic logic to its ultimate conclusion, it is God – the absolute whole – which is absolutely free. Paradoxically, for Hegel, freedom does not come through a world of separate and autonomous individuals, but rather through increasing participation in the whole. Hence, Hegel attempts to synthesize the apparently contradictory messages of the Enlightenment – progress moves toward greater social order versus progress moves toward greater individual freedom.

The contemporary social and political writer Francis Fukuyama in his book *The End of History and the Last Man* argues that it is Hegel's emphasis on the growth of human freedom that is central to his philosophy of historical progress. Although Hegel sees human freedom as only truly realized in the context of the state, according to Fukuyama, Hegel's vision of the ideal state toward which humanity is moving is the liberal democratic state which truly supports human freedom. Again, although freedom is realized in a social context, that social context at an individual level involves the "reciprocal recognition" among citizens of each other's individuality, value, and self-determination. (We all agree to respect each other.) Just as Hegel sees a general direction to time in the realization of the Absolute Spirit, Hegel also sees a general pattern to human history – a **"Universal History"** – that has as its goal and trajectory individual freedom. To quote from Hegel, "The history of the world is none other than the progress of the consciousness of freedom." According to Fukuyama's interpretation of Hegel, history is a struggle and conflict against authoritarian masters and oppressors and the "end of history" is a realization and fulfillment of true individual freedom and the reciprocal recognition among humans of individual value and self-determination.[131]

Generally, though, Hegel is known for emphasizing the state above the individual and the whole above the parts. It is odd then that Hegel did not support the eventual formation of a global government above national states. Kant had argued for such a world organization, but Hegel felt that the nation state had an absolute ethical right to self-determination. As the twentieth-century philosopher Bertrand Russell asks, isn't a global organization a greater whole than the national state, and hence, shouldn't national states find their identity and meaning within a global government, if we were to follow Hegel's logic? Instead Hegel argues that nation states have a "moral" right and obligation in certain circumstances to wage war on each other, in the name of progress.[132] Hegel believed that certain national states at different times achieve a leadership role in human affairs, thus apparently justifying external aggression if it serves progress. As Nisbet points out, nineteenth-century German philosophers of progress often emphasized "power" and the glorification of the state, and thus supported both internal suppression of individual freedom and external expansion and conquest to realize national ideals in the name of progress.

As an overall assessment of Hegel and his philosophy, I think that the dialectical theory of change, and in particular, the idea that change involves an oscillatory process between integration and diversification, captures an important dimension of reality. Polarization and synthesis appear to be pervasive and reciprocal processes in history. At the human and social levels, ideas do seem to often instigate oppositional reactions, which in turn provoke attempted syntheses. Further, the incompleteness of natural realities and their interdependency and interconnection into the whole also seem valid points regarding how the universe is organized. Hegel quite rightly rejects the notion that reality is composed of absolutely distinct and separate entities. Everything does depend on everything else and things interpenetrate, nothing being complete unto itself. If there is a flaw in Hegel's thinking on this point, it is that he sides too much with a holistic perspective. Everything may be interdependent, but following a Taoist logic, everything also has a dimension of distinctiveness and individuality. Hegel's extreme holism is particularly apparent in his notion of an absolute, all encompassing God or Spirit. Although Hegel's dialectic implies that everything is incomplete, possessing inner contradiction, and in a state of becoming (rather than being), he proposes that there is an Absolute which is perfect, complete, and consequently beyond time, or more precisely, at the end of history and time. This combination of unending becoming with an absolute and finalized being is itself a contradiction - perhaps in need of a further dialectic. But Hegel anchors his whole philosophical system to his notion of the Absolute. It is, though, not at all clear whether perfection is a viable idea or whether time (or human history) will come to an end. Is there an absolute whole? Can there even be an absolute whole? Everything we have learned about nature and human history seems to imply incompleteness and never-ending becoming, rather than some final resolution and perfect state. Underneath Hegel's modern abstract philosophical system is the ancient Christian view (or perhaps we should say Zoroastrian view) that time is moving toward completion and resolution in a perfect and complete Godhead.

Hegel and his philosophy would, in fact, instigate numerous and varied counter-reactions and criticisms, thus fulfilling the prediction of his own philosophical system. Although Hegel wished to have the final word in his all-encompassing scheme of thought and theory of the Absolute, he instead became another thesis provoking its own antithesis. In the minds of many, Hegel was either incomplete or in error.

One writer who was both influenced by Hegel and yet critical of him was Karl Marx (1818 - 1883). Marx adopted Hegel's dialectical theory of change, viewing history as an ongoing conflict of opposing forces that progressively lead to higher more advanced syntheses. He also saw an overall direction to history and a final ideal state toward which humanity was moving; that is, like Hegel, Marx subscribed to the ideas of a "universal history" and an "end to history."[133] Further, Marx was a champion of human freedom (as well as human equality), and like Hegel, saw freedom as something realized in a social context. Where he differed from Hegel was, as Marx put it, "standing Hegel on his head," and arguing for a materialist philosophy of history rather than an idealist one. If in Hegel, mind moves matter, in Marx, matter moves mind. Additionally, Marx was more action-oriented in his philosophy, not being content to simply understand human reality, but desiring to concretely influence the course of events. As Marx stated it, "The philosophers have only interpreted the world in various ways: the point however is to change it."[134] Hence, whereas Hegel created a grand metaphysical system of thought, Marx created a philosophical "call to action" that would impact billions of people in the century ahead.

In order to set the historical context for explaining Marx's vision of progress it is important to briefly summarize the Industrial Revolution that was sweeping across Europe and America during the eighteenth and nineteenth centuries. This overview of the Industrial Revolution will also help us to understand the rise of Romanticism which will be described in the next section.

As discussed in the previous sections of the chapter, beginning in England in the seventeenth century, modern factories emerged that accelerated the growth of production throughout the modern West. Fueled by technological inventions, such as the steam engine, and ongoing discoveries in the sciences, such as in chemistry, physical mechanics, and a bit later, the study of electricity, and organized in terms of new principles of efficiency, division of labor, and management in industrial production, the Industrial Revolution generated a great upsurge in the manufacturing of material goods and appeared to many to be creating a world of plenty for citizens of the modern world. As noted in the earlier discussion on science, Newtonian physics provided a new model or metaphor for the organization and operation of human society – the machine and in particular, the mechanical clock. Inspired by this new mechanistic metaphor and the ongoing success of the physical sciences in bringing order and intelligibility to the world of nature, the world of industry became a key element in the new vision of the modern world emerging in eighteenth-century Europe. Many visionaries in the eighteenth and nineteenth centuries, such as the Lunar Society in England, saw great value in science and associated technological developments, espousing a "pro-machine" philosophy and being very optimistic about the future of progress supported through scientific and industrial advancements. The Lunar Society, in fact, inspired by the promises of the physical sciences, contributed many practical and industrial developments and realized the importance of marketing in stimulating the growth of industrial economy. Also the philosophy and practice of capitalism provided another key ingredient in the growth of the industrial economy, producing ongoing innovation through competition among businesses and the creation of wealth. As one other important factor, the Protestant ethic, so argued the sociologist Max Weber, supported a philosophy of diligence and hard work as the road to happiness and personal fulfillment, providing a growing working class of people who would toil for long hours in factories believing in the value of what they were doing. In general, the Industrial Revolution provided a new image of the future, built on material production, a capitalist economy, hard work, and the accumulation of its bounty in the form of increasing material possessions and financial wealth. In the process, a consumer society was being born.

Yet, as soon became apparent, there were various problems associated with this new vision and way of life. Factories, first appearing in rural areas, moved into the cities when steam replaced water as the primary source of power, and not only were multitudes of people displaced from their villages to work in the urban factories, but huge inner city working class slums arose to house all these workers and their families. These new working class urban areas suffered from poor sanitation, crowding, pollution, and in general abysmal living conditions. Wages were usually poor and children, along with adults, were recruited into the workforce where they were expected to work impossibly long work days and where they suffered from disease, infection, depression, and often death as a consequence. Work in factories was invariably monotonous and mechanical (the dark side of the metaphor of the smoothly running machine), and the new factory worker had become nothing more than a "hired hand" with no say or power over the quality and operation of his working environment. To add insult to injury, while the working class lived a poor and dark existence, the capitalist owners of business and industry were accruing huge amounts of wealth at the expense of their employees.[135]

It is in the context of this industrial and capitalist world of the nineteenth century that Marx developed his ideas about history, the nature of progress, and critique of modern life. Nisbet provides a concise description of the essentials of Marx's theory of history and progress. According to Marx, history involves an ongoing conflict of social classes, between the "haves" and the "have nots." There are stages to this class struggle, each stage achieving a higher level of human equality and logical consistency; that is, by Marx's criteria of progress, history advances through conflict and resolution. The historical process of repeated conflicts and resolutions is inevitable according to Marx; there is an overall law and natural direction to history. The ultimate end point of human history will be a utopian state of equality among all people, a resolution of all class conflicts, and the elimination of capitalism and competition. There will also be an end to personal ownership and private property in this utopian state. This ideal state – a "**communist**" state – will be realized through centralized control that serves the collective will of the masses.[136] Though Marx stresses the "scientific" dimension to his thinking, attempting to describe and extrapolate on the facts of history, his philosophy contains a strong moral element as well.[137] He views the lawful culmination of history as a morally ideal state, with the elimination of exploitation, human misery, and inequality.

When Marx described his philosophy as materialist, he meant that the foundation of human identity and human society is its physical economy. It is the triad of natural resources, means of production, and means of distribution that defines the economic foundation of a society and supports all its higher psychological, social, and cultural functions. In this sense, Marx is an economic determinist who believes that it is the economy that determines and controls other aspects of a society. For Marx, since the rich capitalists controlled the economy, they controlled all other aspects of the world.

Further, for Marx the distinctive quality of humans is their capacity to make things – to produce – hence he labels our species *"homo faber"* (man the maker). Literally we are what we make. What is basic to humans is their mode of action – their physical behavior - and Marx describes human behavior as an interactive process with the surrounding physical environment. Humans are physical beings who interact with and manipulate a physical environment, making physical things and often exchanging these things with each other. We are material beings making and distributing material things in a material world. For Marx, all the major forms of knowledge and consciousness (for example, science, art, philosophy, and religion) emerge out of this physical foundation of matter and action and ultimately serve and find their value in the physical world. For example, the value of knowledge lies in its consequences for action and the creation of material things. Literally, for Marx, knowledge is practical power. [138]

According to Marx, the ideology and values of a society are a product of economic power. Those who control the means of production are the most powerful class within a society and, to justify their position and right of power, determine the predominant belief systems and ideals of that society. These central ideas of the society legitimize those in power. Since social power and ideological supremacy are based on economic power, the conflict of social classes within a society is over who controls the means of production and material power.[139] History is a struggle for material power, and what is true and what is right is determined by who possesses material power.

Marx believed that the capitalists controlled the means of production in modern Europe and consequently possessed all the social power. Further, according to Marx, the capitalists had unfairly accrued the vast surplus of material wealth generated by this economic system. Workers were exploited and forced to engage in long hours of hard physical labor that fed the pocketbooks of the wealthy capitalists. Hence, whereas many early philosophers of the Enlightenment saw capitalism as leading to a better life for all, Marx saw capitalism as a form of oppression that created a wealthy class and impoverished the worker class. There was not enough distributive justice in the capitalist system – resulting in a society of haves and have nots. To achieve equality among all humans, which meant, among other things, equal social power, the capitalist system for Marx had to be overthrown.

Marx also believed that capitalism "alienated" the workers from both what they produced, as well as their true human identities. Workers produce what capitalists determine they will produce, and workers do not keep the products of their industry. Workers do not find a creative outlet in production and the products they make are sold to others. If we are what we make, then in a capitalist system, what we make is not of our own choosing and does not even belong to us in the end. It belongs to the capitalists.[140] Capitalism robs us of our identities and freedom of self-determination.

Marx, in general, is critical of the commercialism, consumerism, and monetization of human life that is associated with capitalism. Capitalism leads to the triumph of the economy over all other aspects of social life. Everything

353

produced becomes commerce with a marketable value. The monetary value of things becomes the defining criteria of the worth of man's creations. Life becomes organized around the production of commodities. In fact, individual human beings become commodities who sell their skills and labor for a price, that is, for wages. Individual well being gets defined in terms of the consumption of goods and the overall health and quality of a society is judged in terms of level of production and consumption.[141]

In many important ways Marx is viewed as anticipating contemporary critiques of the capitalist economic system and the social-psychological problems that it generates. Yet if Marx is prescient in his analysis of the effects of capitalism, he is also deeply rooted in the past in his vision of an ideal society. The central human values he supports, which he believes capitalism does not provide for, are social harmony, individual happiness, freedom and self-realization, and human equality. These values, as Wendell Bell points out, are common ideals identified in many earlier utopian visions.[142]

Bell views Marx as a utopian thinker, who describes an ideal society – in fact a "perfect society" – in the future and presents arguments for the desirability of this ideal society and even proposals for how to go about achieving it. As with other utopian thinkers, Marx provides a critique of the world that he lives within and outlines a utopian solution explicitly formulated around eliminating the perceived flaws and problems of his world. The good and the bad are reciprocally defined.

In Marx, as with many other earlier futurists and utopian thinkers, there is a conflation of predicting the future with identifying what is preferable or desirable in the future. It is one thing to make predictions of what will happen in the future – it is quite another thing to identify what one would hope or prefer to happen in the future. Marx clearly makes a variety of predictions about the future. At the most general level he predicts the rise of socialism and the collapse of capitalism. But he also sees this future as morally desirable – the world will improve with this change from capitalism to socialism.

He believes that what will be is what is preferable because he thinks that there is a natural progressive process at work within history. Good is going to triumph. This is the same general mindset that we found both in religious views, such as Christianity, and secular views, such as in Spencer and other theorists of natural progress. This is the same point of view we find in Hegel. The world is necessarily, due to either God or the laws of nature, getting better. Both the philosophers of the Enlightenment as well as the religious thinkers before attempted to derive or justify an "ought" from an "is." [143]

Also, as found in other writers of the period, there seems to be a conflation or blurring of natural necessity and individual choice in Marx. Marx speaks as if the eventual rise of socialism is a necessary consequence of the flow of historical events. He sees the future as determined and does not seem to acknowledge or believe in the uncertainty of the future. Again, both religious and secular thinkers in the West often described history and the future in such a deterministic fashion. Yet Marx also presents a "call to action", arguing that workers should rebel against the oppressor and exploitive capitalist system – that is he speaks as if individual choice and action matter. This same kind of argument can be found in other theorists of progress. The necessary direction of progress is identified – as a global or cosmic force at work – and people are advised to jump on the bandwagon and help to facilitate this process. But in the long run, it really doesn't make any difference if, as these theorists also posit, human society and the universe as a whole is heading that way regardless of what we do or don't do. Whether we decide to be good Christians or not, God will triumph in the end. If one believes in natural or supernatural necessity, then choice doesn't really mean anything in determining the overall course of events. (You do have a choice though in whether you want to be on the "winning" side or the "losing" side, but what kind of a choice is that?)

Thus Marx is a good example of a contradiction that exists within Enlightenment philosophy and the theory of secular progress. Enlightenment philosophy stressed the importance of freedom, yet this same philosophy also embraced the deterministic model of nature provided through science. But how can there be freedom in any true sense of the word, if life is determined. There is no real power or significance to choice unless the future is open to different possibilities.

Marx, like Hegel, believes in perfection and thinks that perfection will be achieved sometime in the future; he also believes in an "end to history." Again, this view reflects an ancient mindset to be found in mythic and religious thinking. Zoroastrianism and Christianity both envisioned an ideal perfect state achieved at the end of history and the end of time. Yet one can question both the idea of perfection, as well as the idea of an end to history and time. How can one legitimately argue, with any credibility or certainty, that there is some ideal human state or social reality that can not be improved upon? Further, just as in Hegel, Marx offers a dynamic theory of history with ongoing change across the ages and then brings the whole process to an end in a perfect social state. He combines a theory of ongoing becoming with a static end.

Finally, Marxist thinking leads to another problem that also shows up in Hegel. If there is some ideal reality toward which human society is headed, then it could be argued that whatever means are necessary to get there should be implemented. The ends are used to justify the means. (Recall Machiavelli.) The authority of both Marx and Hegel has been used to justify war, violence, and oppression as necessary means toward some desirable end. In the case of Hegel, the supremacy of the ideal state, as envisioned to be developing in Germany, was used to justify German aggression and war against "inferior" and less advanced states. In the case of Marx, the promised equality and universal happiness to be realized in a communist state was used to justify violent rebellion and subsequent repression and control of citizens in Russia. In both cases, the inhumane and violent means were justified in terms of ideal ends. Once again, as since time immemorial, violence and war are connected with future consciousness.

As Wendell Bell points out, the value in Marx lies in his comprehensive and telling critique of the flaws of capitalism and the humanitarian ideals he proposes that somehow need to be addressed in human society. Although the communist experiment seems to have failed in the Soviet Union, it does not necessarily follow that capitalism is a morally superior or perfect system. Part of the ongoing critique of modernization has centered around the numerous problems that capitalism seems to generate, for example, excessive commercialization and consumerism and an ever growing unequal distribution of power and wealth. Let us now turn to another wave of thinking that emerged in the nineteenth century and produced perhaps the strongest and most powerful critique of modernization, Enlightenment philosophy, and the rise of capitalism yet to come - Romanticism.

Romanticism

"The world is too much with us; late and soon,
Getting and spending, we lay waste our powers:
Little we see in Nature that is ours;
We have given our hearts away, a sordid boon!"

William Wordsworth

"The world is my idea....The world is my will."

Arthur Schopenhauer

Though Marx and Hegel modified, if not critiqued, certain aspects of the theory of secular progress, both believed that the general idea of progress accurately described the flow of historical time and provided a guiding theme for understanding the future of humanity. Not everyone though in the modern West was sympathetic to the philosophy of progress. Criticisms of the philosophy of the Enlightenment, modernism, and secular progress, go back to the eighteenth century at least. Barely had the modern age been born when it came under attack.

The strongest attack on the theory of secular progress arose in the nineteenth century in the philosophy of **Romanticism**. Romanticism contradicted almost all of the central principles of modernism, science, and rationalism. Recall the distinction between Apollonian and Dionysian modes of consciousness. Enlightenment philosophy, with its emphasis on reason, falls into the Apollonian mindset; as Wilson refers to Enlightenment philosophy, it was "bloodless," focusing on rationality, form, and function. The opposite of the Apollonian mindset, the Dionysian, provided the impetus behind philosophical Romanticism which recoiled not only against modernism, science, and reason, but also against capitalism, industrialism, and the general optimism of the period. Romanticism provided a much different approach and attitude toward the future, as well as the past, than the rationalism and instrumentalism of modernism and the Enlightenment.

Richard Tarnas, in his *The Passion of the Western Mind*, states that two streams of thinking emerged out of the Renaissance – the rational and the romantic. According to Tarnas there were, however, some commonalities between these two ways of thinking: Both were Promethean, challenging the sovereignty of the gods and tradition; both embraced a humanist perspective, setting man in the context of nature; and both had classical origins, in particular, the Apollonian and Dionysian mindsets of ancient Greece.[144]

Even if the origins of Romanticism go back to the Renaissance and classical Greece, it was the increasing influence and ubiquitous presence of modernism, science, and Enlightenment philosophy that instigated the full and intense expression of Romanticism in the nineteenth century. Even if there is a common cultural root, modern Romanticism vehemently attacked and rejected modern Western rationalism. Modern Romanticism set out to dethrone rational Enlightenment and everything associated with it from its position of cultural power. This critique and opposition – this antithesis to reason and progress - created a deep intellectual and cultural schism in modern Western society, or as Watson refers to it, the "modern incoherence," which is still with us up to the present day.[145]

Whereas the Enlightenment emphasized reason, Romanticism embraced emotion, passion, "sensibility", and the a-rational or irrational. Romanticists examined the "dark side" of humanity and not just humanity's higher aspirations and abilities. In Romanticism we see the beginnings of the exploration of the unconscious. Whereas Newton's vision of a clockwork universe inspired a mechanistic and machine model of nature and even human society among secular modernists, romantic philosophers saw nature as alive, inspirited, and organic. Romanticists often reveled in rare nature, in opposition to the constraints and refinements of civilization. As we have seen many philosophers of secular progress championed the importance of order; romantic philosophers embraced chaos, turbulence, the strange, and the macabre. Beauty and the aesthetic, for the Romanticists, took precedence over the utilitarian values of capitalism, industrialism, and technology. Romanticists, often of a more literary than philosophical or scientific bent, saw the dramatic in life; they valued inspiration, imagination, creativity, revelation, and mystery. Whereas science searched for grand abstractions and universal knowledge, the Romanticists valued uniqueness and diversity. Whereas the Enlightenment searched for scientific certainty, the Romanticists embraced uncertainty. The Enlightenment wished to rid humanity of superstition and the supernatural; the Romanticists reasserted the value of ancient myths and the mystical. If the Enlightenment, science, and the Industrial Revolution emphasized understanding and transforming the external world, the Romanticists turned inward, delving into the subjective and the deep inner self. In general, Romanticism brought back into the human equation, in great force, the affective, primordial, subjective, and concrete dimensions of humanity that, in their minds, had been repressed and rejected by the rational and modernized world.[146]

The Romanticists, in numerous ways, questioned the secular and rational ideal of progress. First, they feared that science was more Faustian (a deal with the devil motivated by human ego and vanity) than Promethean. Instead of Bacon's notion of conquering nature through science and technology, many Romanticists wanted to return to a purer harmony and unity with nature. If science wanted to detach itself from nature, adopting an objectivist stance on reality, Romanticists wanted to immerse themselves in nature. Second, following from the first point, Romanticists saw modern civilization as de-humanizing and alienating. Not only was modern humanity cut off from nature through living in cities, but with the scientific emphasis on objectivity, humanity was also cut off from the inner or subjective aspect of reality. Romanticism emphasized the subjective side of human existence and rejected the Enlightenment ideal of a single objective truth. Third, as we have seen, progress and civilization bring in many ways increasing constraint and regimentation. The individual is consequently suffocated in the name of progress. Although philosophers of the Enlightenment often championed the ideals of freedom and self-expression, the Romanticists saw the results of increasing modernization as producing the opposite effect. The Romanticists firmly believed in the value of the individual, which they thought was being undercut in the new modern world order. As noted earlier, the Enlightenment emphasized the apparently contradictory themes of individualism and order; the Romanticists resolved this contradiction by elevating individualism and rejecting social order. Fourth, for the Romanticists, capitalism, industrialism, and consumerism were turning humans into machines - cogs in the wheel of progress and production - who lose themselves in things at the expense of human feeling and human intimacy. The early twentieth-century historian and philosopher, Oswald Spengler, who was strongly influenced by Romanticism, argued that the mechanistic and the commercial - two central themes of secular progress - were incompatible with humanism and were producing a "decline of the West." Echoing a view that would run through Romantic philosophy, Spengler argued that the West needed to reassert the value of the hero over the trader.[147] The highly influential Romantic poet Lord Byron (1788 - 1824) elevated the artist as hero to a central position in Romantic philosophy. In short, to the Romanticists, a philosophy of progress built on rationality, objectivity, mechanization, and efficiency is not progress at all - it is regressive. It ignores the human heart, destroys spontaneity, kills unique individualism, and isolates humans in the unnatural constructions of technology, industry, and urbanization. As Max Weber, the late nineteenth-century sociologist and economist stated, modernity created a "bureaucratization of the human spirit" and placed the human being in an "Iron Cage."[148]

Although the Romanticists were critical of the secular theory of progress and valued the mythical traditions of the past, both Western and Eastern, they were not so much dismissive of the future as simply offering an alternative vision of tomorrow. As one of their general points, emotion rather than dispassionate reason needs to guide the future. In more concrete and personalized terms, the Romanticist replaced the scientist with the artist as the central guiding archetype and consequently replaced science with art as the critical mode of consciousness and knowledge for experiencing and understanding life. If objective truth was the ultimate goal of science, Romanticists elevated beauty to center stage instead. Just as the search for truth had been for many early scientists an effort to read the "mind of God," the creation of beauty became the spiritual quest for the Romanticists. Through the novel, poetry, and the visual and musical arts, the Romanticists created an alternative picture of the world to that of science, and defined a different set of ideals to strive for in creating a better world.

Another important central theme in Romanticism, mentioned above, is creativity. Watson, in his discussion of Romanticism, which he describes as "the great reversal of values," identifies the elevation of creativity as pivotal to Romantic philosophy and art. For the Romanticists, "man" is fundamentally a creative being, who invents both the individual self and values. There is no true self or definitive set of values.

361

Ultimately what is central in human life is created rather than discovered, and thus lies outside the scope of science. Life is art - life is will. The centrality of human will and self-realization, as opposed to reason and objectivity, are clearly apparent in the Romantic philosophies of Fichte, Schopenhauer, and Nietzsche (see below). Hence, it is the creative artist, often solitary and alienated from mass conformist human society, struggling to realize his or her unique vision, that epitomizes the Romantic ideal.[149] It is through the Romantic symphonies of Beethoven that we can experience the meaning, struggle, and direction of life, rather than through the rationalist philosophies of Descartes and Kant.

As noted above, the Romantic reaction produced a highly polarized dichotomy in thinking in Western culture. In the earlier past, we have seen various other oppositions of thought, such as faith versus reason, left versus right brain, and religion versus science. In modern times Romanticism reasserts, with a vengeance, the Apollonian versus Dionysian. Romanticism looks at the subjective and is associated with the study of the humanities; rational Enlightenment emphasizes the objective and supports the study of science. Perhaps, as J. T. Fraser argues, science versus art and the humanities reflects a fundamental difference in temperament among humans, with different people preferring one mode of consciousness over the other. There have also been individuals who have attempted syntheses of these two modes of thinking. Even if the conflict of the Romantic and the rational runs back through much of recorded history and even if it reflects a fundamental difference in human temperaments, it would seem, following the logic of Hegel, that any viable and comprehensive approach to the future needs to find a way to synthesize or bring into balance these two ways of looking at life - it needs to heal the "modern incoherence."

At this point, I want to examine more closely three philosophers of the Romantic era who were both highly influential and who express in unique and significant ways the philosophy of Romanticism. These three philosophers are Arthur Schopenhauer, Søren Kierkegaard, and Friedrich Nietzsche. All three of them, in the true spirit of Romanticism, attacked what they perceived as the excessive rationalism of the modern West.

Arthur Schopenhauer (1788 – 1860) was a somewhat younger contemporary of Hegel and highly critical of him, accusing Hegel of producing "the craziest mystifying nonsense." Schopenhauer followed Kant in arguing that all human experience and knowledge was ultimately subjective and consequently humans could not possess true knowledge of the objective world – the "thing-in-itself." Thus Schopenhauer rejects the Enlightenment and scientific aspiration for objective truth – he thinks it is impossible. Schopenhauer's emphasis on the centrality of subjectivity aligns him with Romantic philosophy, which as noted above, emphasized the subjective over the objective.

But Schopenhauer paradoxically does attempt to look beyond subjective experience and explore the nature of ultimate reality. If the conscious world is nothing but "ideas" – that is creations of the mind – then the ultimate ground of being is what Schopenhauer referred to as "the Will." It is on this point that he quite explicitly rejects rationalism and Hegel in particular. Rationalism, to recall, sees reason as the road to knowledge, and pushing the argument even further, Hegel sees reality as ultimately rational. For Hegel, reality can be understood through reason because reality is rationally structured. Schopenhauer, on the other hand, does not believe that ultimate reality is rational at all. Rather, reality is fundamentally a primordial force – a will – a wanting and desire. The "Will" is a-rational: it is not a "Logos" but an energetic impetus and motive of self-assertion and gratification. It is force rather than form. It is "Will" that drives reality – that creates it in its need for growth and expression. Everything is a manifestation and sublimation of "Will." There are two primary expressions of "Will" – procreation and destruction, that is, "Will" creates "becoming" and "passing away."

In some ways, "Will" sounds like the Hindu god *Shiva*, and that may be no accident since Schopenhauer studied Eastern religion and was definitely influenced by it. (Romanticism was strongly affected by the "Oriental Renaissance" - the rediscovery and renewed appreciation of Eastern culture in nineteenth-century Europe.[150]) Schopenhauer believed that there was really just a single all encompassing "Will" that moved everything. Individuation was actually an illusion of subjectivity - ultimately, as in Hinduism, all is One. The conscious sense of separation does not really exist. Individual self-determination and freedom is also illusory since all action is really an expression of the universal "Will."

Thus for Schopenhauer past and future are ultimately the same - of creation and destruction; all of time is simply the never-ending expression of the universal "Will." According to Schopenhauer, "History shows on every side only the same thing under different forms...". Because of this vision of reality and time Schopenhauer is often seen as a pessimist. He viewed all the various progressive and uplifting philosophies of both past and future, with very few exceptions, as being unrealistic and Pollyannaish. As is often the case, people who are seen as pessimistic view themselves as simply realistic - as did Schopenhauer - and mostly everyone else as engaging in delusory wish-fulfillment.[151]

If Marx turned Hegel on his head, the legacy of Schopenhauer is, to turn both Hegel and the Enlightenment inside out. Reason does not rule reality - there is no "Logos" to either discover or emulate. Reality is a primal motive force that creates and destroys and humans are nothing but expressions of this force of will. The emphasis on will over reason would become a central theme in later Romantic philosophers, notably in Nietzsche, who studied Schopenhauer. The idea of will as fundamental to human reality was also connected with the Romantic ideals of self-assertion, self-expression, and self-creativity, both at the individual and the national levels.

The Danish Christian philosopher Søren Kierkegaard (1813 - 1855) also attacked the rationalism of Hegelian philosophy, as well as the rationalism inherent in both science and Christianity. For Kierkegaard, the universe can not be understood or adequately experienced from a rational point of view, and in fact, he advocated that belief in God, contrary to any presumed proofs or historical evidence for the existence of God, must be approached as a **"leap of faith."** Contrary to the philosophy of the Enlightenment, reason, can not be an ultimate foundation for either action or belief.

For Kierkegaard, it is the concrete lived experience of the individual which is of primary importance. Attempts to encapsulate the universe in an abstract universal system of thought, such as in Hegel, or science for that matter, totally miss the basic fact that life is not experienced from an abstract or general point of view, but from a unique and individual point of view. Philosophy must address life as we find it; hence, individuality and subjectivity should be our starting point rather than abstract universality. Kierkegaard develops his philosophy from the point of view of the subjective lived experience of the individual.

Pivotal to Kierkegaard's philosophy is the theme of individual freedom in the face of the uncertainty of existence. Humans are decision makers; we are always faced with various choices - "either-or" situations. Life is different possibilities branching out in front of us, and we have to make choices regarding which paths to follow. The future is irreducibly a set of "either-or's." Further, for Kierkegaard all decisions are based on values. But values contain a subjective element - values are acts of choice as well. There is no set of universal standards for determining what values or decisions are best. According to Kierkegaard, values are based upon what he refers to as "subjective truths." It is the inner subjective sense of what feels right or true that determines whether a value is embraced or rejected. Consequently, reason can not be the ultimate guide in life, since we can not, through reason, determine what is best - we must experience the subjective validity of a value. Since there are no guarantees in any of this, choosing values and making decisions involves an element of faith. The future is uncertain choices based on faith.

In fact, Kierkegaard sees the self as an act of choice. As self-reflective beings, we can determine what kind of person we choose to be. Kierkegaard uses the expression "authentic self" to refer to the type of self that is freely chosen by the individual. He also uses the expression "knight of faith" to refer to those people who realize that they are free to determine their own destiny. Kierkegaard believed that modernity, through the social and economic forces of mass conformity, was destroying individuality (a criticism we also saw in Marx) and though the promise of the Enlightenment was increased freedom, modernity was producing the opposite effect.

Authenticity in life is also based on the subjective discovery of death. Although everyone possesses the objective knowledge from early on in life that he or she will die, it requires a courageous and intentional effort of consciousness to really feel the inevitability of one's death – to vividly imagine a time when one no longer exists. Only by feeling one's death does one truly appreciate one's life. Again it is the lived or felt experience of life – a subjective reality – that is of paramount importance, and it is upon this inner foundation that the creation of an authentic freely chosen life is built.[152]

Kierkegaard is often seen as the father of **existentialism**, the twentieth- century philosophy that highlighted the dimension of freedom in human existence. Existentialism also emphasizes the subjective dimension of human reality. Both these themes can be found strongly expressed in the philosophy of Kierkegaard. Whereas many nineteenth-century philosophers, such as Marx, Hegel, Comte, and Spencer, saw the future in terms of some universal law of progress that was moving humanity forward in time, Kierkegaard saw uncertainty and individual choice when he looked toward the future. The future is a choice rather than a natural inevitability. There is no certainty in what is to come and we must all, according to Kierkegaard, not only live with that fact, but embrace it as critical to our psychological well-being. Kierkegaard is a philosopher of courage. Kierkegaard reasserts the importance of "either-or" thinking, after Hegel had, as Kierkegaard asserted, obscured the distinction by arguing that everything contains its contradiction. Life is choices - life is risks. There are many possible futures - not just one.

Also, Kierkegaard, in true Romantic fashion, reasserts the significance of individuality and subjectivity after science and the Enlightenment had attempted to turn reality into abstract general truths. Although Enlightenment philosophers argued for the importance of individuality and freedom, they were caught in a contradiction, for they also embraced scientific determinism and the quest for general laws of nature, including laws of the human mind and human society. Kierkegaard would have nothing of this, instead arguing for an extreme philosophy of freedom in the face of an uncertain future. The second philosopher of the nineteenth century who is also seen as a main inspirational figure within existentialism, Friedrich Nietzsche, would take the philosophy of individualism and freedom even further.

Peter Watson, in his intellectual history of the twentieth century *The Modern Mind*, begins his discussion of Nietzsche with the following statement: "There is no question that the figure of Nietzsche looms over twentieth-century thought."[153] The German philosopher Friedrich Nietzsche (1844 - 1900) was one of the strongest spokesmen of Romanticism. Nietzsche attacked the supremacy of reason and the value of modernism. He argued for the importance of passion, and believed that modernism inhibited the creative and higher qualities of humanity. He rejected the Enlightenment ideals of absolute and eternal truth, absolute values, and pure objectivity. Instead, much like a psychologist, he astutely revealed and dissected the subjectivism and relativism inherent in the human mind and human cultures. He was no less critical of Christianity than modernism and argued that moral systems, secular or religious, were ways to control and maintain power over the masses. Power, creativity, and individualist expression were the key ideals and concepts within his philosophy.[154]

Inspired by Schopenhauer's belief in the primacy of will, Nietzsche argues that life, thought, and culture are all manifestations of the **"will to power."** The "will to power," which can also be understood as the urge for individual freedom and self-expression, is the life affirming force in reality. Nietzsche, in true Hegelian fashion, views human history as a clash or conflict between those individuals who possess and embrace the "will to power" and those people (the masses and the poor in life) who do not possess such inner vitality and drive. Nietzsche sees all the higher elements of human culture as having been created by those who possess and exercise the "will to power." Such individuals are the artists, warriors, and conquerors within human history. Such individuals create their own values, rather than submitting to the conformist values of the masses. In effect, Nietzsche, like other Romanticists before him, elevates the "heroic" archetype to center stage in human history. Moreover, he sees this archetype as providing the guiding light for humanity as we move into the future. It is not reason or faith, or social harmony and order that will be our salvation – it is the individual life affirming expression of the "will to power" that will create a better world. As he states, "I teach the No to all that makes weak – that exhausts. I teach the Yes to all that strengthens, that stores up strength, that justifies the feeling of strength."[155]

Founded on the ideals of power, strength, and life affirmation, Nietzsche sees his philosophy as optimistic in contrast to Schopenhauer's pessimism. He believes that he sees what is good and positive in life and he prophesizes a philosophy of hope for the future built upon the eventual realization of his ideals. Yet in contrast to this philosophy of hope for the future, he is highly critical of the most important features of the modern world. There is a very pronounced juxtaposition and contrast in his writings between the innumerable failings of what is real and, in his mind, what is ideal. Highly influential, he is the great critic and dark shadow over the promises and reality of modernity. In examining in more detail his philosophy, I begin with those key metaphysical, epistemological, and ethical ideas he supports, all of which are connected with his theory of the "will to power," and from there move to his most important and basic criticisms of modernism. All of his criticisms make sense given the basic philosophical views that he supports.

Beginning with his metaphysics and theory of reality, Nietzsche generally supports a Dionysian model of existence. One could also say that he follows Heraclitus. Reality is turbulent, filled with "sound and fury," in flux and conflict. To quote, Nietzsche's world is "...a monster of energy, without beginning, without end...a play of forces and waves of forces, at the same time one and many...eternally changing, eternally flooding back, with...an ebb and flow of its forms...". This view is totally at odds with the Newtonian image of an orderly and harmonious universe. Nietzsche rejects the scientific notion of cause and effect as being an accurate description of change, instead arguing that change is due to a struggle of power between the different entities in nature. At times he does acknowledge that there is a balance in nature between order and chaos, and he admires classical art for often synthesizing the two poles of reality, but in general, he emphasizes the chaotic and Dionysian pole of existence. Further, he feels that we should embrace and participate in this "Heraclitian fire" for it is life affirming and the source of all creativity.[156]

Ethics should serve creativity and self-expression - it should strengthen rather than weaken and constrain. Values should be life affirmative. To use a modern expression, values should "self-empower." Consequently, there is an important sense in which Nietzsche argues for an "individualist" ethics - that is, an ethics created and affirmed by the individual that serves the individual. Russell argues that ultimately Nietzsche values war and pride, and elevates the warrior above the thinker-philosopher as the ideal human. Values should serve the "will to power." At times Nietzsche speaks as if "might makes right" - that those who are strongest (possessing the greatest "will to power") deserve to determine their own destiny as well as the destiny of others.

Yet, Russell also notes that Nietzsche values art, literature, and creativity. Nietzsche, in fact, believes that ethics should actually be subordinated to art and aesthetics. For Nietzsche, "art represents the highest task and the truly metaphysical activity of this life."[157] In this regard, Nietzsche is a true Romanticist elevating beauty above the good, or more precisely turning beauty into the ultimate good.

Nietzsche is seen as championing a perspectivist theory of knowledge, but he also argued for a motivational theory of knowledge. All human beliefs are relative to some perspective and are expressions of some motive. He argued that there were many different valid ways of looking at reality. There are always multiple perspectives – multiple points of view – and we should cultivate the habit of trying to see things from different perspectives to gain a better understanding of reality. There is no absolute or single view that is the ultimate or best truth. Further, knowing is always an act of creativity and inventing – it is always subjective. Beliefs are relative to time and place, and we should steer clear of attempts to make universal statements. He states,"...facts are precisely what there are not, only interpretations". Knowing (or believing), in fact, serves the "will to power" – there are no unbiased, dispassionate statements of facts. Assertions of fact or truth always serve personal ends – there is always a motive behind a belief. As he argues, "Ultimately, man finds in things nothing but what he himself has imported into them: the finding is called science."[158] In fact, Nietzsche contends that beliefs should be evaluated in terms of how they serve life. In this sense he is a pragmatist, abandoning the notion of absolute truth and instead looking for the functional value or benefit of a belief. In this philosophy of knowledge, Nietzsche dethrones both rationalism and empiricism, arguing that the "will to power" both determines what we believe and is the final criteria for deciding what we should believe.

Founded on these basic philosophical beliefs, Nietzsche articulates a multi-faceted and extensive critique of modern society. At a general level, he argues that modern society has fallen into a state of nihilism. Modern Western humans have lost faith and hope in both traditional Christianity and the promises of the Enlightenment and the theory of secular progress. In his mind, neither "world-view" delivered what it promised. Christian religion did not realize a heavenly and moral paradise on earth and secularization, science, and capitalism did not bring happiness, material abundance, and self-fulfillment to all people. Both worldviews, in Nietzsche's mind, were too absolutist, grandiose, and universal, presenting visions that presumably explained everything and provided a path of life for everyone. Because Westerners expected too much – thinking that there were eternal truths and absolute values – they became disappointed, frustrated, and disillusioned when these excessive and unrealistic expectations were not realized.

Further, he believed that modern society had become increasingly fragmented and that individuality and creativity had been repressed. In his mind, the West had lost social spirit and cohesion, as well as vitality in its people. Instead of having a sense of community, Westerners had become isolated and separated; equally, instead of a population of unique creative souls, most Westerners were conformists, unconscious "sheep" adopting a "herd" mentality. We all walked alone, in mindless uniformity, in a state of apathy.

Nietzsche, though reporting on the generalized nihilism of his time, was not in his own mind either a nihilist or a pessimist. He believed that the psychological and moral collapse of the West, which he predicted as spreading and deepening in the immediate future, would eventually turn itself around and a new heroic age would emerge. Nietzsche admired the classical civilizations of Rome and Greece, as well as Renaissance Europe, and he believed that a new age resonant with these earlier societies, one that once again affirmed life, creativity, and positive human values, would arise. [159]

Based on both his perspectivist theory of knowledge and his philosophy of "will to power," Nietzsche was especially critical of both authority and conformity, which are really two sides of the same coin. Nietzsche is well known for his pronouncement that "God is dead," by which he meant, at the very least, that modern Western humanity had lost faith and belief in God; (on this point he was simply reporting on the increasing secularization and religious skepticism of the West.) But Nietzsche was not just reporting on a sociological fact; he believed it was just as well that God was dead. Nietzsche saw Christianity as enforcing a "slave morality" on its followers. The good Christian was supposed to be obedient to the will of God, as described in the *Bible* and enforced by the leaders of the Church. Such an approach to life was anathema to Nietzsche's belief that humans should create their own values and not conform to some absolutist authority. Recall, from above, that Nietzsche believed all absolutist claims to knowledge and value are epistemologically in error. In Nietzsche's mind, there can be no God – no absolute source and authority on knowledge and the good. People who follow God are slaves and have abandoned or forsaken life.[160]

But Nietzsche was not just critical of Christianity in this regard. He saw conformity and submission to authority arising in many different aspects of modern life. Humans had become slaves to industry and capitalism, reduced to "industrious ants" serving the rhythm, tempo, and logic of machines and the goals of capitalists. Instead of embracing individual creativity, capitalist society valued money and utility. Nietzsche saw conformity in the social roles people adopt in following convention, again abandoning individual self-expression for a herd mentality. He also saw modern nationalism as contributing to the loss of individuality. According to Nietzsche, both the state and the machinery of capitalism destroy culture and produce mediocrity. Contrary to Marx, he did not think that what was needed was a social movement and rebellion against the oppressiveness of modernity, which would have just been another form of herd mentality, but rather individual transcendence. Finally, Nietzsche even saw science and reason as tyrants in that the philosophy of the Enlightenment presented science and reason as the absolute source of truth. Although the Enlightenment promised liberation from the authority and repressiveness of the church, superstition, and royalty, it simply created a new form of singular authority. Not that Nietzsche was opposed to science and reason, but based on his perspectivist theory of knowledge, he was simply against the idea that science and reason provides the absolute and only truth. He saw the presumed objectivity of science as a pretense.[161]

In general, Nietzsche was critical of all metaphysical schemes of thought, secular or religious. Aside from objecting to their absolutist claims of knowledge, he saw such systems of thought as life denying. Beginning with Plato's disenchantment with the world of time and aspirations toward an eternal realm, the history of metaphysics up through Hegel is a history of abandoning what is real for an idealized fictive realm. As Nietzsche states, "It was suffering and incapacity that created all after-worlds..." Life should rather be embraced, in all its turmoil, struggle, and internal contradictions.[162]

Because of his Dionysian philosophy and resonance with the chaos of life, Nietzsche did not shirk from self-contradiction. In fact, he saw as one central Apollonian myth the idea of a singular consistent self. Just as there are multiple perspectives on reality, there are multiple voices within an individual. Descartes's notion of a single rational subject entails a denial of the richness of mind and consciousness. The human mind shows the same chaos and diversity as the world. We are a multiplicity of drives and ideas.[163]

Nietzsche would replace modern man with a new vision of the ideal human. He believed that modern man, a victim of nihilism and conformity, was doomed. Instead Nietzsche argued for (or prophesized) the emergence of the **"overman"** who embodied those qualities that Nietzsche saw as life-affirming. In his famous work *Thus Spoke Zarathustra* Nietzsche describes the overman and his vision of the future of humanity. In announcing the death of God, the overman fully realizes the "will to power;" he is a "warrior of culture," a lover of dance and laughter, an artist and philosopher, and "aristocrat of the spirit."[164] This futurist image of the superior human though combines both pre-modern and modern ideals, for although Nietzsche was a critic of modernity, his emphasis on individual freedom and self-expression and the rejection of authority, especially religious authority, is a modern concept and his idealized love of the artist and warrior derives from classical thinking and his vision of the self-assertive conquerors of old.

Nietzsche's influence has been significant. There are interpretations of his philosophy that are positive, noting his accurate analysis of the problems of modernity and the nature of the human mind. But there are critics of Nietzsche as well who see his philosophy as providing a justification for war, conquest, inhumanity to man, elitism, and racial supremacy and prejudice. Bertrand Russell, one especially strong critic, sees Nietzsche's philosophy as an expression of a "lust for power" based on an excessively negative image of humanity, a lack of empathy, and a sense of deep fear. Rather than being life affirming, Nietzsche, according to Russell, is life denying.[165] But perhaps most importantly, Russell argues that a philosophical ethics can not be exclusively built upon self-expression, for humans are social creatures and we can not simply disregard the rights and the feelings of others, even if they are weak and poor of spirit, a central insight that Russell sees totally lacking in Nietzsche.

Perhaps the Romanticists were correct in their assessment that modernity and reason had not freed humanity but simply replaced one set of constraints with another. Yet the aspiration toward freedom and self-expression is clearly a modern ideal, and if the Romanticists embraced this ideal, then they were as much children of modernity as the philosophers of the Enlightenment. The strength and value of the Romantic perspective is to bring some balance to modern visions of progress and the good life. A philosophy of the future must speak to the heart as well as to reason. It must balance the Dionysian and the Apollonian. Individuality cannot be sacrificed in the name of social order and progress. There is chaos and uncertainty in life, as well as order and predictability. Life should be approached as a drama and a work of art, as well as through abstract theory. It is important to cultivate optimism about the future, but the optimism must be realistic. Contrary to the elevated visions of the eighteenth and nineteenth century, the dark side of humanity did not disappear. In the twentieth century it came back with a vengeance.

Darwin's Theory of Evolution

"There is nowhere anything lasting, neither outside me, nor within me, but only incessant change. I nowhere know of any being, not even my own. There is no being."

Johan Fichte

"Hence we may look with some confidence to a secure future of great length. And as natural selection works solely by and for the good of each being, all corporeal and mental endowments will tend to progress toward perfection."

Charles Darwin

In this final section of the chapter I describe the development of the theory of evolution from the seventeenth through the late nineteenth centuries. Although the theory of evolution is specifically associated with Charles Darwin (1809 - 1882) and his epochal work *On the Origin of Species by Natural Selection, or the Preservation of Favoured Races in the Struggle for Life* (1859), the theory emerged over a period of roughly two centuries and involved the contributions and discoveries of many significant scientists and writers. The evidence for evolution accumulated across many diverse disciplines of study prior to Darwin. Darwin put the pieces together and added the final central element - the idea of **natural selection**, which Daniel Dennett refers to as "the single best idea that anyone has ever had" - but the scientific concept of evolution had been growing, in modern times at least, since the time of Descartes and Leibniz.[166]

Still it is Darwin and his specific formulation of evolution that created the great public stir and controversy, raising the question of the origin and nature of the human species. Regardless of whether Dennett's view that natural selection is the "best idea" ever is true, Darwin's theory of evolution by natural selection has been the most influential scientific idea to emerge in the modern West, and next to the *Bible*, *On the Origin of Species* is probably the most important and influential book written in Western civilization. Darwin and his theory of evolution transformed the basic picture of nature provided by science, redefined the nature of humanity, overturned almost all metaphysical views in traditional world religions regarding the mechanism of creation and the origins of order, and laid the groundwork for a new view of time and the future. The philosophical as well as scientific implications of Darwin's theory impact many of the most fundamental issues of life and existence.[167] The Darwinian revolution in thought is still in progress today – its implications so profound and deep – that the ripples of the wave set in motion over one hundred and fifty years ago are still traveling outward, instigating counter-reactions and further elaborations across all spheres of human life.

Although the idea of evolution and the specific theory created by Darwin to account for the evolutionary process focuses upon natural and human history, its implications for the future are highly significant. As noted earlier, understanding history helps us to understand the future. Darwin, in fact, along with many other writers and scientists interested in evolution, drew a variety of conclusions concerning the future based on evolutionary ideas. What Darwin did was to identify a pattern to change in the flow of natural time and explain the natural causes of this temporal pattern. Evolution enriched and transformed the nature of historical consciousness and future consciousness. I examine in this section some of the more noteworthy evolutionary ideas on the nature of change and the future.

To set the stage, the modern story of evolution begins in the writings of two seventeenth-century writers, John Ray (1627 – 1705) and Thomas Burnet (1635 – 1715), who attempted to combine and synthesize the new ideas and methods of science with Christian doctrine regarding the history and nature of life and the planet Earth.

In *The Wisdom of God Manifested in the Works of Creation* (1691) Ray systematically describes in great detail a static vision of nature, covering life, the earth, and the heavens above. He assumed, as did most other writers of his time, that the forms of nature, from living species to planets, had been designed and created by God and, further, that the forms of nature had not changed since the original creation. Ray viewed the heavens and the earth as existing in a state of harmony providing stability and support for the existence of life, in particular human life. The heavens and the earth were created for us by God. The constancy of nature reflects the perfection inherent in God's creation. Perfection also shows up in the harmonious relationship of life and physical nature. Each form of life is uniquely adapted and fitted to its surroundings.

Ray was inspired by Newton and the promise of science. Newton had demonstrated that physical nature was governed by a set of scientific laws which gave nature a regularity and uniformity within all its processes. Ray applied this vision to all the diverse forms of nature. He believed that constant and pervasive laws governed biological and physical reality. But he also believed, like Newton, that these laws of nature had been created by God for a purpose. The stability and harmony of nature existed to support the existence of humans. Hence, Ray combined, as did Newton, a mechanistic view of nature where laws determine the processes within nature, with a teleological theory that implied that all these deterministic laws were created for a purpose. Like Newton, he saw no incompatibility or contradiction between these two views of reality, mechanistic and scientific and teleological and religious. Ray supported "experimental philosophy" and the scientific search for knowledge and saw himself, in this regard, as serving the will of God.[168]

Thomas Burnet also wrote on a large scale; his four-volume *The Sacred History of the Earth,* published during the 1680's, covered the entire past history of the earth as well as its predicted future leading to the end of time. Burnet, like Ray, attempted to integrate Biblical ideas regarding creation with Newtonian science and saw absolutely no contradiction between science and Christian religion. As did Newton and Ray, Burnet believed that the universe and the Earth had been created by God; (all three rejected the possibility that the universe and the earth were eternal). Burnet also thought that although the Earth and the heavens were shaped out of a primordial chaos by God, the Earth was initially created in a state of perfection and that, following Newton, God had also established the laws of nature at the beginning.

But given these basic assumptions and starting points, Burnet goes in a different direction than Ray. Burnet describes a transformative history for the Earth; according to Burnet, the Earth has not been always the same. Burnet describes the Earth in the beginning as existing in a state of perfection – a paradise which contains the Garden of Eden and its original sole human inhabitants, Adam and Eve. Yet due to the repeated sins of humanity throughout early history, as well as the pre-ordained plans of God, Burnet contends, following the storyline in the Old Testament, that the Earth was subjected to a great flood or deluge which transformed its original perfection into the ragged, cavernous, and mountainous disarray of continents and islands existing in the present day. For Burnet, the Earth is now a "cracked ruin" – a poor resemblance to its initial perfect state. In essence, Burnet invokes the *Bible*, with its ideas on human sin and disobedience and the great flood, to explain contemporary geography.

Burnet, like most Christians of his time, believed in the literal truth of the *Bible*, which included the story of the flood and Noah's ark. Burnet also believed, based on the famous calculation of Bishop Usher (using personal chronologies described in the Old Testament), that the Earth was approximately only six thousand years old. His future predictions of the Earth also derived from the *Bible*. He foresaw a coming conflagration that would return the Earth to a state of perfection and paradise where Christ would rule over the earth as prophesized in the

Bible. In this regard, note the similarities with St. Augustine, as well as with Christian millennialism.

In this comprehensive description of past and future, Burnet combines linear and cyclic conceptions of time. Although there is a history to the Earth, involving change and a direction to the change, there is also a cycle to the entire process. The Earth begins in perfection, degenerates through destruction, and then once again rises to perfection in the end. The paradise and state of perfection at the end of time though is different than at the beginning. A story and narrative has unfolded – mankind falls but is redeemed and Christ rules over the Earth in the grand conclusion and finale of history. Geology is clearly mixed together with theology.

Although Burnet sees a restoration of perfection in the future – an optimistic vision regarding where we are heading – he attributes this restoration to the will of God. When Burnet looks at nature as such, he sees everything as decaying and wearing down. There is no intrinsic restorative power in nature – it is only God who creates or restores order according to Burnet. Yet because he also attempts to follow Newton and the new principles of science, Burnet thinks that the laws of nature directly determine the course of events within the world. Burnet combines God and the laws of nature by arguing that God created the laws such that these laws will produce the very historical changes intended by God and described in the *Bible*. The laws of nature – created by God – produce historical change and lead to both the deluge and decay and the eventual restoration to perfection. Again, as with Ray, there is an attempt to combine teleological and mechanistic views of change and find a way to reconcile the idea that God directs history with the scientific idea that the laws of nature determine history.[169] These efforts to combine science and religion on the part of Burnet and Ray are representative of popular thinking during their time and even into the following century. According to Watson, up to the end of the eighteenth century, the main concern of geologists and other investigators of natural history was to find ways to reconcile the *Bible*, and in particular the book of *Genesis*, with the findings of geological and natural science.[170]

In the writings of Ray and Burnet, the stage is set for the ongoing tension and struggle that shows up throughout the historical development of evolutionary thought over the next two centuries. Writers and researchers repeatedly attempted to find ways to make the accumulating scientific evidence consistent with the Biblical story of creation and, in general, make scientific ideas consistent with religious ideas, but the evidence would not fit and piece by piece the Biblical and religious ideas lost credibility and were replaced by a new way of thinking. This shift in thinking, as noted above, was the result of numerous discoveries and new ideas contributed by many individuals before Darwin. Darwin did not overturn the Biblical story of creation, many people did, and it was not that most of these researchers and writers did not want to preserve the Biblical story – far from it – they kept trying to preserve whatever elements of it they could, but the evidence just kept forcing them further and further away from the Biblical view.

A case in point is the issue of fossils. The famous scientist Robert Hooke (1635 – 1703), a contemporary and rival of Isaac Newton, argued that fossil bones of animals that no longer seemed to exist in the modern world appeared to indicate that once there had lived animals on the earth that had since become extinct. Hooke's hypothesis though was rejected by most early scientists, since it seemed to contradict the Biblical story that God had made all the various life forms at the beginning of creation and that since then, there had not been any significant change. Noah presumably did take two members of every species on the arc. The idea of extinction ran against the popular view that God's initial creation was perfect – if it was, then why would some species go extinct? It was not until a hundred years after Hooke had proposed his ideas on extinction that popular opinion began to shift. So many different fossil skeletons had been uncovered by then that did not match any existing animal species that it seemed increasingly irrational to deny the evidence.[171]

If evidence grew in the eighteenth and nineteenth centuries for biological change on the earth, what new facts and ideas emerged during this period regarding change in the heavens above? To recall, Plato viewed the heavens as both perfect and stable, and Christianity inherited this notion of a stable and harmonious heaven. Galileo found himself in trouble with the church in arguing that the heavens were neither perfect nor stable. To recall, Descartes hypothesized that the universe had evolved through the interaction of physical forces and Liebniz also believed in cosmic evolution (as well as biological extinction and evolution), but the Platonic-Christian view of a stable perfect heaven, supported by Newton's idea of stable laws and a static universe, was the dominant view through the seventeenth and early eighteenth centuries. But following the lead of Galileo, the astronomer William Herschel (1738 - 1822) (the discoverer of the planet Uranus) began to accumulate mounting evidence, based on improved telescopic devices, that there were fundamental changes occurring in the heavens. Through detailed observations of stars and gaseous nebulae throughout the sky, Herschel hypothesized that the stars were not static or permanent but rather formed out of nebulae. Creation was not something that happened once and for all at the beginning of time, but was ongoing. As with Galileo, who first saw the craters of our moon and observed the moons of Saturn and Jupiter revolving around these giant planets, improved observations and astronomical technology led to a change in thinking from a stable heaven to a changing one.[172]

Midway through the eighteenth century, the general theory of cosmic evolution reasserted itself in the writings of Immanuel Kant. In his book *Universal Natural History and Theory of the Heavens*, Kant argued in a vein similar to Descartes and Leibniz that evolution was a general process occurring through the universe. Kant believed that through the influence of the laws of nature, order and structure emerged out of amorphous and chaotic beginnings. Kant saw his explanation of cosmic order as following from Newtonian ideas, specifically invoking natural laws to explain natural phenomena, but Kant's vision of the heavens was decidedly anti-Newtonian in describing the cosmos as dynamical and evolutionary, rather than static and stable. Yet, Kant still wanted to preserve a place for God and argued that the evolutionary and creative effect of natural laws was evidence for a divine intelligence behind the creation of these natural laws.[173] Kant saw cosmic evolution, as well as social historical evolution, as teleological, being guided by the plans and purposes of God.[174] Thus Kant typifies a common view that would appear and reappear over the next centuries. Argue for the reality of evolution and the determinism of natural laws, but assume that the laws of nature were created by God and the resultant evolutionary process was God's intention – evolution is the mechanism of God's creation.

The mathematician Pierre-Simon LaPlace (1749 – 1827) however, found no evidence of God within the emerging sciences of astronomy and astrophysics. LaPlace adopted Herschel's idea of "nebular condensation" to explain the formation of the solar system and, in general, saw the heavens as undergoing change due to the inherent forces and laws of nature. To recall, LaPlace supported the philosophy of determinism, arguing that given complete and detailed knowledge of the universe, it would be possible, in principle, to predict the entire future history of the cosmos. LaPlace believed that natural events followed from natural laws and could see no convincing reason to add any additional factors (e.g., the intentions of God) into the picture if the laws of nature provided a complete explanation. He argued, "If we trace the history of progress of the human mind, and of its errors, we shall observe final causes perpetually receding, according as the boundaries of our knowledge are extended."[175] Final causes, which are central to teleological explanations, would include divine purposes or intentions behind the processes of nature. LaPlace saw final causes as unnecessary and a sign of ignorance. Before natural laws were identified, primitive humans explained natural occurrences as due to the intentions and actions of deities, but what explanatory function do such deistic beliefs serve if the processes of nature can be completely predicted from natural laws? Hence, in finding natural laws which explain the processes of nature, it seems gratuitous and unnecessary to continue to include God in the equations.

Thus we see in LaPlace the recognition that if natural laws completely explain natural phenomena, then spiritual and religious ideas about nature serve no function. Instead of trying to combine the scientific and the religious, LaPlace jettisons the religious since it seems superfluous. This realization repeatedly occurs in researchers and scientists in the coming century, as more and more of nature became intelligible through natural laws. Repeatedly, in one case after another, the need to postulate a teleological or supernatural force behind some natural phenomenon seemed unnecessary. When Darwin addressed the origin of species, he was able to provide an explanation in terms of natural law that seemed to make the idea of a supernatural creator superfluous.

If change became an increasingly obvious fact regarding the heavens, a similar realization steadily emerged regarding the Earth and the history of humankind. Although, as described earlier, pre-modern humanity was not totally oblivious to historical change, the depth, richness, and scope of humanity's understanding of the past were significantly limited. Furthermore, the Biblical account of creation, as noted above, was interpreted to mean that the variety of species had been fixed at creation. Also, based on the famous calculation of Bishop Usher (1581-1656), who argued that the earth was created in 4004 BC, most Western Christians believed that the world was relatively young and that humankind had been around since the original creation. Historical and scientific research beginning in the eighteenth century would drastically expand and alter these ancient and traditional views.

Understanding the modern development of the study of history is highly relevant to understanding the theory of evolution, for as the biologist Kenneth Miller points out, evolution, in fact, is a theory and description of history.[176] As the study of natural and cultural history developed in modern times an immense amount of evidence began to accumulate regarding the richness and vast reaches of the past. Relics, fossils, and other records and effects of the past have over the last few centuries steadily been unearthed, filling in more details and progressively revealing earlier and earlier beginnings for both natural and cultural phenomenon.[177] One of the great achievements of modern times is the discovery and dramatically heightened awareness of the vast legacy of the past and the breadth and depth of historical change. Temporal consciousness has jumped forward in the last few centuries.[178] The increasing sense of historical change and all the various forms of evidence supporting this enhanced understanding of the past provided Darwin with a foundation on which to build his theory of evolution. Evolution provided a systematic description and explanation for biological change that was connected to and grounded upon the emerging historical story of natural change.

Critical to the foundation for the theory of evolution was the discovery of **"deep time."** James Hutton (1726 - 1797), who is frequently identified as the father of modern scientific geology, published in 1795 his famous work *Theory of the Earth*, "the earliest treatise which can be considered a geological synthesis rather than an imaginative exercise," and described in this book a vision of the history of the earth that extended back in time indefinitely for millions upon millions of years. It is important to note that at the time Hutton published this book, the historical accuracy of the Great Flood as well as the entire story of creation as described in *Genesis* was not questioned. Yet Hutton's ideas would challenge all of this. He is famous for the statement describing the vast panorama of earthly time, "...we find no vestige of a beginning, - no prospect of an end." According to Hutton, the earth was very old - much older than most everyone had supposed.[179]

Through a detailed examination of geological strata and forms of deposits around the world, Hutton provided a scientific explanation for how the surface of the earth was transformed over time. He argued that geological change involved two complimentary processes, destructive deterioration and creative restoration. (Note the age old theme of creation and destruction in explaining change.) Contrary to the idea found in Burnet, that nature left to itself deteriorates over time, Hutton presented a great deal of evidence that nature also rebuilds and that new geological structures emerge while older ones wear away. Likening his model of the Earth to the Newtonian idea of nature as a perfect machine, Hutton viewed earthly time as an endless cycle of creation and decay – of becoming and passing away.

Hutton described this cyclical process entirely in terms of natural forces, without invoking any divine guidance or intervention – as had Burnet – and found such natural forces as sufficient to explain the present geological conditions of the Earth. Further, he contended that all these natural forces that existed throughout the history of the earth were the same forces we see at work in the world today. There was nothing unusual or miraculous about change in the past – the laws and forces of nature at work on the earth are constant and uniform through time. Thus Hutton's position is seen as a supreme example of the general theory of **uniformitarianism**. The Biblical view of the past invoked unique and catastrophic events, such as the flood, in describing the history of the Earth as well as the cosmos and, during Hutton's time, this perspective on history was identified as **catastrophism**. Hutton rejected catastrophism because he believed that natural change was due to uniform and constant natural forces – in the spirit of Newtonian science which described natural change as due to universal and constant laws.

Because Hutton supported a cyclical theory of time regarding the history of the Earth, he saw the earth as a stable system. Though there is constant change, the changes balance out and, to use modern scientific terminology, the earth exists in a state of "dynamic equilibrium," stability being maintained through balanced and cyclic oppositional processes. Further, Hutton saw life and the natural world as existing in a state of harmony with each other. Life fits the environment and the environment fits life – there is a complementarity here as well. Thus we have a perfect system, dynamical yet stable, where forces balance and all the parts fit together into a harmonious whole.

Because Hutton basically argued that regarding the earth, there is "nothing new under the sun," Stephen Jay Gould contends that Hutton had no real conception of history. What exists today is what existed in the past – there is no real change. So it is rather ironical that the man who opened up the vastness of time on the earth – who "discovered" deep time – ended up rejecting the idea that there had been any significant changes across the wide expanse of historical time. This extreme uniformitarianism regarding the past would, however, dramatically change in the century ahead.[180]

It is important to see the relevance of Hutton's ideas to both the theory of evolution as well as to our thinking about the future. Hutton transformed our understanding of time – time was vast and the events in time could be accounted for through natural law and natural forces. Just as the past extends backwards much further than previously supposed, the future stretches out ahead of us "with no prospect of an end." Just as science was demonstrating that the natural world as observed in the present could be explained through natural laws and natural forces, Hutton proposed that both the past and the future could be understood in a similar fashion. Further, the state of the present could be explained in terms of forces at work throughout the past. What Hutton provided for evolutionary theorists, including Darwin, was a much expanded panorama of time in which natural forces could produce fundamental change in the world. Again, with a degree of irony, though Hutton saw the forms and forces of nature as uniform across time, it was his discovery of deep time that opened the possibility that given a sufficient amount of time big changes could occur in the biological realm.

Because Hutton adopted a cyclical vision of earthly time with no real historical change, within the confines of this model there is no explanation of origins or beginnings. Extending backwards into the past, the world simply exists just as it does today. But how did this world and the natural laws that govern it come into existence? Or are the world and its laws eternal? And how does Hutton account for the perfect harmony between life and the natural world? On these questions, in spite of Hutton's efforts to be scientific and empirical regarding the study of the earth, he ultimately resorts to a supernatural or divine explanation for the origin of the earth and the harmony of nature. So although Hutton does attempt to provide a naturalistic explanation for geological structures and geological processes, he does not extend this naturalistic perspective to explaining how the forms or laws of nature came about. In this regard, he ends up with a view similar to Newton's; natural laws explain how the world operates and how it is ordered, but God is invoked to explain the origin and formation of nature. Both Hutton and Newton lived in a divinely created static world.

Though Hutton believed in a stable albeit dynamic and cyclic Earth, the accumulating number of animal and plant fossils seemed to imply something different, as least regarding the history of life. To set the stage for the transformation in thinking that occurred surrounding the meaning of fossils, let us begin with Carolus Linnaeus (1707 - 1778) who founded the modern system of biological classification. Linnaeus believed that there existed a natural order and arrangement of living species. He extensively studied the distinctive anatomical features of living forms and attempted to arrange and classify living forms based on observable similarities and differences. But like other early scientists, Linnaeus combined the methods of science with religious concepts as well. For Linnaeus, each species was a distinct and stable entity created by God at the beginning of time. Again noting the adaptive fit of living forms to their environment, Linnaeus invoked God to explain this harmony and concordance. Yet, Linnaeus was aware of the growing collection of fossils that seemed to be of species that no longer existed on the earth, but he had no clear way to account for their existence that did not upset his vision of a stable and perfect set of living forms created by God.

One particular point regarding the classification system created by Linnaeus that would open a Pandora's box in the century ahead was that he grouped humanity together with the apes as a common biological group. According to Linnaeus, no anatomical distinctive feature could be found that clearly separated humans from apes. In his time, this point created a controversy, since the common opinion in the West was that humanity was special and clearly distinct from the rest of the animal kingdom. What would be eventually suggested, because of this anatomical affinity of human and apes, was that somehow humans and apes were biologically related.[181]

Linnaeus's contemporary Comte de Buffon (1707 - 1788) thought differently about both fossils and the relationship among the various living species. Buffon believed, just as had Hooke a century before, that fossils were evidence for extinction; thus the perfect order of living forms was not so perfect. Further, Buffon took a naturalist point of view regarding history and believed that natural forces, uniform throughout time, produced any changes that had occurred throughout history. Hence, Buffon thought that the extinction of living forms throughout history was due to natural forces. This idea would be adopted by Darwin a century later. In presenting this argument, Buffon distinguishes between the idea of uniformitarianism, which is that the same laws and forces have been in operation throughout the history of the earth, and the idea that there have been no significant changes in natural, and in particular, biological forms throughout history. Buffon accepted the first idea, but not the second one. In his mind, contrary to Hutton's uniformitarianism, constant laws and forces can, over time, produce changes in the specific forms that populated the natural and biological world. This is a critical insight, also central to Darwin's theory of evolution.

Buffon also anticipated Darwin on the cause of extinction. Specifically, Buffon saw the natural world as undergoing changes through time, and hypothesized that different forms of life may have become extinct because they could not survive within changing environmental conditions. Buffon not only acknowledged the reality of biological extinction, he also entertained the possibility that new forms of life could arise within nature as well.

Contrary to Linnaeus, Buffon did not see the biological world as a set of clearly distinct and separate species, but rather as one great whole where differences among living forms involved variations and gradations rather than clearly distinct and separate groups. In Buffon's mind all of life was connected and across the vast domain of living forms there were innumerable commonalities of structure and function. Life does not appear to be a set of separately created forms, but a huge family of interrelated beings. Again, this idea anticipates Darwin and the evolutionary view that living forms are all connected via common descent. Buffon did not propose a theory of evolution – for one thing he was not aware of the vast stretch of deep time that Hutton was to uncover in the decades ahead – but Buffon did grasp the notion of biological change and biological interconnectedness, two key elements in Darwin's theory.[182]

The founder of modern paleontology, Georges Cuvier (1769 – 1832), extensively studied, catalogued, and classified existing fossils, including recently uncovered mastodon and mammoth skeletons. Based on the pervasive fossil evidence that had accumulated by the late eighteenth century, Cuvier championed the idea that there had been significant mass extinctions in the past. Cuvier had a powerful influence on the scientific community because of the large amount of evidence and detailed analysis he presented on fossil remains. After Cuvier it no longer seemed realistic to deny extinction as a fundamental feature within the natural history of the past.

Cuvier believed that the scientific evidence pointed to a series of pervasive or "catastrophic" upheavals in both biological and geological history. He opposed the uniformitarianism doctrine that natural change in the past had been slow and steady. Catastrophic geological changes produced mass extinctions of innumerable life forms which according to Cuvier, were followed by creative outbursts of new life forms. To some degree, more recent geological and fossil evidence has supported Cuvier's theory of catastrophism, at least in so far as there is strong evidence for mass biological extinctions coupled together with significant geological and meteorological changes. Cuvier, though, believed that the creation of new life forms was connected with the hand of God and not due to naturalistic forces, and he did not think that new life forms evolved from earlier life forms.[183]

One important discovery during the time of Cuvier that was crucial in understanding the history of the earth was made by the British "surveyor and self-made engineer," William Smith (1769 - 1839). What Smith realized in his examination of geological strata was that different strata were associated with distinctive sets of different biological fossils. Each geological layer or strata, as a record of the history of the earth, was uniquely connected with a biological record of past living forms. Paleontology and historical geology come together in a unified image of the past. Cuvier was aware of this highly significant connection between geological and biological records of the past in formulating his views on the history of the earth.[184]

Another significant development that contributed to the new way of thinking about history and time was the scientific work going on in archeology and cultural history. The Biblical account of creation fitted together explanations of the origin of the physical universe, biology, and the history of humankind. Just as investigations in geology and paleontology were unsettling Biblical views on the history of the physical and biological realms, the scientific study of human history was producing some challenging facts and conclusions as well. By the early decades of the eighteenth century, archeology had unearthed ancient relics and repositories of long vanished cities and cultures that seemed to indicate a longer, more ancient history for humanity than what was described in the *Bible*. Additionally, there appeared to be important pieces missing, as indicated by archeological findings, in the

Biblical telling of human history. In fact, these scientific and historical investigations instigated a rise in Biblical criticism regarding the accuracy of the *Bible* as a record of the past. Moreover, for centuries many people had been aware of primitive stone tools that had been uncovered throughout Europe which seemed to have been made by humans from some much earlier era. As the study and collection of these fossil tools advanced, it became apparent that they showed a developmental progression in style and sophistication. There were some writers who suggested that there had been humans prior to Adam and, based on the evidence and following an "evolutionary" logic, argued that humankind had progressed from exceedingly primitive beginnings to our present civilized state. Hence, while geology and paleontology were creating a history of nature that appeared to contradict Biblical accounts - especially in terms of the deep expanse of time in the past - archeology was putting together a history of humanity that also challenged the *Bible* and again seemed to require a much more expanded view of the past. Further, these new views of geology, life, and the history of humanity seemed to fit together into a coherent whole. A new story of history was emerging.[185]

Moving on to the next important individual in this history of evolutionary thought, Jean de Lamarck (1744 - 1829) is frequently identified as the person who proposed a mistaken and scientifically untenable theory of evolution based on the idea that offspring inherit the "acquired characteristics" of their parents. For example, if parents exercise their physical bodies and increase their muscular strength through this process, Lamarck believed that this acquired increased muscular strength would be passed on to their offspring. This idea of inherited acquired characteristics runs counter to modern genetic theory - according to contemporary scientific thinking, the experiences and activities of parents do not affect the genes that they pass on to their offspring. Lamarck had no conception or knowledge of the nature and operation of genes and the role of genetics in evolution.

Although Lamarck's theory of evolution through acquired characteristics was generally rejected by the scientific community in the decades to come, it was Lamarck who first proposed a comprehensive theory of how the evolution of all of life from simple beginnings could have occurred through entirely natural forces. Lamarck may have been wrong regarding the mechanism that produced evolution, but he clearly articulated a general description of evolution and provided the first naturalistic explanation of how it happened. This is a highly significant shift in thinking, for almost all modern scientific writers prior to Lamarck invoked some type of divine or supernatural force in explaining some aspect of life and natural change.

A fundamental assumption in Lamarck's theory of evolution is that he saw species as transformative rather than static or stable. For Lamarck, the fossil evidence seemed to indicate that the forms of life had changed throughout history. Species are mutable. It is only from our very limited time perspective that life and the environment seem stable, but the geological and fossil evidence tells a different story. Life is adapted to the environment and will remain stable if the environment remains stable, but the environment is not always stable. If the environment undergoes change, life will change with it.

Lamarck believed that present life forms evolved from earlier life forms and that there was a general overall direction in nature from simpler life forms to more complex, organized, and intelligent life forms. When the environment undergoes changes, life is challenged to adapt to these changes, and living forms extend their capacities and abilities in order to continue to exist. The results of such adaptive efforts get passed on to offspring. This theory of adaptive evolution through acquired characteristics was applied by Lamarck to humans as well. He argued that humans arose out of apes - apes that were pushed into having to expand their intelligence in order to continue to survive.

The idea that there is a direction to change is basic to Christian and other teleological views of time - the direction is set by God and determined by God's purpose. Yet, Lamarck excludes God from his theory of biological evolution and attempts to explain evolution in purely naturalistic terms. Evolution is due to ongoing adaptation. He still believes, however, that there is a direction to evolution – a direction that results from the forces of nature. To recall, supporters of the idea of progress, such as Spencer, Marx, and Comte, believed that there was a direction to history that could be understood and explained through natural, as opposed to supernatural, forces. Yet in the decades ahead, as evolution became the accepted explanation of biological change, the debate would arise over whether there was a progressive direction to evolution. Lamarck believed that there was such a natural direction to evolution, a progressive direction defined in terms of increasing complexity and intelligence. He believed that increasing complexity and intelligence was a result of having to adapt to a changing environment. But adaptive success may not imply increasing complexity – there are innumerable simple organisms that are adaptively very successful, having existed for millions of years. How is humanity somehow more adaptively successful than bacteria, sharks, or crocodiles? Is there a direction to evolution and is this direction somehow progressive? Can we understand the idea of progress in nature without the idea of God? These questions would be discussed and debated in the centuries ahead.[186]

Charles Darwin was aware of Lamarck's theory of evolution, as well as similar evolutionary speculations published by his grandfather Erasmus Darwin (1731 – 1802). Indeed, by the 1830's most stratigraphers believed that the fossil record showed improvement or progress in living forms across the different geological strata. They recognized that fossils of increasingly complex organisms appear as we move from older to more recent geological strata. In the 1840's the three great periods of geological history, Paleozoic, Mesozoic, and Cenozoic had been named and identified, each successive period corresponding to the emergence and flourishing of, respectively, fish and invertebrates, reptiles, and finally mammals. Further, the ideas of struggle, competition, biological divergence, and even evolution were "in the air," as topics of discussion and debate. In general, evolution had become a popular idea before Darwin created his theory of how evolution took place.[187] Two writers though would provide the final critical inspiration for Darwin.

Thomas Malthus (1766 – 1834) in *An Essay on the Principle of Population* (1798) argued that species populations left unchecked always increase geometrically, whereas available food supplies do not show a corresponding increase. Hence, the number of members of a species is always kept in check because there is not enough food to go around to feed a geometrically increasing population. Many members of each generation starve to death. The implication of Malthus's thesis is that there is ongoing competition in each generation for food and other necessary resources. This implication of a necessary element of competition in nature would provide one of the essential components to Darwin's theory of natural selection. Darwin, in fact, credited Malthus with providing the key inspirational element in formulating his theory of evolution.[188]

The second person who strongly impacted Darwin's thinking was Charles Lyell (1797 – 1875) the founder of modern geology. In his *Principles of Geology* (1830-1833) Lyell adopted and defended Hutton's theory of uniformitarianism and added new evidence and arguments to the idea that the constant and moderate forces of nature had produced all geological change throughout history. Lyell opposed catastrophism and was highly critical of Lamarck's theory of evolution. Contrary to the popular view of the day, Lyell saw no direction toward improvement in the fossil evidence. Instead he believed that each geological epoch brought with it a set of living forms specifically created and adapted to their particular environment. According to Lyell, natural history was cyclical and there was no overall direction across cycles.

What Darwin got from Lyell was a powerful sense of the vastness of natural history and the belief that constant and moderate natural forces, given sufficient time, could produce significant effects. Like Lyell, he accepted uniformitarianism. Darwin also found in Lyell a friend and intellectual colleague who provided a critical ear for his ideas. Lyell and Darwin corresponded extensively as Darwin was writing *The Origin of Species.* Yet, even after the publication of the ground-breaking work, Lyell remained unconvinced of either evolution or Darwin's explanation in terms of natural selection. While Lyell eventually did accept evolution, but he did not believe that natural selection could explain the emergence of new species.[189]

Gould argues that Lyell conflated and confused the different meanings of uniformitarianism. Like Hutton, Lyell believed that the laws of nature were constant through history and that such laws were sufficient to explain geological change. But Lyell inferred from this premise of constancy of laws that the conditions of the earth did not dramatically change throughout history (history was cyclical) and that the rate of change was constant throughout history. Lyell associated catastrophism – abrupt and significant change – with teleological and Biblical thinking about history, but as Gould correctly points out, catastrophic change need not have supernatural causes. The collision of a meteor or comet with the earth, which presumably occurred coincident with the extinction of the dinosaurs, was a natural event that produced fundamental and pervasive ecological changes. Lyell believed that the fossil record, which seemed to indicate abrupt and pronounced changes in biological populations, reflected incomplete evidence and that with accumulating evidence the suggestion of sudden big changes would disappear. Darwin also took this the position that there were no catastrophic changes in ecosystems, an idea that came to be referred to as gradualism. But Lyell also believed that if there were uniform laws at work, then there couldn't be any pronounced and directional changes across even larger periods of time.[190] Darwin did not accept this meaning of uniformitarianism; constant laws producing gradual changes, in Darwin's mind, could produce big changes over sufficient time. The question for Darwin was how.

Darwin's theory of evolution through natural selection is the essence of simplicity. Thomas Huxley, perhaps Darwin's strongest advocate, remarked that when he read Darwin's explanation of evolution it struck him how obvious the explanation was and Huxley wondered why he hadn't thought of it himself. Yet, even if the theory is simple, the implications of the theory are enormous. There have been numerous and varied interpretations of the meaning and significance of Darwin's theory.

As the historian of science, John Green, notes in *The Death of Adam*, all the elements of Darwin's theory were in place by 1818. Evolution had emerged as a popular scientific idea in both astronomy/cosmology and paleontology. Hutton had demonstrated that the earth was very old – much older than previously supposed – and that natural laws could account for present geological conditions. Cuvier had clearly demonstrated the factual reality of extinction and provided the methodological principles for reading the fossil record. Buffon and others had noted the ongoing variability in species, as well as suggesting that extinction may be a result of the ongoing struggle for survival. Finally, Malthus had shown that reproduction rates invariably exceed the resources needed for survival. The idea had even been suggested that natural selection of more favorable variations could be used to explain the evolution of the different races of humanity.[191] As noted earlier, what Darwin accomplished was putting together the accumulated pieces of research and theory produced by others. Darwin found the key to creating his theoretical synthesis in the idea of natural selection.

Darwin's argument is that life forms exhibit variation (perhaps random variation) in offspring produced in each generation; that given the limited food and resources in the environment too many offspring are produced for all to survive; and therefore there is natural competition over resources among the members of a species. Because there is variability among the species, some members will possess greater abilities for finding resources and staying alive. Those members possessing these favorable traits will survive and pass on those favorable traits or abilities to their offspring. Favorable traits steadily accrue and magnify over successive generations due to the ongoing process of natural selection of those members of a species better able to survive. Given sufficient time this ongoing selective process produces the evolution of new species and eliminates various species that are not able survive.[192]

As the noted philosopher and social thinker George Herbert Mead (1863 - 1931) pointed out, what Darwin provided in his theory of evolution was a general explanation, in terms of a fundamental natural law, for the great variety of different existing species. Species or biological forms were not created individually (for example, by God), but rather a universal and ubiquitous principle was responsible for all species. Form was not assumed as a given, but rather, form was the result of a dynamical process in nature. In essence, instead of simply stating that God made the forms of nature (without any further explication) or that forms are eternal (as in Plato), Darwin argued that at least biological forms had evolved, and he provided a real, understandable explanation for how this happens.[193]

The argument is often made that Darwin's theory implies that evolution occurs by chance. Such an interpretation of Darwin is at best a half-truth and it is exceedingly misleading. As just noted, Darwin provided an explanation of evolution through natural law - the antithesis of randomness or chance. It is true that reproductive variation may be random or due to chance, but the law of natural selection goes to work on these variations, perpetuating those members of a species that are most adaptive or capable in the environment. The conditions of the environment are clearly not random and ultimately it is the resources, opportunities, and dangers of the structured environment that select which members of a species survive. The more accurate and complete description of Darwin's theory of natural selection is that it combines chance and natural law providing an alternative explanation for species to that of divine creation.[194]

Clearly Darwin saw order and lawfulness in the evolution of life; what Darwin did not see was purpose or intelligent design. Evolution occurs through natural selection and there is no overall purposeful direction to this process; there is no need to postulate a guiding force or intelligence behind the process. There is no plan or goal to the evolution of life – natural selection explains why life evolves.[195] Natural selection also explains why living forms seem so well adapted to their environment – if a living form is maladapted, it dies, and due to the ongoing competition among members of a species, only those most capable pass on their traits through inheritance. Adaptation or the harmonious relation of life to the environment is selected for. There is no need to postulate an intelligent or purposeful force that creates life forms that are fitted to the environment.

Darwin's abandonment of purposeful design or direction in his theory instigated a great deal of controversy. As described in this history of evolutionary thought, most scientists and philosophers prior to Darwin accepted the Christian idea that there was purpose and intelligence behind both creation and natural change. It seemed to Darwin's critics that his theory of evolution made God unnecessary. Also in jettisoning the idea of purposeful direction, the future no longer seemed certain or secure. The future, within a Darwinian universe, was not being directed or guided toward some divinely determined end. Critics of Darwin found this loss of purposeful and divine direction to time abhorrent and totally disconcerting. What was the point – what was the meaning of it all, if there was no divine purpose guiding natural change?

Though Darwin was a scientist in the tradition of Newton, believing that there were natural laws to explain the order and processes of nature, Darwin's theory of evolution constitutes a real break with one fundamental assumption in Newtonian science. Newton believed that order was imposed on nature, directly through the laws of nature but ultimately through the hand of God who created the laws. Newtonian science is "top-down" – order comes from above and is imposed on nature below. Darwin turns this view upside down. As noted above, biological forms are not given or created by God; the forms of life arise through natural forces, and in particular, natural selection. As Tanner Edis states "...Darwinian evolution...radically undermines the whole top-down universe, situating creativity squarely in the material world."[196] For Darwin, biological order arises from the bottom-up. In essence, Darwin abandons the idea that order in nature requires an intelligent higher form producing that order. This bottom – up evolutionary perspective would permeate out through much of modern science in the coming century.

Another way to state this last point is that Darwin attempted to explain the complexity and organization of the natural world without assuming that something just as complex already exists which gave the world its complexity and organization. For Darwin, complexity arises in life from simple beginnings; complexity is not imposed on life by a complex Creator. The direction in evolution is from the simple to the complex, and Darwin believed that he had discovered a naturalistic mechanism that would explain how something more complex could arise out of something simpler.

Although Darwin did not see purpose or intelligent design in the evolution of life, he did believe that evolution lead to improvement – that is he saw evolution as progressive. Accordingly, he believed that natural selection led to improvement or progress. As Nisbet points out, Darwin often used the words "progress" and "evolution" interchangeably. Darwin spoke of "higher" and "lower" life forms, equating lower with simpler, older, and more primitive species. Higher life forms were more complex, and hence, similar to Lamarck, Darwin used increasing complexity as a criterion for defining evolutionary progress. Darwin also saw in animal and human evolution both increasing intelligence and increasing moral capacity. He believed that the future would see humanity further evolving in both intelligence and morality. So although there was no apparent purposeful direction to the evolution of life, there seemed to be for Darwin a naturalistic direction to evolution – a direction that was progressive. This direction, though, emerged out of nature.[197]

Another central point in Darwin's theory that would generate intense debate and controversy concerned the origin of humanity. Christianity saw humankind as a special creation of God. Humans were clearly different from animals, not only possessing rational intelligence and a moral sense, but also, according to Christianity, an immortal immaterial soul. It was evident to many of Darwin's critics, even with the publication of *On the Origin of Species,* in which Darwin does not directly address the issue, that the theory of evolution implied a totally different explanation of the origin of humans. Darwin removed all doubt concerning his view of humanity with the publication in 1871 of *The Descent of Man and Selection in Relation to Sex.* As he states in this book, "The main conclusion here arrived at, and now held by many naturalists who are well competent to form a sound judgment, is that man is descended from some less highly organised form." To recall, Lamarck had argued that humans were descended from apes, and there were many other scientists and philosophers, Darwin notes, who also held such a view. What Darwin provides in *The Descent of Man* is an explanation of the origin and development of humanity in terms of his particular theory of evolution through natural selection.

From a Darwinian perspective, humanity is not distinct from the rest of life. Humanity is part of nature and is connected with the rest of life through common descent. Our ancestry can be traced back to the simplest of creatures that populated the world millions of years ago. Humanity shows great similarity and commonality with other species, and in particular primate and mammal species. Placing humankind within nature – rather than separate from nature – was a real blow to humanity's ego and totally contradicted Christian doctrine regarding the special creation of Adam and Eve. Further, many Christians worried that if Darwin was right, then what happens to the idea of a human soul?[198]

Darwin is particularly concerned in *The Descent of Man* to demonstrate that not only are there innumerable anatomical and physiological commonalities between humans and other life forms, but there are clear connections between humans and animals, especially higher animals, regarding intelligence and morality – two of the presumed distinguishing characteristics of humans. Darwin's argument is that intelligence and morality evolved in degrees from lower animals up through higher animals and eventually humans. Adopting his "gradualist" position on evolution, he wanted to demonstrate that nothing "catastrophic" or special occurred in the emergence of humans. For Darwin the difference between humans and other animals, even regarding intellectual and moral abilities, is one of degree.

Further, following a line of thinking that in fact stretches all the way back to the ancient Greeks, but had become increasingly championed by various writers in his day, Darwin argued that humankind and human civilization gradually evolved from a state of primitive barbarism. Inspired by the study of apes and newly discovered primitive cultures around the world, a variety of scientists and scholars in the late eighteenth century, including Buffon, Jean Jacques Rosseau (1712 – 1778), Lord Monboddo (1714 – 1799), and Lamarck, all argued that human history should not be seen as a fall from perfection (the Garden of Eden or Golden Age myths), but rather as a rise from savagery. As I noted earlier, archeology had begun to uncover significant evidence that humankind had progressed from the primitive in the distant past to the more advanced over time. (The first Neanderthal skull was unearthed in 1856.[199]) Darwin embraced this theory of cultural history but combined it with biological history and his theory of evolution into one grand scheme of human evolution. For Darwin, the entire bio-social history of humanity is one of steady progression upward from the simple and the primitive to the complex, intelligent, and increasingly civilized. Herbert Spencer, for one, who both influenced Darwin as well as being influenced by him, took a similar view, and saw cultural development as a continuation and further elaboration of biological evolution. For Darwin and Spencer, as well as many other scientists and philosophers of the day who were influenced by them, evolution became an all embracing theory which explained the entire history of humanity and provided a conceptual framework for understanding the future of humankind.[200]

The promise of science, from its beginnings in Galileo, Kepler, and Newton had been to explain all of nature in terms of natural laws. Yet, as I have described in my history of modern science, philosophy, and evolutionary thinking, God and various supernatural forces were often included in both theories of nature and theories of humanity. Even Descartes who championed the scientific method and argued for an evolutionary explanation of the development of the astronomical and stellar systems, believed that God had created the universe and that God had given humankind a special "immaterial" mind (or soul) that was exempt from the deterministic laws of nature. But the philosophical movement toward viewing everything in naturalistic and secular terms steadily gained strength throughout the eighteenth and nineteenth centuries, and with the publications of both *On the Origin of Species* and *The Descent of Man*, humankind was assimilated into the scientific and naturalistic model and the role of God was pushed further out of the picture. For writers such as Spencer, who embraced the idea of evolution, evolution provided a naturalistic concept that comprehensively explained the history and present conditions of all of nature, including humanity.

Eliminating divine purpose and special creation in the grand scheme of things dramatically altered how humanity viewed itself in the context of the cosmos. As Tarnas notes, humankind "...was not God's noble creation with a divine destiny, but nature's experiment with an uncertain destiny." *Homo sapiens* are simply "a highly successful animal." According to Tarnas, this insight was both liberating and alienating, for divine purpose

402

gave humans a sense of meaning and security, as well as a yoke and constraint on behavior and thinking. Although evolution was embraced by philosophers of progress as a naturalistic justification and foundation for their belief in increasing improvement in humankind and human society, Tarnas argues that Darwin's theory of evolution also undercut the optimism of the Enlightenment. Not only did the theory of evolution seem to imply that Christianity, as well as other religious doctrines, was nothing but an "anthropocentric delusion," but given the dethronement of humanity from a special position within the cosmos, there was no longer any guarantee or promise for the indefinite success of the species. In the ongoing competitive reality of nature, who is to say if humankind will survive? No one is watching out for us. The future of humanity is uncertain. Further, culture and ethics can no longer be seen as having some higher, divine origin or justification – both civilization and morality are expressions of the evolutionary process, part of nature rather than being divinely ordained or created. Hence, the ideal future defined in terms of ethical standards is a creation of the human mind.[201]

Tarnas's interpretation of the effect of Darwinian thinking on humanity's sense of purpose and direction brings us back to the issue of progress and how it connects with Darwin's theory of evolution. The issue is complex and controversial. Tarnas argues that the theory of evolution through natural selection undermines both religious and secular ideas regarding the inevitable progress of humankind. Within a Darwinian context, there is no guarantee that humans will continue to progress, let alone survive. There is clearly no purpose or plan to evolution, thus undercutting teleological or religious ideas of progress, such as in St. Augustine and later Christianity. Yet Darwin, as noted above, saw evolution through natural selection as leading to progress and improvement. I have already described earlier how nineteenth-century writers such as Marx, Comte, and Spencer all believed that there was a natural law of progress in human history, and with the emergence of Darwin's theory, many philosophers of progress, Spencer being one noteworthy example, embraced the evolutionary perspective as providing a scientific explanation and grounding for the reality and inevitability of progress. This coupling of the ideas of progress and evolution became known as **"Social Darwinism"** – a very powerful intellectual force in the late nineteenth and early twentieth centuries. But Social Darwinism has been criticized on both logical and moral grounds.[202]

Let us try to disentangle and clarify the connection between evolution and the theory of progress. First, it should be noted that both Judeo-Christian and modern secular theories of progress adopted a linear and progressive view of time. Darwin's theory of evolution through natural selection also presented a linear and progressive view of time, at least in so far as Darwin believed that evolution led to increasing complexity and intelligence in the biological realm. The difference among these three views is that the Judeo-Christian vision of the world is teleological and supernaturally directed, whereas the secular theory of progress and Darwin's theory of evolution do not postulate any supernatural force producing and guiding progress.

Second, when Darwin first published *On the Origin of Species* he clearly appeared to be arguing that competition among members of a species led to evolution. Since by implication it was those members of a species best adapted to the environment that survived and reproduced, Spencer suggested the phrase "survival of the fittest" to concisely describe the Darwinian principle of natural selection and Darwin in later editions of *On the Origin of Species* included this phrase to describe his theory. Philosophers of progress, such as Spencer, viewed this competitive reading of Darwin as a justification for the economic and social reality of competitiveness in the modern European world. To recall, Adam Smith had made the idea of competition central to his economic theory of capitalism. Hence, defenders of capitalism embraced Darwin's theory as a justification for their economic system, and in general, competition in all aspects of life was presumably vindicated by Darwin's theory of evolution. This is the core

404

belief underlying Social Darwinism. Competition is how nature works and competition produces progress, hence competition is good. In fact, the twentieth- century historian J. B. Bury argued that the idea of progress "evolved" into the idea of evolution - evolution presumably providing a scientific justification and explanation for progress built on competition.

But there have been critics of this whole line of thinking connecting evolution with secular progress. Thomas Huxley (1825 - 1895), the great defender of Darwin's theory, took issue with Spencer on whether evolution, a scientific theory, provided a justification for a social or moral philosophy or way of life. How, Huxley argued, does one derive an ethics from statements of fact? "Let us understand, once for all, that the ethical progress of society depends, not on imitating the cosmic process, still less in running from it, but in combating it." What theorists of progress often conflate is the idea that there is a direction through history (however defined) with the idea that this direction is ethically a good thing. Religious views, such as in Zoroastrianism, Judaism, and Christianity, connected the direction of history with moral advance because God, the source of what is good, is orchestrating the process of history. Secular views of progress, which identified values such as rationality, freedom, wealth, and the well being of humanity as the criteria and goals of progress, provided non-religious ethical ideals to strive for. Progress could be defined relative to such ideals and it could be argued that history was moving toward the increasing realization of these ideals. But it can't be simply assumed that a direction to history means things are getting better in some ethical sense. Does survival of the fittest somehow translate into an ethical prescription? As Green puts it, survival is "a brute fact, not a moral victory."[203]

Moreover, for critics of Social Darwinism, the ethical implications drawn from a competitive model of evolution seem to be inhumane. Social Darwinism downplayed cooperation, nurturance, compassion, and community in favor of competition and individualism. It seemed to support a "law of the jungle" morality and philosophy. Also Social Darwinism provided an

405

ethically suspect and self-serving justification for the authority and privileges of those who possessed social and economic power. Social Darwinism supported the status quo and biological racism. Europeans could feel superior to other cultures and other races based on the idea that they were more advanced on the evolutionary scale, and rich capitalists could feel morally exonerated and superior to the poor and weak because they had earned their positions of power through the natural law of competition. Darwin's theory of evolution was interpreted to mean that nature rewarded individual competition and therefore competition was a good thing and those in positions of power were "superior" human beings. Further, the Social Darwinist emphasis on competition and "survival of the fittest" seemed to fit with Hegelian philosophy and its emphasis on conflict as a driving force in progress, as well as Nietzschean philosophy, with its emphasis on individualism and the will to power, thus providing a justification for war, conquest, and the subjugation (or elimination) of those not strong or fit enough to defend themselves.[204]

Yet in spite of such criticisms of connecting Darwin's theory of evolution with some type of moral or prescriptive theory of progress, it is clear that Darwin did believe in a naturalistic conception of progress. Evolution produces increasing complexity and intelligence and in this sense, evolution is progressive. Whether this direction to evolution is morally good or bad can be debated, and further, which moral ideals should be included in an ethical theory of progress can also be debated, but Darwin's theory does provide a factual hypothesis regarding the natural dimension of evolutionary progress. In this regard, Darwin's theory of evolution aligns with those theorists of progress, such as Smith, Marx, Spencer, and even Hegel, who believed that there was a natural and inevitable direction to time (a natural law of progress). One could say though, quite justifiably, that Darwin's description and explanation of the evolutionary process is unequivocally the most empirically corroborated theory of natural progress to ever have been advanced.

Darwin's ideas on evolution exerted a significant and continued influence on intellectual and popular thinking throughout the latter decades of the nineteenth century and into the twentieth century. Evolution became a general mindset that was increasingly applied to all aspects of natural and human reality. In the twentieth century other areas of science, besides biology and paleontology, including physics, cosmology, psychology, ecology, and anthropology were all influenced by the evolutionary framework. From an evolutionary perspective, nature is a dynamic and changing reality in which forms evolve in complexity rather than being created in their present state at the beginning of time. Order emerges from the bottom up in nature rather than being imposed from above by God.[205]

In order to understand the impact of evolutionary thinking on subsequent science, philosophy, and general intellectual thought, it is important to distinguish three different though related ideas. The first idea is that biological evolution has occurred within natural history. The second idea is Darwin's theory that biological evolution is due to natural selection. The final idea is the more general concept that all or most natural forms are transformative and evolve in complexity over time.

Darwin explained the increase in complexity in biological forms across time through natural selection, but to recall, Lamarck attempted to explain increasing biological complexity through inheritance of acquired characteristics. Both Lamarck and Darwin agreed that biological evolution had occurred – they differed in their theoretical explanations of the cause behind it happening. Darwin, in fact, though emphasizing the principle of natural selection (which was unique to his theory of evolution), acknowledged that there were perhaps other causative factors at work in evolution – even inheritance of acquired characteristics. It is often argued that evolution is a theory, rather than a fact, but as science writers such as Kenneth Miller and Richard Morris point out, this argument is mistaken and confused. The immense amount of fossil evidence collected over the last two centuries demonstrates the general fact of biological evolution; biological evolution is a general fact not a theory. Evolution can be "seen" in the fossil record. Darwin's specific explanation through natural selection is indeed a theory, yet even here, Darwin's theory is the most scientifically supported and substantiated of any theory of biological evolution.[206]

The phenomenon of biological evolution should be distinguished from the more general idea of natural and cosmic evolution. Over the last century and a half, since the publication of *On the Origin of Species*, many different aspects of nature have been re-conceptualized in evolutionary terms. It is not just that biological forms evolve, but all other forms in nature from atoms to molecules, planets, solar systems, and galaxies appear to have evolved or emerged from simpler beginnings. Further, building upon the early ideas of Descartes, Herschel, Leibniz, and Kant, contemporary cosmologists describe the entire history of the universe in evolutionary terms, in the sense that order and complexity has developed over time out of chaos and simplicity. Whatever mechanisms are behind natural and cosmic evolution (and there are theories), the general scientific consensus has grown that all of nature, including humanity, is a result of an evolutionary process.

As one general point regarding both biological and cosmological evolution, even if Darwin's theory of natural selection turns out to be only part of the story of how natural forms evolve, Darwin's belief that evolution can be explained through natural causes and natural laws has emerged as the guiding principle in contemporary science. Supernatural explanations of evolution have increasingly been pushed out of the picture. There continue to be explanations offered of both evolution and the origin of species that postulate some type of purposeful or "intelligent design" or creator. Such explanations, though, are no longer very popular in contemporary science and from a scientific point of view are highly problematic for a variety of reasons.[207]

Evolution has had a strong impact on philosophy as well as science. Many of the great philosophers of the late nineteenth and early twentieth centuries, such as Charles Sanders Peirce

(1839 – 1914) and Alfred North Whitehead (1861 – 1947), attempted to create comprehensive evolutionary theories of reality. Such theories argued in a broad vein that the universe was "process" and "change" rather than a collection of static entities, harkening back to Heraclitus, and that there was a direction to change – an evolutionary direction that involved such factors as the ongoing creation of novelty, increasing complexity and intelligence, and the emergence of mind and consciousness. Spencer, for one, saw mind itself as an evolutionary phenomenon – adaptively advancing in its capacities across time.[208]

Thus, given its influence on social and anthropological thinking, theories of progress, academic philosophy, and many diverse areas of science, it is clear that the impact of evolutionary thought has been immense. Regardless of whether natural selection turns out to be the total answer to biological evolution, Darwin's theory of evolution instigated and inspired a monumental and pervasive transformation across most areas of human research and thinking. As the historian Peter Watson so aptly puts it, "Evolution is the story of us all."[209]

One final issue to consider regarding evolutionary theory is the relative significance of competition and cooperation. As noted above, Social Darwinists embraced the idea of competition within Darwin's theory, but Darwin had other thoughts about evolution, especially as expressed in *The Descent of Man*, that revolve around such concepts as love, sympathy, mutuality, and cooperation. To recall, Darwin saw social institutions and morality in humans as a consequence of evolution though natural selection. He believed, though, that integral to human morality and social organizations was a highly developed capacity for concern and caring among humans. Behaviors, feelings, and modes of thinking connected with cooperation and mutual affection would be highly advantageous for the survival of the group. A cooperative group is much more efficient in facing the challenges of life than a non-cooperative group. Hence groups of humans showing greater cooperation and caring would survive, passing on the traits connected with cooperative ethics in its individuals, whereas groups of humans and the individuals in these groups not showing these traits as strongly would falter and fail. Those emotions and moral principles that bind humans together would be selected for within the evolutionary process. Beginning with Darwin, but continuing to the present day in writers such as E. O. Wilson and Michael Shermer, the argument has been presented that cooperative and caring ethics in humans is a consequence of evolutionary forces at work in our history.[210]

The evolutionary writer David Loye goes so far as to argue that Darwin in *The Descent of Man* emphasized cooperation and caring much more than individual competition and survival of the fittest. Further, according to Loye, Darwin saw the central driving force in the future evolution of humans as morality rather than cut-throat self-centered competition. Loye accuses earlier writers and thinkers of excessively emphasizing the competitive theme in evolutionary theory in order to scientifically justify the competitive and individualist behavior and philosophy of modern Western society. Further, Loye states that Darwin is not fixated on a biological level in understanding past or future human evolution; Darwin also discusses topics connected with mind, education, and intelligence extensively in *The Descent of Man*.[211]

The debate continues to the present in contemporary evolutionary thought: What are the relative roles of competition and cooperation in the evolution of life?[212] Although Darwin is more strongly associated with the theme of competition, especially within Social Darwinism, he saw the significance of both factors in understanding past and future human evolution. In earlier portions of this book, I described the related dual themes in theories of progress of individual freedom and social order: in Hegel, of opposition and synthesis, in Empedocles, of love and hate, in Bloom, of conquest and reciprocity, and in mythology, of the nurturing mother and violent hunter. There seems to be a common theme - a common debate - that takes various forms in understanding the forces that produce change in nature and in human history.

In summary, although there are numerous issues surrounding evolution, many of which Darwin himself was aware in his day, again the general point should be highlighted that the historical phenomenon of evolution, both biological and cosmological, is supported by a vast wealth of factual evidence. There are debates as to whether evolution is always gradual or whether there are sudden spurts at times. There are debates over whether natural selection can explain all of evolution or whether there is some other mechanism or cause involved. There is the issue above regarding the relative importance of cooperation and competition. There are debates over the connection between evolution and progress. But through a series of discoveries and insights beginning in the seventeenth century, and culminating in Darwin's extensive research and grand theoretical synthesis, the general phenomenon of evolution has become increasingly apparent to scientists, philosophers, and other students of nature. Creationism and other theories of divine intervention – in particular the theory that natural forms were instantaneously created sometime in the past – though highly resistant to the discoveries of evolutionary science, seem increasingly dubious. As Kenneth Miller states "It is high time that we grew up and left the Garden."[213]

Evolution not only overturned static creationism as expressed in Christianity, as well as teleological and anthropocentric views that saw the universe as purposefully created for the benefit of man, it also, perhaps most fundamentally, overturned the Platonic-Newtonian static image of the universe.[214] In the nineteenth and twentieth centuries, the pivotal theory that has influenced the course of most of science has been the theory of evolution. In a second major wave of scientific thinking, the concept of dynamic evolutionary change replaced Newton's stable and harmonious machine as the central idea in science. In the process of this deep and pervasive transformation in science, Newton's clockwork model of the universe was overturned, nature and reality were redefined, concepts of progress and time significantly changed, and the origin and development of the universe and humanity was opened to scientific study and debate. As a result, contemporary science views reality, time, order, and the future in predominantly evolutionary and dynamical terms and the entities and laws of nature no longer look so permanent. This evolutionary transformation drew its fundamental inspiration from Darwin.

References

[1] Durant, Will, and Durant, Ariel *The Story of Civilization VII: The Age of Reason Begins*. New York: Simon and Schuster, 1961; Nisbet, Robert *History of the Idea of Progress*. New Brunswick: Transaction Publishers, 1994.

[2] Best, Steven and Kellner, Douglas *The Postmodern Turn*. New York: The Guilford Press, 1997, Page 18.

[3] Anderson, Walter Truett *All Connected Now: Life in the First Global Civilization*. Boulder; Westview Press, 200, Page 32.

[4] Shlain, Leonard *The Alphabet Versus the Goddess: The Conflict Between Word and Image*. New York: Penguin Arkana, 1998.

[5] Ray, Paul and Anderson, Sherry *The Cultural Creatives: How 50 Million People are Changing the World*. New York: Three Rivers Press, 2000, Pages 70 - 71.

[6] Nisbet, Robert, 1994, Pages 79 - 81.

[7] Watson, Peter *Ideas: A History of Thought and Invention from Fire to Freud*. New York: HarperCollins Publishers, 2005, Pages 319, 331 – 333, 742.

[8] Christian, David *Maps of Time: An Introduction to Big History*. Berkeley, CA: University of California Press, 2004, Pages 342 – 352.

[9] Christian, David, 2004, Pages 354 - 364.

[10] Watson, Peter, 2005, Page 315.

[11] Polak, Frederik *The Image of the Future*. Abridged Edition by Elise Boulding. Amsterdam: Elsevier Scientific Publishing Company, 1973, Pages 70 – 73; Watson, Peter, 2005, Page 333.

[12] Watson, Peter, 2005, Page 355; Polak, Frederik, 1973, Pages 68 – 70.

[13] Watson, Peter, 2005, Pages 339 – 362.

[14] Watson, Peter, 2005, Pages 363 – 369.

[15] Watson, Peter, 2005, Page 369.

[16] Watson, Peter, 2005, Pages 370 – 371.

[17] Wikipedia – Robert Grosseteste - http://en.wikipedia.org/wiki/Grosseteste; Wikipedia – Roger Bacon - http://en.wikipedia.org/wiki/Roger_Bacon; Watson, Peter, 2005, Page 376.

[18] Watson, Peter, 2005, Pages 331, 370.

[19] Christian, David, 2004, Pages 335 – 342, 367 – 376.

[20] Tuckman, Barbara *A Distant Mirror: The Calamitous 14th Century*. New York: Ballantine Books, 1978.

[21] Shlain, Leonard, 1998, Page 309.

[22] Watson, Peter, 2005, Pages 320 – 326.

[23] Shlain, Leonard, 1998, Chapter Twenty-Nine.

[24] Watson, Peter, 2005, Pages 392 – 394.

[25] Watson, Peter, 2005, Pages 377 – 388.

[26] Watson, Peter, 2005, Page 396; Wikipedia – Petrarch - http://en.wikipedia.org/wiki/Petrarch.

[27] Watson, Peter, 2005, Pages 400 – 401; Wikipedia – Erasmus - http://en.wikipedia.org/wiki/Erasmus .

[28] Watson, Peter, 2005, Pages 397 – 398, 409 – 410.

[29] Watson, Peter, 2005, Pages 394 – 395, 403, 411 – 412.

[30] Wikipedia – Leonardo da Vinci - http://en.wikipedia.org/wiki/Leonardo_da_Vinci; Lombardo, Thomas *The Reciprocity of Perceiver and Environment: The Evolution of James J. Gibson's Ecological Psychology*. Hillsdale, NJ: Lawrence Erlbaum Associates, 1987, Pages 50 – 53.

[31] Nisbet, Robert, 1994, Pages 101 – 110; Morris, Richard *Time's Arrows: Scientific Attitudes Toward Time*. New York: Touchstone, 1986, Pages 65 – 67.

[32] Watson, Peter, 2005, Pages 460 – 461.

[33] Nisbet, Robert, 1994, Pages 118 - 124; Shlain, Leonard, 1998, Chapter Thirty.

[34] Shlain, Leonard, 1998, Chapter 30; Watson, Peter, 2005, Pages 462 – 469.

[35] Watson, Peter, 2005, Chapter Twenty-One.

[36] Christian, David, 2004, Pages 380 – 401.

[37] Whitehead, Alfred North *Science and the Modern World*. New York: The Free Press, 1925; Hall, A. R. *The Scientific Revolution 1500 – 1800: The Formation of the Modern Scientific Attitude*. Boston: The Beacon Press, 1954; Singer, Charles *A Short History of Scientific Ideas to 1900*. London: Oxford University Press, 1959; Cronin, Vincent *The View From Planet Earth: Man Looks at the Cosmos*. New York: Quill, 1981, Chapters Five to Eight; Ferris, Timothy *Coming of Age in the Milky Way*. New York: William Morrow and Company, 1988, Chapters Four, Five, and Six.

[38] Watson, Peter, 2005, Pages 474 – 475.

[39] Watson, Peter, 2005, Pages 476 – 477, 516 - 517; Wikipedia – Galileo Galilei - http://en.wikipedia.org/wiki/Galileo.

[40] Wilson, E.O. *Consilience: The Unity of Knowledge*. New York: Alfred A. Knopf, 1998, Pages 24 – 31; Nisbet, Robert, 1994, Pages 112 – 117; Tarnas, Richard *The Passion of the Western Mind: Understanding the Ideas that have Shaped Our World View*. New York: Ballantine, 1991, Pages 272 – 281.

[41] Wilson, E. O., 1998, Pages 24 – 30; Wilson, E.O. "Back from Chaos" *The Atlantic Monthly*, March, 1998b.

[42] Oliver, Martyn *History of Philosophy*. New York: MetroBooks, 1997, Pages 66 - 67.

[43] Watson, Peter, 2005, Pages 489 – 490.

[44] Wilson, E. O., 1998b.

[45] Oliver, Martyn, 1997, Pages 66 – 67; Solomon, Robert *The Big Questions: A Short Introduction to Philosophy*. 6th Ed. Orlando, Florida: Harcourt College Publishers, 2002, Pages 19 – 21; Russell, Bertrand *A History of*

Western Philosophy. New York: Simon and Schuster, 1945, Pages 541 - 545.

[46] Wilson, E. O., 1998; Wilson, E.O., 1998b.

[47] Nisbet, Robert, 1994, Pages 112 - 115.

[48] Smith, Norman Kemp *Descartes Philosophical Writings*. New York: The Modern Library, 1958, Pages 91 - 144, 161 - 248; Smith, T.V. and Grene, Marjorie *From Descartes to Locke*. Chicago: The University of Chicago Press, 1957, Chapter One; Solomon, Robert, 2002, Pages 168 - 170; Russell, Bertrand, 1945, Pages 557 - 568.

[49] Watson, Peter, 2005, Pages 491, 515 - 517.

[50] Wilson, E. O., 1998b.

[51] Morris, Richard, 1986, Pages 69 - 72.

[52] Christian, David, 2004, Page 431.

[53] Watson, Peter, 2005, Pages 739 - 740.

[54] Fraser, J. T. *Time, the Familiar Stranger*. Redmond, Washington: Tempus, 1987, Page 41.

[55] Fraser, J. T., 1987, Page 40.

[56] Watson, Peter, 2005, Pages 481 - 483.

[57] Smolin, Lee *The Life of the Cosmos*. Oxford: Oxford University Press, 1997, Pages 142 - 143.

[58] Berman, Morris *The Reinchantment of the World*. New York: Bantam, 1981, Chapters Three and Four.

[59] Ackoff, Russell "From Mechanistic to Social Systematic Thinking" *Systems Thinking in Action Conference*, Pegasus Communications, Inc., 1993; Capra, Fritjof *The Turning Point*. New York: Bantam, 1983, Chapter Two.

[60] Watson, Peter, 2005, Pages 484 - 489.

[61] Watson, Peter, 2005, Pages 492 - 495.

[62] Randall, John *Aristotle*. New York: Columbia University Press, 1960.

[63] Goerner, Sally *Chaos and the Evolving Ecological Universe*. Luxembourg: Gordon and Breach, 1994, Chapter One.

[64] Whitehead, Alfred North, 1925; Smolin, Lee, 1997.

[65] Koestler, Arthur *The Act of Creation*. New York: Dell, 1964, Chapter Six.

[66] Nisbet, Robert, 1994, Pages 4 - 5.

[67] Nisbet, Robert, 1994, Page 317.

[68] Smolin, Lee, 1997, Chapters Sixteen to Eighteen.

[69] Nisbet, Robert, 1994, Pages 156 - 159.

[70] Nisbet, Robert, 1994, Pages 124 - 139.

[71] Watson, Peter, 2005, Pages 497 - 498.

[72] Watson, Peter, 2005, Pages 501 - 502.

[73] Watson, Peter, 2005, Pages 503 - 505.

[74] Smith, T.V. and Grene, Marjorie, 1957, Pages 339 - 454; Watson, Peter, 2005, Page 509.

[75] Elwes, R.H.E. (Ed.) *Spinoza: On the Improvement of the Understanding, The Ethics, and Correspondence.* New York: Dover Publications, Inc., 1955; Damasio, Antonio *Looking for Spinoza: Joy, Sorrow, and the Feeling Brain.* Orlando, Florida: Harcourt, Inc., 2003; Watson, Peter, 2005, Pages 505 – 507.

[76] Wikipedia – Baruch Spinoza - http://en.wikipedia.org/wiki/Spinoza.

[77] Watson, Peter, 2005, Pages 509 – 527; Wikipedia – Voltaire - http://en.wikipedia.org/wiki/Voltaire.

[78] Chappell, V. C. *The Philosophy of David Hume.* New York: The Modern Library, 1963; Kant, Immanuel *Critique of Pure Reason.* Amherst, New York: Prometheus Books, 1990; Russell, Bertrand, 1945, Pages 659 – 674, 701 – 718; Solomon, Robert, 2002, Pages 170 – 175; Tarnas, Richard, 1991, Pages 336 – 354.

[79] Nisbet, Robert, 1994, Pages 179 - 186.

[80] Nisbet, Robert, 1994, Pages 160 – 167; Watson, Peter, 2005, Pages 507 – 509.

[81] Nisbet, Robert, 1994, Pages 134 – 138.

[82] Watson, Peter, 2005, Pages 532, 546 – 547.

[83] Condorcet's book was published after his death.

[84] Nisbet, Robert, 1994, Pages 206 - 212; Bell, Wendell *Foundations of Future Studies: Human Science for a New Era.* Vol. II. New Brunswick: Transactions Publishers, 1997, Pages 32 - 38; Wilson, E. O., 1998, Pages 15 – 22; Wilson, E. O., 1998b.

[85] Watson, Peter, 2005, Pages 547 – 548.

[86] D'Souza, Dinesh *The Virtue of Prosperity: Finding Values in an Age of Techno-Affluence.* New York: The Free Press, 2000, Pages 169 - 183.

[87] Watson, Peter, 2005, Pages 550 – 571.

[88] Christian, David, 2004, Pages 406 – 432.

[89] Nisbet, Robert, 1994, Pages 187 - 193.

[90] Watson, Peter, 2005, Pages 541 – 543.

[91] Wikipedia – Thomas Paine - http://en.wikipedia.org/wiki/Thomas_Paine; Wikipedia – Thomas Jefferson - http://en.wikipedia.org/wiki/Thomas_Jefferson; Watson, Peter, 2005, Pages 575 – 585.

[92] Cornish, Edward "Futurists" in Kurian, George Thomas, and Molitor, Graham T.T. (Ed.) *Encyclopedia of the Future.* New York: Simon and Schuster Macmillan, 1996.

[93] Miller, Kenneth *Finding Darwin's God: A Scientist's Search for Common Ground between God and Evolution.* New York: Perennial, 1999, Pages 165 - 195.

[94] Bell, Wendell, Vol. II, 1997, Chapter One; Wilson, E. O., 1998, Chapter Eleven; Wilson, E.O. "The Biological Basis of Morality" *The Atlantic Monthly,* April, 1998c.

[95] Nisbet, Robert, 1994, Pages 153 – 156, 212 – 216; Watson, Peter, 2005, Pages 546 – 548.

[96] Nisbet, Robert, 1994, Pages 229 – 236, 251 - 258.

[97] Watson, Peter, 2005, Pages 548 - 549, 648 - 653.

[98] Postrel, Virginia "Technocracy R.I.P." *Wired*, January, 1998.

[99] Nisbet, Robert, 1994, Chapter Seven.

[100] Watson, Peter, 2005, Page 549.

[101] Christian, David, 2004, Page 393.

[102] Christian, David, 2004, Pages 427 - 430.

[103] Nisbet, Robert, 1994, Pages 220 - 223; Watson, Peter, 2005, Page 548.

[104] Bell, Wendell, Vol. II, 1997, Pages 7 - 14.

[105] Nisbet, Robert, 1994, Pages 112 - 115.

[106] Bell, Wendell, Vol. II, 1997, Chapter One; Clute, John *Science Fiction: The Illustrated Encyclopedia*. London: Doarling Kindersley, 1995, Pages 34 - 35.

[107] Wagar, W. Warren "Utopias, Futures, and H.G. Wells' Open Conspiracy" in Didsbury, Howard F. (Ed.) *Frontiers of the 21st Century: Prelude to the New Millennium*. Bethesda, Maryland: World Future Society, 1999.

[108] Bell, Wendell, Vol. II, 1997, Chapter One.

[109] Nisbet, Robert, 1994, Page 239.

[110] Nisbet, Robert, 1994, Chapters Six and Seven.

[111] Nisbet, Robert, 1994, Pages 258 - 267.

[112] Wilson, E. O., 1998b.

[113] Christian, David, 2004, Pages 432 - 439.

[114] Best, Steven and Kellner, Douglas, 1997, Page 18.

[115] Ray, Paul and Anderson, Sherry, 2000, Pages 25 - 30, 70 - 78.

[116] Best, Steven and Kellner, Douglas, 1997, Pages 202.

[117] Goerner, Sally, 1994; Prigogine, Ilya *The End of Certainty: Time, Chaos, and the New Laws of Nature*. New York: The Free Press, 1997.

[118] Wilson, E. O., 1998, Chapter One.

[119] Nisbet, Robert, 1994, Pages 275 - 286; Tarnas, Richard, 1991, Pages 379 - 383.

[120] Russell, Bertrand, 1945, Chapter Twenty-Two.

[121] Palmer, Donald *Looking at Philosophy: The Unbearable Heaviness of Philosophy Made Lighter*. 3rd Ed. Boston: McGraw Hill, 2001, Pages 224 - 232, 243.

[122] Russell, Bertrand, 1945, Pages 741 - 742.

[123] Adams, Fred and Laughlin, Greg *The Five Ages of the Universe: Inside the Physics of Eternity*. New York: The Free Press, 1999, Pages xvii - xxi; Wright, Robert *Nonzero: The Logic of Human Destiny*. New York: Pantheon Books, 2000, Chapter Two.

[124] Solomon, Robert, 2002, Pages 86 - 87, 144 - 145.

[125] Fukuyama, Francis *The End of History and the Last Man*. New York: The Free Press, 1992, Pages 62 - 63.

[126] Fukuyama, Francis, 1992, Page 60.

[127] Nisbet, Robert, 1994, Pages 220, 267 - 269.

[128] Russell, Bertrand, 1945, Page 737.

[129] Nisbet, Robert, 1994, Page 269.

[130] Solomon, Robert, 2002, Pages 220 – 221.

[131] Fukuyama, Francis, 1992, Pages 59 – 67, 143 – 156; Palmer, Donald, 2001, Page 231 – 232.

[132] Russell, Bertrand, 1945, Pages 741- 742.

[133] Fukuyama, Francis, 1992, Page 64 – 69.

[134] Palmer, Donald, 2001, Pages 258 – 261.

[135] Watson, Peter, 2005, Pages 550 – 571.

[136] Nisbet, Robert, 1994, Pages 258 – 267.

[137] Russell, Bertrand, 1946, Page 782; Bell, Wendell, 1997, Vol. II, Pages 44 – 63.

[138] Russell, Bertrand, 1946, 782 – 790; Palmer, Donald, Pages 259 – 261.

[139] Tarnas, Richard, 1991, Pages 314, 329.

[140] Palmer, Donald, 2001, Pages 259 – 260.

[141] Best, Steven and Kellner, Douglas, 1997, Pages 50 – 57.

[142] Bell, Wendell, 1997, Vol. II, Pages 48, 63.

[143] Bell, Wendell, 1997, Vol. II, Pages 44 – 63; Russell, Bertrand, 1946, Pages 782 – 790.

[144] Tarnas, Richard, 1991, Page 366.

[145] Watson, Peter, 2005, Pages 610, 733.

[146] Tarnas, Richard, 1991, Pages 366 – 378; Russell, Bertrand, 1946, Pages 675 – 684; Watson, Peter, 2005, Pages 613, 732.

[147] Watson, Peter *The Modern Mind: An Intellectual History of the 20th Century*. New York: HarperCollins Perennial, 2001, Pages 171 – 173.

[148] Watson, Peter, 2005, Page 613; Nisbet, Robert, Pages 318 – 320.

[149] Watson, Peter, 2005, Pages 606 – 623.

[150] Watson, Peter, 2005, Pages 589 – 605.

[151] Palmer, Donald, 2001, Pages 233 – 242; Russell, Bertrand, 1946, Pages 753 – 759; Nisbet, Robert, Pages 319 – 320.

[152] Palmer, Donald, 2001, Pages 242 – 253; Best, Steven and Kellner, Douglas, 1997, Pages 40 – 50.

[153] Watson, Peter, 2001, Page 39.

[154] Solomon, Robert, 2002, Pages 61 - 63, 92, 109, 265, 286 – 289; Best, Steven and Kellner, Douglas, 1997, Pages 57 – 77. .

[155] Palmer, Donald, 2001, Pages 267 – 275; Watson, Peter, 2001, Pages 39 – 40; Tarnas, Richard, 1991, Pages 370 – 371; Best, Steven and Kellner, Douglas, 1997, Page 65.

[156] Best, Steven and Kellner, Douglas, 1997, Pages 59-60, 70; Solomon, Robert, 2002, Page 370.

[157] Best, Steven and Kellner, Douglas, 1997, Page 59; Russell, Bertrand, 1946, Page 760; Solomon, Robert, 2002, Pages 368 – 370; Watson, Peter, 2001, Page 40.

[158] Tarnas, Richard, 1991, Page 370; Best, Steven and Kellner, Douglas, 1997, Pages 67 – 71; Palmer, Donald, 2001, Pages 268 – 269.

[159] Best, Steven and Kellner, Douglas, 1997, Pages 57, 64 – 65, 73.

[160] Solomon, Robert, 2002, Pages 92, 109, 286-289; Palmer, Donald, 2001, Pages 272 - 273.

[161] Best, Steven and Kellner, Douglas, 1997, Pages 60 - 62, 67, 71.

[162] Palmer, Donald, 2001, Page 272; Kaufmann, Walter (Ed.) *The Portable Nietzsche*. New York: The Viking Press, 1954, Page 143.

[163] Best, Steven and Kellner, Douglas, 1997, Pages 62 - 63.

[164] Kaufmann, Walter, 1954; Best, Steven and Kellner, Douglas, 1997, Page 73.

[165] Russell, Bertrand, 1946, Pages 760 - 773.

[166] Darwin, Charles *On the Origin of Species by Natural Selection, or the Preservation of Favoured Races in the Struggle for Life* (1859). New York: New American Library, 1958; Dennett, Daniel C. *Darwin's Dangerous Idea*. New York: Simon and Schuster, 1995, Page 21.

[167] Watson, Peter, 2005, Page 641.

[168] Green, John *The Death of Adam: Evolution and Its Impact on Western Thought*. Ames, Iowa: Iowa State University Press, 1959, Pages 1 - 13.

[169] Gould, Stephen Jay *Time's Arrow Time's Cycle: Myth and Metaphor in the Discovery of Geological Time*. Cambridge: Harvard University Press, 1987, Pages 21 - 51.

[170] Watson, Peter, 2005, Page 14.

[171] Green, John, 1959, Pages 43 - 48, 89 - 127.

[172] Green, John, 1959, Pages 31 - 35.

[173] Green, John, 1959, Pages 28 - 30.

[174] Watson, Peter, 2005, Page 548.

[175] Green, John, 1959, Pages 36 - 37.

[176] Miller, Kenneth, 1999, Pages 36 - 37.

[177] Watson, Peter, 2001, Pages 249 - 255.

[178] Tarnas, Richard, 1991, Pages 330 - 331.

[179] Gould, Stephen, 1987, Pages 1 - 3; Watson, Peter, 2005, Pages 631 - 633.

[180] Green, John, 1959, Pages 76 - 81; Gould, Stephen, 1987, Pages 61 - 97.

[181] Green, John, 1959, Pages 131 - 137.

[182] Green, John, 1959, Pages 55 -59, 138 - 155.

[183] Green, John, 1959, Pages 106 -109, 116 -119, 123 -125.

[184] Watson, Peter, 2005, Page 633; Green, John, 1959, Pages 86 - 87.

[185] Watson, Peter, 2005, Pages 13 - 17, 624 - 629.

[186] Green, John, 1959, Pages 155 - 166.

[187] Green, John, 1959, Pages 166 - 169; Gould, Stephen, 1987, Page 158; Watson, Peter, 2005, Pages 637 - 640.

[188] Nisbet, Robert, 1994, Pages 216 - 220; Green, John, 1959, Pages 258 - 260.

[189] Green, John, 1959, Pages 249 - 260; Gould, Stephen, 1987, Pages 99 - 179.

[190] Gould, Stephen, 1987, Pages 117 - 126.

[191] Green, John, 1959, Pages 244 – 247.

[192] Green, John, 1959, Pages 264 – 265; Miller, Kenneth, 1999, Pages 7 – 10.

[193] Mead, George Herbert *Movements of Thought in the Nineteenth Century*. Moore, Merritt (Ed.) Chicago: University of Chicago Press, 1936, Pages 157 – 160; Green, John, 1959, Pages 270 – 271; Morris, Richard *The Evolutionists: The Struggle for Darwin's Soul*. New York: W.H. Freeman and Company, 2001, Pages 52 – 54.

[194] Edis, Tanner *The Ghost in the Universe: God in Light of Modern Science*. Amherst, New York: Prometheus Books, 2002, Chapter Two; Watson, Peter, 2001, Page 115; Miller, Kenneth, 1999, Pages 233 – 239; Green, John, 1959, Pages 301 – 305.

[195] Fraser, J. T., 1987, Page 132.

[196] Edis, Tanner, 2002, Page 55.

[197] Nisbet, Robert, 1994, Page 175; Green, John, 1959, Pages 297 – 298, 327 - 328.

[198] Watson, Peter, 2005, Page 644.

[199] Watson, Peter, 2005, Page 628.

[200] Green, John, 1959, Pages 201 – 219, 320 – 335.

[201] Tarnas, Richard, 1991, Pages 326 – 327.

[202] Watson, Peter, 2001, Pages 41 – 43, 245 – 246; Morris, Richard, 1986, Pages 80 - 85; Watson, Peter, 2005, Pages 674 – 676.

[203] Green, John, 1959, Page 335.

[204] Shlain, Leonard, 1998, Page 379; Watson, Peter, 2001, Pages 40 – 41.

[205] Wilson, E. O. *Sociobiology: The New Synthesis*. Cambridge, MA: Harvard University Press, 1975; Gribbin, John *Genesis: The Origins of Man and the Universe*. New York: Delta, 1981; Prigogine, Ilya and Stengers, Isabelle *Order out of Chaos: Man's New Dialogue with Nature*. New York: Bantam, 1984; Davies, Paul *The Cosmic Blueprint: New Discoveries in Nature's Creative Ability to Order the Universe*. New York: Simon and Schuster, 1988; Smolin, Lee, 1997; Loye, David *The Evolutionary Outrider: The Impact of the Human Agent on Evolution*. Westport, CT: Praeger, 1998; Sahtouris, Elisabet *EarthDance: Living Systems in Evolution*. Lincoln, Nebraska: IUniverse Press, 2000; Wright, Robert, 2000; Watson, Peter, 2001, Pages 245 – 255; Morowitz, Harold *The Emergence of Everything: How the World Became Complex*. Oxford: Oxford University Press, 2002; Loye, David (Ed.) *The Great Adventure: Toward a Fully Human Theory of Evolution*. Albany, New York: State University of New York Press, 2004.

[206] Morris, Richard, 2001, Pages 45 – 46; Miller, Kenneth, 1999, Page 53 – 56.

[207] Miller, Kenneth, 1999, Chapters Two to Six; Edis, Tanner, Chapter Two. But also see Gardner, James *Biocosm: The New Scientific Theory of Evolution: Intelligent Life is the Architect of the Universe*. Makawao, Maui, Hawaii: Inner Ocean, 2003 for a naturalistic theory of intelligent design.

[208] Green, John, 1959, Pages 305 – 306; Whitehead, Alfred North *Process and Reality: An Essay in Cosmology*. New York: Harper and Row, 1929.

[209] Watson, Peter, 2001, Page 772.

[210] Wilson, E.O., 1998c; Shermer, Michael *The Science of Good and Evil*. New York: Times Books, 2004.

[211] Loye, David "Darwin, Maslow, and the Fully Human Theory of Evolution" in Loye, David (Ed.) *The Great Adventure: Toward a Fully Human Theory of Evolution*. Albany, New York: State University of New York Press, 2004. (b)

[212] Margulis, Lynn *Symbiosis in Cell Evolution*. 2nd Ed. New York: W. H. Freeman, 1993; Margulis, Lynn "Gaia is a Tough Bitch" in Brockman, John *The Third Culture*. New York: Touchstone, 1995; Sahtouris, Elisabet, 2000.

[213] Miller, Kenneth, 1999, Page 56.

[214] Watson, Peter, 2005, Page 641.

Conclusion and Summary

"To understand and change the present condition of our species, we must gain insight into the past. If we do not, we cannot exert a lasting influence on the future."

Leonard Shlain

Darwin's theory of evolution would have a significant impact on American philosophy and science. The Darwinian emphasis on adaptation, competition, and the challenges of survival, coupled together with Darwin's progressive vision of the history of life and humanity, resonated with the pioneer spirit and pragmatic optimism of America. Many of the most influential American thinkers at the turn of the century, such as Charles Pierce, William James (1842 – 1910), and John Dewey (1859 - 1952), enthusiastically applied evolutionary ideas and themes to philosophy, psychology, and education. As one significant illustration, in considering the question of the nature of consciousness, James asked what adaptive value consciousness must serve such that it would have evolved in the human species. Dewey applied evolutionary thinking to ethics, arguing that ethical standards, rather being static and unchanging, are dynamic and evolve over time.[1]

James and Pierce are particularly noted for creating the philosophy of **pragmatism**, which highlighted the functional and practical benefit of ideas. For example, the pragmatic theory of truth equated truth with utility or its "cash value."

Does an idea or principle work, producing intended beneficial effects and results? Does an idea or theory produce accurate concrete predictions? Grounded in the American emphasis on making and doing – on constructive and creative action – pragmatic philosophy was concerned with how thought served action. Pragmatism both expressed and reinforced the action-oriented philosophy and approach to the future of the American people.[2]

Dewey, who along with James, helped to create "functional psychology" which focused on the functions and benefits of psychological processes, had a big impact on American education. American higher education, in its early formative years, reflected a combination of religious and secular ideas, emphasizing moral and character development through both reason and religious values and ideals. But as Watson recounts, American universities became increasingly secular after being founded by religious organizations. Dewey saw the American educational system as excessively authoritarian in its approach to learning, reinforcing conformity and passive learning in students – in essence, following a factory and European lecture model of education. Dewey believed that education needed to align better with the democratic way of life in America and should encourage individuality and active participation in the learning process.[3]

Given these action-oriented, progressive, and individualistic themes in American thought and philosophy, it is not surprising that a great wave of prolific invention swept across the country between 1870 and 1900. Much of the spirit and momentum of the technological revolution that would transform human life in the twentieth century was spearheaded by the "can-do" creative inventiveness of the American people that blossomed and realized its potential in the last decades of the nineteenth century. If Europe dominated the world economically, militarily, and scientifically in the nineteenth century, the balance of power began to shift more toward America. In the process a uniquely Americanized version and vision of a positive future developed that attracted millions upon millions of immigrants hoping to realize their dreams in the "land of opportunity" (clearly a futurist image). America, in fact, became the new "promised land."

The emergence of the theory of evolution also provoked another wave of conflict and tension between the scientific-secular world and Western religion. As Watson states, Darwin was not that strongly motivated to criticize or attack Church doctrine, although he was well aware of the apparent contradictions between his explanation of the origin of life and man and the story of creation in *Genesis*. But defenders of Christian cosmology and history did realize the severity of the challenge of evolutionary thinking to traditional Christian views and repeatedly attempted to defend the Christian theory of creation against evolution. Furthermore, even if Darwin was relatively non-confrontational, popular writers on science and evolution were openly and decidedly critical in their treatment of religious doctrine, attacking the factual accuracy of the Bible, questioning all types of supernatural beliefs, and in particular, finding great fault in the authoritarian and dogmatic attitude of the Church and religious belief systems. The reaction of the Catholic Church to this new wave of scientific and philosophical criticisms was similar to previous episodes involving perceived threats to Christian thinking – Church leaders became more entrenched, reasserting their authority, and in fact, creating the doctrine of "papal infallibility" as an ultimate defense and justification of the truth of Christianity. Of special significance, the Church rejected the philosophy of evolution and change, arguing for the reality of "eternal truths" presumably contained within Christian doctrine. Watson contends that the Catholic Church retreated backwards to Middle Age thinking, re-asserting the philosophical and theological authority of St. Thomas Aquinas and, in general, back-peddling away from modernism and the contemporary world.[4]

If the Catholic Church seemed to be running away from modernity into the past and erecting barricades against the advances of science, the theory of secular progress, especially in so far as it was associated with European expansionism, had its problems and critics as well. The results of European colonialism were mixed at best; although European countries, such as Great Britain and France, in the name of progress brought new social and political ideals and modern products and technologies to non-Western cultures, there was the growing realization that the West was culturally arrogant and elitist in attempting to impose its values on other people and societies. Non-Western people around the world were being used to support the economic and material growth of the West and were being robbed of self-governance, as their indigenous cultures and ways of life were being repressed, if not destroyed, by the West. Perhaps the Euro-centric vision of the future was not for everybody; perhaps this vision could only be realized at the expense of others.[5]

As we have seen, there were also powerful negative reactions from within Europe against the ideals of secular progress, capitalism, science and industry, and the modern way of life as benefiting those who lived within this social system. Interestingly, the very success of modernism had a negative effect on many Westerners. The increasing speed, energy, innovation, and creativity associated with accelerative technological, as well as cultural change made modern life confusing and bewildering to many individuals. The rejection of the past – of heritage and tradition – left many people feeling lost without a mental anchor. The decline in religious faith in the nineteenth century robbed humanity of a sense of cosmic purpose. Freedom and the rejection of higher religious authority brought with it anxiety and uncertainty in the face of the future. To many it seemed that the world was falling apart.[6]

Hence, as humanity entered the twentieth century, many social and technological forces and many different and often contradictory ideas and belief systems were having a multi-faceted and mixed impact on the ever transforming nature and make-up of future consciousness. There was a strong vein of optimism, both in Europe and America, associated with the rationality and apparent benefits of science and technology. Economies were growing, wealth was accumulating, standards of living were on an upswing – capitalism and consumerism were increasingly powerful forces in the world. In general, Watson describes the course of events in the eighteenth and nineteenth centuries as involving a fundamental transition of authority from the soul to the scientific experiment.[7]

Yet, in spite its dogma and entrenchment in the past, organized religions were still extremely influential in determining the beliefs and values people had concerning the future. The vast bulk of people in the world still believed in miracles, an afterlife, reincarnation, and God. Furthermore, the Romantic vision of life, as expressed through the arts, literature, and the humanities, provided still another popular alternative to science, technology, and abstract rationalism. Growing out of the writings of the Romantic philosophers Kierkegaard and Nietzsche, the philosophy of existentialism grew and flourished in the decades both before and after the turn of the century, first through the novels of Fyodor Dostoevsky (1821 – 1881) and then later in the twentieth century through the writings of Sartre, Camus, and Heidegger, among others. Dostoevsky, in such books as *Notes from the Underground*, *Crime and Punishment*, and *The Brothers Karamasov*, delved into the complex, contradictory, and mysterious nature of the human self, human passion, madness, and good and evil. This turning inward toward the tumultuous and uncertain depths of the human soul and human desire contrasted sharply with the externally focused, clear vision of science and rational Enlightenment. Dostoevsky's fascination with the irrational and with the undercurrents of the human mind anticipated twentieth-century Freudian depth psychology and Freud's interest in the primordial forces of the human mind. In fact, the irrational and the unconscious would emerge as powerful themes in psychology, philosophy, and the arts early in the twentieth century – perhaps in resonance with the sense of madness and loss of order connected with the modern way of life.[8]

427

In general, various concerns over the excessively rationalistic, optimistic, and Euro-centric vision of secular progress were beginning to intensify, ultimately leading to a loss of faith in progress early in the twentieth century.[9] Fear and uncertainty regarding the future entered the twentieth century along with hope and conviction. Both the Dionysian and the Apollonian crossed over from one century into the next.

Finally, the seeds of revolution, social upheaval, and war were taking root in Europe at the end of the nineteenth century; the legacy of Marx would provide the ideological justification for the Communist Revolution in Russia – as an alternative to capitalism. Moreover, the ideas of Nietzsche and Hegel, among other German philosophers, would fuel and justify growing German nationalism[10] and a philosophy of cultural superiority and military conquest that would explode in the coming decades ahead. As in the past, at least in the minds of some individuals, the violent and dominance driven streak within human nature was connected with future consciousness.

┼══╾══┼

In summary, my starting point in this book was to examine the complex and holistic nature of future consciousness, looking at all its fundamental dimensions, including perception, behavior, thought, emotion, motivation, self, and ethics, and their interconnections. I systematically described the many different cognitive processes involved in future consciousness, including planning, imagination, foresight, and decision making and I also explored in depth the motivational and emotional aspects of future consciousness, such as fear, depression, hope, optimism, and pessimism. From various angles, I examined the relationship between consciousness of the past and consciousness of the future and the tension between the desire for stability and the desire for growth and change. Throughout the opening chapter, I emphasized the importance of future consciousness in everyday human life and how its enhancement or impoverishment respectively impacts mental health and mental illness. I also explored the philosophical and cosmic dimensions of future consciousness. I closed the opening chapter with a discussion of the contemporary transformation and the need for the continued evolution of future consciousness, providing summary lists of the multiple benefits of future consciousness and the multiple ways to enhance it.

In the next chapter, I traced the development of human psychology and the various capacities of future consciousness from their biological and primordial foundations through prehistory up to the emergence of agriculture and urbanization. Looking at its evolution further helped us to understand and explain the nature and innumerable functions of future consciousness. I connected the evolution of future consciousness to basic psychological processes and issues of life, such as tool making, hunting, mating, reproduction, sexuality, child-rearing, art, death, the self, choice and free will, the creation of "thinking spaces," culture and social organization, war and conquest, agriculture, trade, and the manipulation the environment. All these foundational elements of life and mind are still with us in our contemporary world, as fundamental undercurrents and components of human existence. Much of future consciousness is built upon these basic concerns and issues of life. First introduced in the opening chapter, a key theme throughout the chapter was the principle of reciprocity as a way to understand the various complementarities and interdependencies of psychology, society, and life.

In the third chapter, I discussed the importance of mythic and religious thinking in the development of future consciousness. Myth provided the first, and perhaps most fundamental, archetypes of human existence, the goddess and the hunter, as well as the first theories of the past and the future. Looking at myth and religion helped to elucidate the conceptual and thematic heritage and global richness of future consciousness. I explained the importance of the narrative and religious prophecies in mythic future consciousness and discussed in detail the shift from female to male religions and visions, and the significance of this shift in affecting ideas about the future. I also highlighted the various gods and goddesses within ancient religions and how these deities personified, in the minds of believers, the saga and drama of time and the future. Additionally, I described various significant themes in mythic views of life, time, and the future, including order and chaos, death and rebirth, polytheism and monotheism, and the soul and reincarnation. I described dualistic and teleological views of reality and the relevance of such philosophies to thinking about the future; I looked at the importance of Taoism and Zoroastrianism regarding theories of time and the basic contrast between cyclic and linear views, respectively highlighting balance and harmony versus dominance and conflict. In general, I outlined the various connections between religious and philosophical cosmologies and theories of time. I repeatedly noted commonalities and historical connections among the different myths and religions, as well as fundamental differences, many of which anticipate and prefigure modern issues and disagreements regarding the nature of reality, good and evil, and the future.

I discussed the Axial Age, considered by many historians to be a fundamental turning point in the evolution of the human mind; in this context, I described how the abstract theoretic mode of understanding developed in both religion and ancient philosophy. In the context of Greek literature, mythology, and philosophy, I examined the issue of self-determination versus destiny and the rule of the gods, and described how both the Apollonian and the Dionysian perspectives on life arose within ancient Greece. I looked at the importance of Plato, the dualist, rationalist, and mystic, and Aristotle, the naturalist and logician, and how their ideas would affect later thinking in science, philosophy, and religion.

I described the evolution of monotheism, the historical narrative of Judaism, worldly and other-worldly images of the future, and the emergence of the idea of a Messiah, which served as the starting point for the birth of Christianity. As I argued, Christianity combined many oppositional themes, bringing together diverse ideas and influences in the history of religion and myth. I reviewed the dichotomy between the peaceful and inner vision of the future espoused by Jesus and Buddha and the violent and warlike vision of the future contained in Zoroastrianism, Judaism, and *The Revelation*. Finally, I examined the rise of Islam and the attempted synthesis of faith and science within this religious system; further I described the numerous common elements in Christianity and Islam regarding their views on reality and the future, and considered why these two major world religions have had such a conflictual history.

In the final chapter I focused on the rise of modernism and the theory of secular progress in the West, which would eventually have a great impact throughout the rest of the world as well. I described the growth of individualism, the economic way of life and model of the future, and the escalating attack of secular thinking on European religious authoritarianism. Much of modern history, from the Renaissance onward, can be seen as an ongoing conflict between religion and secularism concerning the future, although along the way there have been repeated attempts to synthesize the two perspectives. I described the shift in the center of gravity from Asia and the Middle East to Europe that took place during the High Middle Ages, the Renaissance, and the Age of Exploration. Next I examined in depth the Scientific Revolution, with its emphasis on reason, empirical observation, and experimentation. These scientific methods of inquiry into nature provided still another challenge to Western religious authority. Modern science brought with it the philosophy of determinism, the mechanistic/machine model of nature, and the theory of natural laws, all these ideas serving as a foundation for a theory of the future and a basis for making predictions about the future.

Next, I described the Age of Enlightenment and the growth of the theory of secular progress and its impact on social, political, and economic thinking. I discussed the themes of democracy, freedom, and capitalism, and the emergence of new secular values for the future. I explained the hypothesized natural law of progress, the theories of history and the future inspired by this idea, and how various individuals proposed that humanity apply scientific principles to guiding, if not controlling the future of human society. In this context I described some of the main features of utopian thought from Sir Thomas More up through the late nineteenth century. In summary, following the lead of Nisbett, I argued that the Enlightenment actually produced two apparently contradictory theories of progress: Increasing individualism and freedom versus increasing social order.

After the Enlightenment, I turned to the ideas of Hegel and Marx, two of the most influential philosophers of the nineteenth century. Both Hegel and Marx articulated general theories of human history, as well as the future. Hegel's philosophical system involved a conceptual synthesis of diverse ideas and themes from the past, emphasizing the centrality of spiritual evolution toward the realization of God. Furthermore, Hegel's theory of the dialectic supported the view that progress inevitably involves conflict and perhaps war and conquest. Before discussing Marx, I described some of the main features of the Industrial Revolution, noting both its positive and negative effects. Marx was one of the great critics of mass industrialization and capitalism and, standing Hegel on his head, argued for a materialist view of history and progress. A utopian thinker, Marx called for a social revolution as a necessary step toward the future realization of a better human society.

After Hegel and Marx, I looked at Romantic philosophy and its critique of the values of Enlightenment philosophy, especially rationalism, industrialism, and social order and control. Romanticism offered a Dionysian view of life and the future, emphasizing beauty, art, passion, will, and creativity. In the case of Kierkegaard, faith, subjectivity, freedom, uncertainty, and extreme individualism were key philosophical ideas. Between Kierkegaard and Nietzsche, although also reinforced by the earlier philosophical conclusions of Kant, the philosophy of relativism became a powerful current in

European thought, challenging the objectivist view of science and the Enlightenment.

Finally I described the historical development of the theory of evolution, which culminated in Darwin's theory of natural selection. I looked at the discovery of deep time and a variety of related findings across the natural and cultural sciences that transformed the Western view of reality into a much more dynamic and expansive vision of history and time. Darwin, building upon these discoveries, provided a scientific theory of the evolution of life and humanity that directly challenged the creationist account in *Genesis*. Darwin's evolutionary vision replaced the static view of nature contained in Newtonian physics and through the principle of natural selection explained the origins of order without recourse to either teleologism or creationism; for Darwin order evolves from the simple to the complex. Although throughout the historical development of evolutionary science from the seventeenth to the nineteenth century, various efforts were made to reconcile and synthesize Western religion with this new emerging view of history and time, the overall result has been tension and conflict between a God-centered static creationism and scientific law-centered dynamic evolutionism. These alternative visions of historical time and natural reality have clear and disparate implications regarding the future of life and humanity. Secular and scientific thinkers inspired by Darwin applied evolutionary ideas, in particular competition and natural selection, to human society, drawing a variety of conclusions regarding present human conditions and how to direct human evolution in the future. This philosophy of Social Darwinism though was criticized on a variety of points. One objection has been that Social Darwinism with its emphasis on competition is based on a biased and limited view of Darwin's thinking. Darwin, in fact, saw moral development and cooperation of central importance in understanding human history, as well as providing a sense of direction for humanity in the future. In conclusion I noted that the contrasting pair of ideas of competition and cooperation has been a major theme throughout human history, perhaps, in fact, going all the way back to the goddess and hunter archetypes of ancient mythology.

In closing, these numerous intellectual and social trends and modes of consciousness expressed throughout human history provide the foundation for contemporary thinking on the future. As can be seen, future consciousness is a complex phenomenon and has manifested itself in varied and often conflicting ways throughout our evolution and social development. These conflicts continue. Clearly, future consciousness has been of pivotal importance in defining the meaning and purpose of human existence for people around the world. As stated in the introduction, our capacity for future consciousness is a uniquely human achievement, though from this historical and psychological study, it should be apparent that it has grown and evolved in steps, spurts, and, at times, detours and regressions across the ages. It is as if the conscious expansion and evolution of our awareness of the future (and reciprocally of the past) has been an ongoing struggle against various interfering and constraining forces throughout our history. Enlightenment, however it is defined, is a battle against the forces of dogmaticism and darkness. I will continue and complete this narrative and drama in a second volume, *Contemporary Futurist Thought*, in which I will pick up many of the themes introduced in this volume and trace their development up to present times.

References

[1] James, William *The Principles of Psychology*. Volumes I and II. New York: Dover Publications, 1890; Watson, Peter *Ideas: A History of Thought and Invention from Fire to Freud*. New York: HarperCollins Publishers, 2005, Pages 695 - 697.

[2] James, William *Pragmatism* (1907). Cleveland: The World Publishing Company, 1955; Watson, Peter, 2005, Pages 688 - 694.

[3] Watson, Peter, 2005, Pages 695 - 700.

[4] Watson, Peter, 2005, Pages 707 - 711.

[5] Watson, Peter, 2005, Pages 678 - 682.

[6] Watson, Peter, 2005, Pages 729 - 730.

[7] Watson, Peter, 2005, Page 740.

[8] Watson, Peter, 2005, Pages 718 - 734.

[9] Nisbet, Robert *History of the Idea of Progress*. New Brunswick: Transaction Publishers, 1994, Pages 317 - 351.

[10] Watson, Peter, 2005, Pages 664 - 671.